S0-AWV-377

HANDBOOK OF INDUSTRIAL RESIDUES

ENVIRONMENTAL ENGINEERING SERIES

The Noyes Series in Environmental Engineering addresses technologies and strategies which will seek to provide cost-effective solutions to environmental needs confronting both municipalities and industry in the future. The topics in the series include not only detailed treatment technologies, but also management approaches needed to meet the changing environmental and economic conditions. We hope the series will prove to be a timely and valuable reference source for professionals working in the environmental protection field.

HANDBOOK OF INDUSTRIAL RESIDUES

Volume 1: Industries and Management Options

Volume 2: Treatment Technology

by

JON C. DYER

Environmental Technology Consultants, Inc.
Springfield, Virginia

and

NICHOLAS A. MIGNONE

International Paper Company
Mobile, Alabama

Environmental Engineering Series

NOYES PUBLICATIONS
Park Ridge, New Jersey, U.S.A.

Library of Congress Catalog Card Number: 82-19082
ISBN: 0-8155-0924-3
Printed in the United States

Published in the United States of America by
Noyes Publications
Mill Road, Park Ridge, New Jersey 07656

10 9 8 7 6 5 4 3 2 1

Library of Congress Cataloging in Publication Data

Dyer, Jon C.
Handbook of industrial residues.

Bibliography: v. 1, p. ; v. 2, p.
Includes index.
Contents: v. 1. Industries and management options -- v. 2. Treatment technology.
1. Factory and trade waste. I. Mignone, Nicholas A. II. Title.
TD897.D93 1983 628.5'4 82-19082
ISBN 0-8155-0924-3 (set)

Preface

The control of industrial residuals within a municipality will be a pressing issue during the 1980s. As industries are required to treat and/or pretreat their process streams, the resulting residuals will have to be handled in an environmentally and economically suitable fashion. This need will bring the municipal and industrial managers together in a problem-solving environment. The development of mutual trust, a sound understanding of the residuals to be handled, and the availability of applicable technologies and management options will be essential to meeting this need.

The intent of this handbook is to provide information, data, and concepts that will prove beneficial to the municipal/industrial managers tasked with handling industrial residuals. The authors of this handbook have been fortunate to be involved in a number of USEPA projects that address this area of concern. The material presented in this handbook contains information and data developed with USEPA funding assistance and technical input. The authors wish to thank the USEPA personnel who have been involved for their assistance, efforts and dedication. Appropriate references are included for pursuit of more detail, if desired.

Contents and Subject Index

VOLUME 2: TREATMENT TECHNOLOGY

List of Tables

VOLUME 1

VOLUME 2

List of Figures

VOLUME 1

VOLUME 2

Volume 1

Industries and Management Options

-1-

INTRODUCTION

OBJECTIVES OF HANDBOOK

The *Handbook of Industrial Residues - Volume 1 and Volume 2* has been prepared to assist publicly owned treatment works (POTWs) and their industrial users solve waste management problems.

This handbook pulls together, under one convenient title, some of the most important information POTWs and industrial users (IUs) need in finding solutions to waste problems.

With the enactment of the Clean Water Act of 1977 (CWA — Public Law (PL 95-217) and the promulgation of General Pretreatment Regulation (40 CFR, 403) by the United States Environmental Protection Agency (USEPA) on June 26, 1978, the control of industrial toxics has become a reality. The regulation went into effect on January 31, 1982, but some portions are being reviewed for potential revision.

Since regulations in this area are changing, it is critical that the reader obtain the most current information by consulting the *Federal Register* and its indices.

The pretreatment of industrial wastewaters will create residuals which will have to be controlled. As higher degrees of effluent quality are obtained, larger volumes of resulting solids will be generated. In addition to municipal waste disposal, municipalities will have to be concerned with the treatment, disposal, and management of industrial residuals generated within the community. This Handbook will assist municipalities in determining the type and extent of the problem in their respective communities and how to approach solving it.

The pretreatment regulations and the program requirements affect both municipalities and indirect industrial dischargers, as well as the

NPDES (National Pollutant Discharge Elimination System) States and EPA itself:

- USEPA develops regulations, standards and guidance for implementing pretreatment programs. EPA Regional Offices are responsible for assisting states and municipalities in developing programs and for enforcing categorical standards directly where applicable. Pretreatment Program Guidance (304-g) is also available through EPA.
- NPDES States - Administer and enforce pretreatment program requirements within their respective states. Where applicable, enforces categorical standards directly.
- POTWs - As required by 40 CFR, Part 403, POTWs develop and implement local pretreatment programs to control industrial discharges into the municipal system and where applicable, enforce categorical standards and installation of technologies.
- IUs - Industrial users pretreat their wastes to meet appropriate standards, perform self-monitoring and reporting, and handle their industrial residuals (sludge) in a lawful manner.

Industrial users may well look to their respective POTWs as a source of information and assistance in meeting their pretreatment responsibilities. The *Handbook of Industrial Residues* brings together current information on residual waste management options and requirements, data on categorical industries regulated by federal pretreatment standards, and pretreatment and sludge management technologies. This Handbook is a source document that can be utilized in a total waste management approach (covered later in this volume) or to assist in providing solutions to work problems. This Handbook will provide the POTW manager with information on what residuals industries will generate, answer questions about problems associated with priority pollutants, and make IUs aware of the types of pretreatment processes and equipment available. This is not a design manual; specific engineering and control technologies cannot be specified on the basis of this Handbook alone.

The Handbook has been divided into two (2) volumes:

- Volume 1 - *Industries and Management Options*
- Volume 2 - *Treatment Technology*

Volume 1 contains information on management options available for dealing with residuals, a process approach to total waste management, and several scenarios which depict the options available to small, medium, and large POTWs in addressing their industrial residuals problem. In addition, *Volume 1* contains available information pertaining to the original 34 categorical industries identified by USEPA to be addressed by Federal categorical pretreatment standards. Information includes industry description, categorization, process description, wastewater characteristics, control and treatment technologies, type of residuals generated and resid-

Table 1: Information Available on Categorical Industries

Categorical industry	General industry description	Industrial categorization	Process description	Wastewater characteristics	Control and treatment technology	Type of residue generated	Residual management options	Toxic categorical standard not expected[a]
Adhesives and Sealants	A[b]	A	A	A	A	I[b]	I[c]	
Aluminum Forming	NA[d]	NA	NA	NA	NA	NA	NA	
Auto and Other Laundries	A	A	A	A	A	I	I	X
Battery Manufacturing	NA	NA	NA	NA	NA	NA	NA	
Coal Mining	A	A	A	A	A	A	A	
Coil Coating	A	A	A	A	A	I	I	
Copper Forming	NA	NA	NA	NA	NA	NA	NA	
Electrical Products	NA	NA	NA	NA	NA	NA	NA	
Electroplating	A	A	A	A	A	A	A	
Explosives Manufacturing	A	A	A	A	A	A	A	
Foundry	A	A	A	A	A	A	A	
Gum and Wood Chemicals Mfg.	A	A	A	A	A	I	I	X
Inorganic Chemical Mfg.	A	A	A	A	A	I	I	
Iron and Steel Mfg.	A	A	A	A	A	A	A	
Leather Tanning & Finishing	A	A	A	A	A	A	A	
Mechanical Products	NA	NA	NA	NA	NA	NA	NA	X
Nonferrous Metals Mgf.	A	A	A	A	A	I	I	
Ore Mining and Dressing	A	A	A	A	A	I	I	
Organic Chemicals	A	A	A	A	A	A	A	
Paint and Ink Formulation	A	A	A	A	A	A	A	
Pesticides Manufacturing	A	A	A	A	A	A	A	
Petroleum Refining	A	A	A	A	A	A	A	
Pharmaceutical Mfg.	A	A	A	A	A	A	A	
Photographic Equip.&Supplies	A	I	A	A	A	I	I	
Plastics Processing	NA	NA	NA	NA	NA	NA	NA	X
Plastics and Synthetic Materials	A	A	A	A	A	I	I	
Porcelain Enameling	A	A	A	A	A	I	I	
Printing and Publishing	NA	NA	NA	NA	NA	NA	NA	X
Pulp, Paper & Paperboard	A	A	A	A	A	A	A	
Rubber	A	A	A	A	A	I	I	X
Soap and Detergent Mfg.	A	A	A	A	A	I	I	
Steam Electric	A	A	A	A	A	A	A	
Textile Mills	A	A	A	A	A	A	A	
Timber Products Processing	A	A	A	A	A	A	A	

a USEPA - Effluent Guidelines Division report by James D. Gallup and Debbie Seal, July 1, 1981, listing paragraph 8 exclusions.
b A - adequate information available to allow discussion.
c I - insufficient data available to allow discussion or draw conclusions.
d NA - data base has not yet been established.

ual management options. The initial data base on these 34 industries is not yet complete and several industries could not be discussed as indicated by the symbol "NA" in Table 1. In addition, information on residual generation and residual management options is insufficient in several other industries and could not be discussed at this time. They are indicated by the symbol "I" in Table 1.

Volume 2 contains discussions and descriptions of possible pretreatment and sludge residual treatment and management technologies applicable to the categorical industries.

–2–

MANAGEMENT OPTIONS

OVERVIEW OF INDUSTRIAL RESIDUALS PROBLEM

One of the most challenging tasks confronting the environmental pollution control profession during the 1980's is the cost-effective, environmentally sound handling of the wastes and residuals from industrial production. During the 1970's and with the Federal Water Pollution Control Act of 1972 (PL 92-500), Congress directed a determined effort to improve the quality of the nation's waterways. To help reach this goal, USEPA and states provide construction grants to build POTWs. POTWs must meet secondary treatment standards of 30 mg/l BOD_5 and 30 mg/l of SS or more restrictive water quality standards as applicable. This higher degree of treatment will result in the generation of greater volumes of wastes and residual from both municipal and industrial sources.

Figure 1 represents an example of the total waste management concept that will become increasingly more recognized and appreciated. The concept of a total waste management approach in dealing with specific local jurisdiction is a logical step in achieving the most cost effective solution to waste management. The POTW manager should and can take a leading role in resolving total waste management problems.

Study and planning costs associated with evaluating the feasibility of total waste management within a community may be grant-eligible under the 201 Construction Grant Program. Eligibility will be determined on a case-by-case basis but in no case will funds be available for construction; only study and planning monies are eligible. To determine eligibility, the POTW should check with its State or managing USEPA construction grant officer. The cost of managing industrial residuals is still the responsibility of industry.

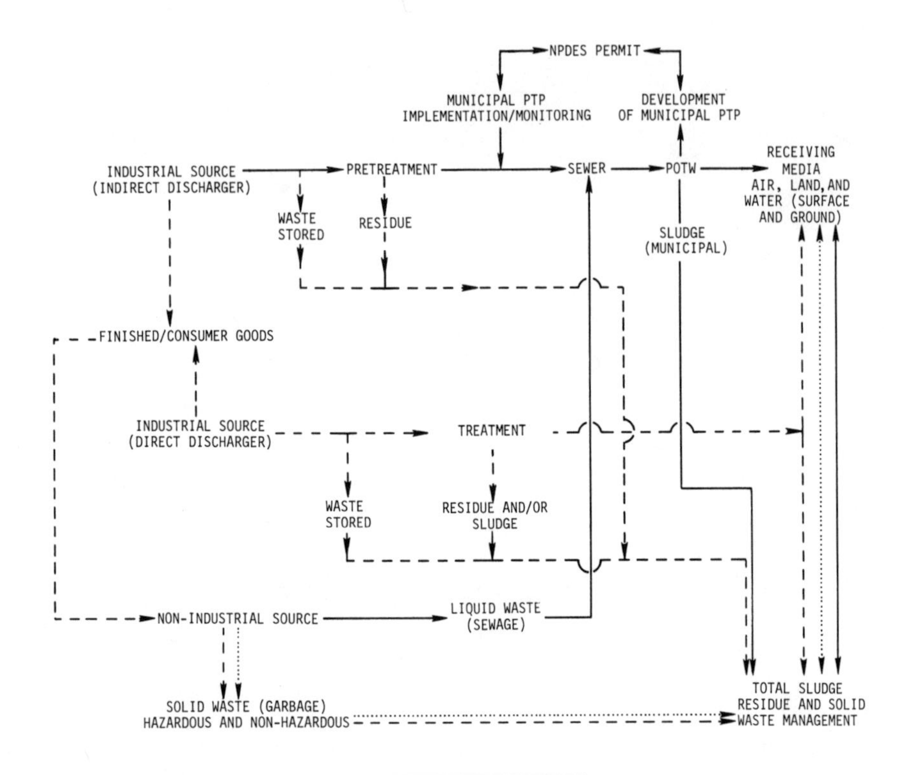

LEVEL I

- planning, design, and construction of POTWs to treat wastes from municipal sources, including industrial wastes discharged into municipal sewers and the treatment and disposal/utilization of POTW sludges;
- development of pretreatment programs that address discharges of industrial wastes into POTWs;
- tracking and planning of facilities for the disposal/utilization of pretreatment residues and stored wastes from these industrial dischargers into the municipal system, in combination with POTW sludges.

LEVEL II

- tracking, planning, design, and construction of facilities for the disposal/utilization of sludges and stored wastes from the treatment of wastes from direct industrial source dischargers;
- tracking, design, and construction of facilities or best management practices (BMPs) for management of nonpoint sources of pollution resulting from various activities (e.g., construction, urban runoff, agriculture, silviculture, mining, etc.).

LEVEL III

- planning, design, and construction of facilities for the treatment and disposal/utilization of other wastes (e.g., municipal solid waste, hazardous waste, radioactive waste, etc.) in combination with Level I or Level II.

Figure 1: Concept of total waste management.

In the 1980's, there will be a national thrust directed at managing waste solids — both municipal and industrial. The CWA and several other major Federal laws affect the evaluation and selection of residuals management systems. This Federal legislation has refocussed the environmental effort. The Federal regulations, coupled with applicable state and local requirements, have a major impact on proper selection of sludge management systems.

As depicted in Figure 1, the sludge generated at the POTW is greatly influenced by the industrial contribution into the municipal system. The nature, strength, and volume of industrial pollutants directly affect particular sludge management systems. Sections 307 and 402 of the CWA authorized regulations (40 CFR, Part 403) for industrial pretreatment. One of the objectives of the pretreatment regulations and standards is to reduce or prevent the introduction of pollutants into POTWs which will contaminate the sludge and thereby impair opportunities for the utilization and disposal of sludge. When industrial pretreatment regulations take effect, the amount of municipal sludges suitable for disposal via more cost-effective options should increase. For example, POTWs involved in landspreading practices should insist on effective pretreatment of appropriate industrial wastes in order to meet regulations designed to protect human health and the environment in a cost-effective manner.

This Handbook will provide guidance for the formulation of one or more decision matrices that will aid in the selection of the most efficient and effective management practices for any number of situations encountered. This decision matrix or decision tree will be the foundation for the planning, development and implementation of total waste management practices.

To utilize the most efficient waste management techniques, knowledge of influent characteristics, potential pollutants, sludges generated, industrial residuals generated and treatment/disposal options must be integrated in the decision matrix.

PROCESS APPROACH TO TOTAL WASTE MANAGEMENT

A process approach is essential in solving the residuals question within a specific community. Many factors need to be considered, evaluated, ranked, and reviewed prior to arriving at a cost-effective and practical solution. Consideration should be given to the development of a public information program and an advisory group. Public and special interest input to residuals management is important in selecting a community supported solution. Table 2 is a detailed outline of factors to be considered in selecting the proper residuals management option.

Table 2: Detailed Outline of Conceptual Approach for Solving a Residuals Management Problem

I. RESIDUALS INVENTORY DEVELOPMENT

- A. Educate and Involve Industrial Community
  - ° Seminars
  - ° Meetings
- B. Develop and Mail Industrial Questionnaire
- C. Use Residuals Manual to Develop Gross Volumes
- D. Follow-up
  - ° Phone
  - ° Site Visits (may require some industrial sampling)
- E. Development of Advisory Group
  - ° Technical
  - ° Legal
  - ° Financial
  - ° Industrial
  - ° Federal
  - ° State
  - ° Local Government
  - ° Civic

II. DEVELOPMENT OF ALTERNATIVES

- A. Categorization of Waste Residues
  - ° Quality
    - ° Hazard Classification
      - ° Known Composition
      - ° Leachate Analysis
    - ° Chemical Classification
  - ° Quantity
    - ° Weight
    - ° Volume
  - ° Geographical Origin
    - ° Quadrant Designation
  - ° Form
    - ° Liquid
      - ° Slurry
      - ° Cake
      - ° Solid
  - ° Generation Source
    - ° Industrial Classification
    - ° Process Classification
  - ° Resource Potential
    - ° Reuse
    - ° Recovery
- B. Management Options for Waste Residuals
  - ° Quantity Reduction
    - ° Thickening
    - ° Concentration
    - ° Conpaction
    - ° Incineration
  - ° Stabilization
    - ° Biological
    - ° Chemical
    - ° Physical
  - ° Ultimate Disposal
    - ° Landfill
      - ° General
      - ° Segregated
    - ° Retention Basins
  - ° Reuse and Recovery
    - ° Internal Consumption
    - ° Supplier Return
    - ° Waste Exchange
    - ° Co-disposal
    - ° Reclamation
  - ° Internal/External Alternatives
  - ° Site Development Alternatives
    - ° Combined Sites
    - ° Segregated Sites
    - ° Centralized Sites
    - ° Noncentralized Sites
- C. Evaluation of Potential Sites for Treatment and Disposal Systems
  - ° Development of Criteria for Preliminary Selection
    - ° Geological and Topographical
    - ° Proximity to Residential and Commercial Development
    - ° Accessibility to Transportation Modes
    - ° Area Requirements for Treatment and Disposal Options
    - ° Relative Availability
    - ° Environmental Considerations
    - ° Ultimate Best Use Considerations
  - ° Application of Criteria for Selection of Potential Sites to Accommodate Screening of Alternative Systems
  - ° Consultation with Advisory Group on Findings

(continued)

Table 2: (continued)

D. Survey and Evaluation of Existing Treatment and Disposal Facilities
- Industrial
- Commercial
  - Within Area
  - Outside Area
- Public
  - Municipal
  - County
  - State

E. Evaluation of Logistic Alternatives
- Transport Modes
  - Roadway
  - Rail
  - River
  - Conduit
- Direct Delivery versus Transfer Stations
- Centralized versus Noncentralized Disposal Destinations
- Combined versus Segregated Disposal Destinations

F. Management and Operation Alternatives
- Local Government
- Local Authority
- Industrial Consortium
- Commercial
  - Lease
  - Ownership

G. Monetary Screening
- Determine Quantity of Units and/or Systems
  - Capital Facilities
  - Operation and Maintenance
- Develop Unit and System Cost Basis
  - Capital Facilities
  - Operation and Maintenance
- Establish Life-Cycle Procedure for Equating Capital Costs and Operation and Maintenance Costs to Common Cost Factors
- Delineate Total Life-Cycle Costs for System Analysis Applications

H. Nonmonetary Screening
- Development of Ranking Methodology for Nonmonetary Factors
  - Environmental Effects
    - Aquatic Biota
    - Terrestrial
    - Wildlife Habitat
    - Cultural Areas
    - Air Pollution
    - Groundwater and Surface Pollution
    - Aesthetic
    - Land Use
    - Noise, Odor and Dust
  - Social Factors
    - Dislocation
    - Employment Changes
  - Contribution to Program Objectives and Goals
  - Energy and Resource Use
    - Energy
    - Land Commitment
    - Chemicals
  - Reliability
    - Frequency of Process Upsets
    - Frequency of Spills
    - Implementation Capabilities
      - Institutional
      - Financial
      - Legal
    - Risk and Functional Sensitivity
      - Acceptability
        - Public
        - Industry

III. SCREEN ALTERNATIVES FOR DETAILED ANALYSIS

A. Select Alternatives for Detail Analysis
B. Utilize Advisory Group
C. Review

(continued)

Table 2: (continued)

IV. DETAIL ANALYSIS OF VIABLE ALTERNATIVES
- A. Development Cost
- B. Capital Cost
- C. Operation Cost
- D. Maintenance Cost
- E. Social Cost

V. DEVELOP FINANCING AND USER COST ESTIMATES

VI. Implementation
- A. Implementation of Selected Alternatives
- B. Management
- C. Industrial Contracts
- D. Summarize Findings
  - ° Review with Advisory Group
- E. Finalize Study
  - ° Public Meeting

Available Options and Selected Scenarios

As indicated in Table 2, a number of solutions are possible in solving the residuals management question. Prior to selecting the approach to be used for a specified location and type of residuals generated and proceeding with an engineering design, serious consideration must be given to the institutional alternatives available. How these alternative fit the specific local situation and conditions is a major factor in developing a successful total waste management approach. The alternatives range from a totally municipal operation to a totally private operation with a range of variations in between.

Basic Alternatives

- *Totally Public* - This option involves an approach where all facilities, equipment, etc. are owned and operated by a public entity — municipality, county, public authority, etc. Financing is through available avenues — municipal bonds, authority bonds, county bonds, etc. — with each industrial user (IU) paying its proportionate share of costs.
- *Totally Private* - This option involves an approach where all facilities, equipment, etc. are owned and operated by a private firm. Financing can be through various programs — conventional loans, small business administration, corporate debentures, etc. Each user — municipal or industrial — pays its proportionate share of treatment/disposal costs.
- *Consortium* - The option involves some combination of public and private ownership and operation of facilities, equipment, etc. A wide range of combinations from public ownership/private operation to private ownership/operation and public control is possible. These combinations need to be evaluated on a case-by-case basis and selected on their specific advantages and disadvantages under actual local conditions.

Figure 2 depicts the array of options available for managing residuals.

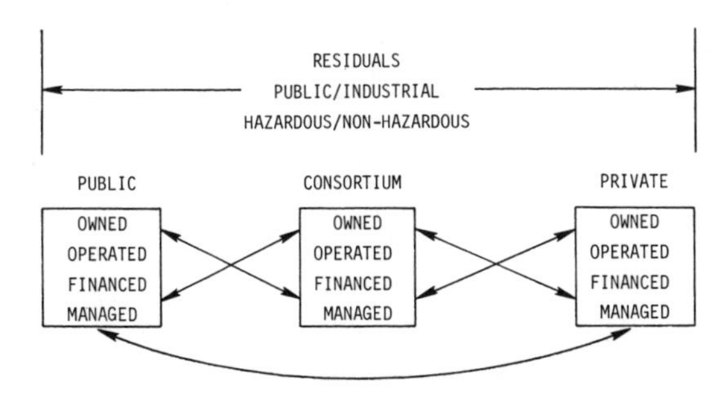

Figure 2: Residuals management options.

The USEPA has produced a guidance document entitled, *Centralized Waste Treatment Alternatives for the Electroplating Industry*, EPA-625/5-81-017. Although this document deals primarily with centralized electroplating and metal finishing wastewater treatment it also details financing options. This information could be applied to the financing questions having to do with total residuals or waste management arrangements as well. The feasibility of establishing a regional industrial wastewater (RWT) facility to serve the 68 industrial plants along the lower reaches of the Monongahela River has been studied; see the report entitled, "The Feasibility of a Regional Industrial Wastewater Treatment Facility", EPA-600/2-80-065.

Additional EPA publications of interest include: "Federal Financial Assistance for Pollution Prevention and Control" and "Loans to Small Businesses under the Federal Water Pollution Control Act". Contact the EPA Economic Assistance Coordinator in the Regional Offices for more information.

SCENARIOS

Four scenarios are presented in this section to illustrate some of the planning recommendations in this manual. The scenarios discussed are based on fictitious municipalities and do not address all the particular problems, solutions, and interactions that may come into play in a real situation. The scenarios are presented with the intent to show the potential benefits — environmentally and economically — to municipalities and their respective industrial users in working together to address the industrial residual problem. There is no intent to develop or promote preconceived ideas and solutions. However, the factors covered in Table 2 should be considered in the planning process. The POTW can play a key role in orchestrating cost-effective solutions and each scenario has been developed with that theme.

This section discusses four different situations: a small town with one major categorical discharger, a small town with one small categorical industry, a city with several categorical and noncategorical dischargers, and a large metropolitan area with a major industrial community.

Scenario #1

POTW Description: Keithville is a small rural community — population approximately 3,400 — located in the southeastern part of the U.S. The existing wastewater treatment facility consists of an Imhoff tank followed by a trickling filter system. Secondary effluent is discharged directly to a small creek. The treatment system has been overloaded for several years and violations of the Town's NPDES permit are frequent. The municipal sludge is dewatered on sand drying beds and disposed of in the county sanitary landfill. The town would like to upgrade the treatment system and land-apply the resulting POTW sludge. The Town has applied to the State and the USEPA for construction grant funding.

Industrial Users: The only industrial user of the POTW is a relatively large non-ferrous metals manufacturing industry. This firm is the major employer in Town and a good working relationship already exists between the plant manager and the Town officials. The POTW regularly receives discharges from the industry that contain heavy metals and oil and grease. Due to the variety of plant operations, the discharge has a varying pH. In addition, batch process tanks are occasionally dumped into the POTW system. The industry is working to comply with applicable Federal pretreatment standards and has asked the Town to work with it. The main question to be resolved is the acceptable — environmentally and economically — disposal of the residuals from the pretreatment processes.

Approach: The Town has elected to develop a local pretreatment program to comply with the General Pretreatment Regulation (40 CFR, 403). The Town manager has asked the Town's consultant to evaluate the impact of the industrial discharge, with pretreatment, on the proposed POTW system and determine if land disposal of the POTW sludge is a viable approach. The industry's consultant and the Town consultant worked together to optimize the cost-effective solution for the industry and the Town. The *Handbook of Industrial Residues* was a basic resource document. Volume 1 provided information on wastewater characteristics, residuals to be generated, and applicable regulations to be addressed in light of available disposal options. Volume 2 provided pretreatment and sludge residual treatment technologies.

Viable Alternatives: The conceptual approach (Table 2) was simplified for this particular situation. The viable alternatives were evaluated and the following decisions/selections were made:

- The POTW will develop and enforce a local pretreatment program.

- The POTW will work with the industry and assist in interfacing with the State.
- The POTW will expand its treatment facility and land-apply the municipal sludge.
- The POTW will charge the industry a user charge for treatment based on actual waste flows, concentrations, etc.
- The industry will install pretreatment, including flow equalization, neutralization, and dewatering.
- The industry will apply to the State for approval of an on-site dewatered sludge land disposal system.
- The industry and the POTW will work together to meet all applicable standards and insure continued compliance.

Scenario #2

POTW Description: Jadfield is a small mid-western town of approximately 8,000 people. The municipal wastewater treatment system consists of a primary clarifier followed by a trickling filter. The facility is being upgraded by the addition of land treatment (slow-rate irrigation) prior to discharge into the local river. The sludge is applied to local farm land. The POTW has experienced periodic upsets in the past which have caused the POTW to violate its NPDES permit. During these upsets, the POTW sludge has had to be disposed of in the sanitary landfill. The state health department is disturbed over this practice. The POTW is developing a local pretreatment program as part of its construction grant responsibilities.

Industrial Users: The POTW accepts discharges from a number of nondomestic sources (meat packing, dairy, etc.) but has only one categorical industry — Diamond Industries, a small (12,000 gallon per day) electroplater. The plater has been in the POTW system for a number of years and has never been required to provide any form of pretreatment. Occasionally, various plating tanks have been batch dumped into the municipal system. The POTW has notified Diamond Industries of the Federal requirements for pretreatment and the Town's responsibility for enforcement. The shop manager is willing to work with the Town, but needs technical assistance and is depending on the POTW manager to help him.

Approach: The Town is committed to develop a viable local pretreatment program and protect the operation of the new municipal wastewater treatment facility. The POTW manager is working with all of the nondomestic dischargers in order to insure a smooth working program. The POTW manager and the shop manager for the plater get together to resolve the problem relative to Diamond Industries' compliance with Federal regulations and the local pretreatment program. The *Handbook of Industrial Residues* is used as a resource document to answer questions as to the regulatory and legal responsibilities and management approach. Volume 1 provides information on the industrial process and the type of waste problems created. Volume 2 identifies treatment and disposal options.

Viable Alternatives: The conceptual approach presented in Table 2 was modified to fit their situation. The following alternatives were analyzed and selected to address the pretreatment needs for Michaelton:

- The POTW pretreatment program will be developed and enforced equitably.
- The land treatment systems will be protected through pretreatment program compliance.
- Initially the industry was considering the elimination of the cadmium line (major contributor of dump problems) because it represented less than 7% of his business and would increase his pretreatment costs. As an alternate to this and overall pretreatment problems, the industry will utilize the Providence Method of pretreating electroplating wastewater. The Providence Method is a basic rinsing system designed to remove the majority of contaminating drag out in a small volume before a flowing rinse which can flow at a fairly high rate to ensure good product quality. By using this method the pollutants in question are separated from the main wastewater flow and are treated or recovered separately.
- The electroplater will install pH control on discharges to the POTW and discontinue batch dumping procedures.
- The POTW sludge will continue to be land applied by local farmers.
- The local pretreatment program and associated ordinance will provide a stable base for future industries that may choose to locate in Jadfield.

Scenario #3

POTW Description: Michaelton is a city of approximately 75,000 people, located in the mid-Atlantic region of the U.S. The municipal wastewater treatment system consists of an activated sludge system followed by mixed-media filtration. The POTW discharges into a river with restrictive water quality standards. The plant is well operated, but has a history of intermittent NPDES permit violations. The sludge treatment system consists of thickening and incineration. Operational problems, due to high strength sidestreams from the sludge system, have caused some permit violations. The City does not have the option of land applying sludges due to high metals content. The existing sanitary landfill is being closed by the State due to hazardous waste dumping by several area waste haulers. The City is concerned and has decided to address waste disposal problems within the local pretreatment program being developed. The POTW manager is progressive and is seriously considering a total waste management approach. The City feels that compliance with the pretreatment program being developed will protect the POTW and insure environmentally acceptable solutions to the waste disposal problem.

Industrial Users: The POTW accepts discharges from several categorical industries — electroplaters, metal finishers, a printing plant, a small pharmaceutical firm and several food-related industries. Residuals currently generated by local industry are collected by local waste haulers. The industrial discharges to the POTW do not seem to have adverse effects on the plant, with the exception of metals contamination of the POTW sludge. A few biological upsets at the POTW were thought to be due to industrial dumping, but the source could not be determined. The industrial community is happy with the labor market, transportation, and other benefits found locally, and is desirous to meet regulations and continue business.

Approach: The City is required by the General Pretreatment Regulations (40 CFR, Part 403) to develop an approvable local pretreatment program. The POTW manager has instructed the City's consultant to conduct an industrial waste survey (IWS) in order to obtain more information on the nondomestic users of the POTW. This information will be useful when using The *Handbook of Industrial Residues*. The POTW manager, with approval from the City fathers, is developing an advisory group to assist in selecting viable alternatives to solve the total waste residual problem. The advisory group is comprised of the POTW manager, three industrial managers, and two representatives from concerned citizen/environmental groups. The technical team utilized the conceptual approach outlined in Table 2 and screened a number of treatment/disposal options. Several options were developed and presented to the advisory group. These options covered the question of both the municipal residuals (sludge/solid waste) and the industrial residuals.

Viable Alternatives: The process approach utilized was reviewed, evaluated, and accepted. A treatment and management option was selected. The final decision of ownership and operation has not been fully decided in regard to the industrial residual option. However, the following direction has been selected:

- All industrial dischargers will meet applicable pretreatment standards.
- The POTW will use a user charge system for recovery of collection and treatment costs for liquid wastes. Rates will be based on actual flows, concentrations, etc.
- The POTW sludge will be acceptable for land application and the City plans to cease incineration of sludges and contract with local farmers for reuse of the municipal sludge.

- The City is working with the State to develop an acceptable sanitary landfill site and will control its use through the new City Ordinance.
- A number of industries have compatible wastes and a form of centralized treatment and dewatering for the metal-type wastes appears to be cost-effective. The location, ownership, management, etc. has not yet been determined. Options available include: joint ownership, city owned and operated, industry owned and city operated, etc.
- The City and industry are working with the State to develop an approved site for a segregated landfill operation which is specific for the metal residuals.
- Collection and hauling of residuals will be handled by local haulers on a competitive bid basis.
- A waste exchange program is being considered with other industries within the geographic region, in order to reduce residuals to be treated.
- The City and the industrial community in conjunction with the advisory group will work together to insure compliance with all regulations, standards, etc.

Scenario #4

POTW Description: Banbury is a midsized northeastern city of approximately 38,000 permanent residents.

The main commercial activities in Banbury center on the State University campus. The treatment plant, which is a twenty year old 4.5 MGD conventional activated sludge system, has experienced moderate odor problems and occasional upsets leading to short term permit limit exceedances. Sludge treatment consists of dewatering and anaerobic digestion. Digested sludge is alternately landfilled or spread on county owned non-agricultural property.

The relatively minor odor problems are intensified because of the plant's proximity to recently developed private apartment housing. A consultant's study of the odor problem concluded that odors generally are intensified simultaneous with treatment upsets. On the basis of the consultant's general knowledge of the industrial makeup of the City, he postulated that slug loads of toxic pollutants originating at the University Hospital or the large commercial laundry serving the University and its adjunct Hospital may be the cause of the observed upsets and odors. In recent years, the odor problems have become a political issue.

Industrial Users: Banbury has no heavy industry and no industries subject to categorical pretreatment standards. The University operates its own wastewater treatment plant serving the campus, but the University Hospital is a discharger to the POTW as is the commercial laundry. Because the plant is less than 5 MGD and there are no categorical industries or problems with sludge disposal, it is a borderline issue whether or not a formal pretreatment program would be required.

Approach: Because the need for a pretreatment program is not well defined in this case, the city fathers contacted the EPA Regional Administrator (RA) for his determination as to the need for a program. In this case, the RA formally exercised his option not to require a program. Nevertheless, because of increasing political pressures associated with plant odors, it was decided to perform a discharge survey at the contributing industries, determine problem pollutants and incorporate limits for those pollutants into its wastewater ordinance.

Viable Alternatives: The approach needed to solve Banbury's industrial waste problem is relatively straightforward. Since odors and occasional slight upsets are the main issues, and were attributed to the Hospital and a laundry by the city's consultant, the following steps have been selected:

- A consultant was engaged to sample at the Hospital and laundry to determine the exact nature of their discharges.
- Based on this sampling program the consultant recommends new, more stringent toxic limits to be incorporated in the ordinance.
- The ordinance is also amended to require periodic self-monitoring by the Hospital and laundry and to provide the right of entry for random spot checks by the POTW.
- The POTW assists the two industries in determining appropriate in-plant controls to eliminate slug loads and generally improve discharge quality to meet the new ordinance limits.

-3-

CATEGORICAL INDUSTRIES

INTRODUCTION

This section contains the available information pertaining to the original 34 categorical industry groups identified by the USEPA to be regulated under promulgated categorical standards. It should be noted that the industrial categorization information contained in this volume is not indicative of changes proposed by the U.S. Environmental Protection Agency. However, all technical information is correct and may be used in making any planning, management or engineering decisions. Proposed changes include: the exclusion of Mechanical Products under paragraph 8(a) of the NRDC Settlement Agreement; the inclusion of the Electroplating Industry under Metal Finishing, and the combination of the Organic Chemicals and the Plastics and Synthetic Materials Industry.

ADHESIVES AND SEALANTS

General Industry Description

Adhesives and sealants are manufactured through the mixing or compounding of various components. This industry can be operated from one-man garage-type to a large industrial complex. The vast majority of adhesive and sealant plants discharge wastewater to a POTW.

Industrial Categorization

This industry has been divided into the following subcategories: animal glue and gelatin, water-based adhesives, solvent-based adhesives with contaminated water, solvent-based adhesives without contaminated water, hot melt thermoplastic adhesives, and dry blend adhesives.

Process Description

Manufacturing processes for all subcategories within the industry are basically the same. Both water base and organic solvent base adhesives are produced by mixing raw materials in mixing tanks under ambient temperatures or heating tank contents with steam. Nonsolvent base adhesives (thermoplastic and dry blend adhesives) are also produced in mixing tanks. Thermoplastic adhesives require heat while dry blends do not. All production processes described above are batch processes. The one exception is animal glue production, which involves hot water applications for the extraction of glue from the raw materials.

Solvents are needed in most adhesives to disperse the binder to a spreadable liquid form. In most wood and paper-bonding adhesives, the solvent is water. In many adhesives based on synthetic resins, rubbers, and natural gums, a variety of organic solvents are required to achieve the necessary solubility and to provide some minimum percentage of base solids. Thermoplastic adhesives and dry blended adhesive materials are composed completely of solids and contain neither water nor solvent-based materials. Polymeric, thermoplastic solids are converted to mobile fluids when subjected to sufficient amounts of heat.

Wastewater Characteristics

Table 3 shows raw wastewater characteristic data. The main source of wastewater for animal glue and gelatin, water-based adhesives, and solvent-based adhesives with contaminated water, is washing of the process vessels and lines.

Table 3: Adhesives and Sealants–Raw Wastewater Characteristics

	Subcategory[a]		
Wastewater parameters[b]	Animal glue and gelatin	Water based adhesives	Solvent based adhesives with contaminated water
BOD	1.2 - 4,800	2.1 - 4,200	13,000
Chromium	10 - 20		
COD	10,000	16,000	22,000
$Nitrogen-NH_3$	16	5 - 20	20
Oil and Grease	400 - 1,500		
pH (units)	9 - 12	9	
TOC	2,600	3.8 - 7,700	4,200
TSS	1.7 - 4,500	2.1 - 4,300	36

[a] There are no pollutants in the wastewater flow for the subcategory "Solvent Based Adhesives without Contaminated Water" and there is no wastewater flow for the subcategories "Hot Melt Thermo-plastic Adhesives" and "Dry Blend Adhesives."

[b] In mg/l, unless otherwise stated.

Control and Treatment Technology

In-Plant Control Technology: Some in-plant controls that are applicable to this industry are: rinse recycle to reduce rinse water volumes, recovery of by-products that can be sold instead of discharged as a waste stream, minimize equipment washouts, and use steam instead of water to reduce wastewater volume.

End-of-Pipe Treatment Technology: Data on end-of-pipe treatment technologies are very limited. Clarification in lagoons will remove 90 percent of the TSS, 60 percent of the oil and grease, and 15 to 20 percent of the BOD and COD. Adding a dissolved air flotation separator increases oil and grease removal but does not improve the other removal efficiencies. The combination of clarification in lagoons followed by ultrafiltration improves COD and oil and grease removal to approximately 70 and 95 percent respectively but does not improve BOD removal. Research is currently being conducted on biological treatment and the use of clarification followed by activated carbon.

Type of Residue Generated

Information on types and characteristics of residues generated by this industry is insufficient and cannot be discussed at this time.

Residual Management Options

Lack of sufficient information on types and characteristics of residues generated does not permit a discussion of specific management options at this time.

ALUMINUM FORMING

Sufficient data is not available on any aspect of this categorical industry.

AUTO AND OTHER LAUNDRIES

General Industry Description

Laundry facilities use a variety of methods to obtain a clean product. With the exception of dry cleaning plants, the industry uses substantial quantities of process water. In 1979, more than 90 percent of all laundries discharged to POTWs. Concentration of pollutants is generally 10 to 20 times more than average domestic wastewaters. It is usually strongly alkaline, highly colored, and contains large quantities of soap or synthetic detergents, soda ash, grease, dirt, and dyes.

Industrial Categorization

This industry has been divided into the following subcategories: industrial laundries; linen supply, power laundries (family and commercial), and diaper service; auto wash establishments; carpet and upholstery cleaning; coin operated laundries, dry cleaning, and laundry and garment service not elsewhere classified; and dry cleaning plants, except rug cleaning.

Process Description

Industrial Laundries: Industrial laundries are located in highly populated areas, and discharge large quantities of high strength wastewater into POTWs. A medium-sized industrial laundry may process between 80,000 to 100,000 pounds of dry wash per week. Articles are subjected to a series of wash and rinse operations to remove oil and grease, and to loosen soil. Some items are dyed and rinsed. Excess water is extracted and the items are dried in a dryer. The wastewater has the appearance of thin oily mud and contains material from towels used by printers, tool and dye makers, filling station attendants, etc. The soil may be in the form of paints, varnishes, lacquer, latex rubber, ketone solvents, inks and dyes. Thus, laundry effluent contains products its customers are using, plus laundry agents, including alkalies, soaps, detergents, bleaches, starches, blueing compounds, fabric softeners, fungicides, petroleum solvents, and enzymes.

Linen Supply, Power Laundries (Family and Commercial), and Diaper Services: This subcategory has the second strongest average waste load. Operations are similar to industrial laundries, except that two sudsing stages are used, with a rinse step between them. In addition, a sour step is utilized in place of the dye step mentioned above. Sour is an acid chemical added at the end of the operation to negate the swelling effect of the alkali. Starch as well as other compounds are added frequently to linen wash loads. Waste characteristics from this subcategory are similar to industrial laundries, except that the strength of the waste is usually lower.

Auto Wash Establishments

Tunnel-Type — The vehicle is pulled through a "tunnel" type area past different operating stations. Operation is generally fully automatic, with operations such as interior cleaning, wiping, and drying performed manually.

Bay-Type — In this coin-operated type of auto wash, the customer parks his car in a bay area, and a wand-type water spray unit is used to soap and rinse the vehicle.

Wastes from both types of auto wash systems contain high amounts of total solids, suspended solids, oil, grease, and BOD.

Carpet and Upholstery Cleaning: In a carpet cleaning operation, the carpet is first beaten to remove dust and dry solids and then wetted with water and a mild, diluted detergent. The carpet then passes through a system of either rollers or brushes which work the detergent into the fiber. A clean water rinse follows. Excess water is squeezed out and the carpet is air dried.

Upholstery cleaning is basically a dry process and no wastewater is produced.

Coin-Operated Laundries, Dry Cleaning Facilities, and Laundry and Garment Services Not Elsewhere Classified: Most coin-operated laundries contain between 25 and 35 machines, each of which uses 25 to 30 gallons of water per washing cycle. An average weekly wastewater volume of 50,000 gallons can be expected from such an operation. Approximately 100 pounds of commercial detergent would be used per week.

Coin-operated dry cleaning is a solvent cleaning process with no process wastewater discharge.

Laundry and garment services not elsewhere classified include Chinese and French hand laundries, facilities where clothes are altered and repaired, and pillow cleaning operations. Since their effluent is small in both volume and contaminant levels, these operations have not been included in this summary.

As a group, the effluent of industries in this subcategory is weaker than domestic sewage and is more easily treated by municipal wastewater treatment plants.

Dry Cleaning Plants Except Rug Cleaning: Perchloroethane is used to remove the dirt from the fabric and then the solvent is recovered and recycled by a filter system.

Soil extracted from the cleaned materials should be disposed of by a scavenger. Wastewaters are not usually generated.

Wastewater Characteristics

Table 4 shows raw wastewater characteristic data.

Control and Treatment Technology

In-Plant Control Technology: Industrial laundries using dry cleaning methods before washing can reduce the oil and grease content of wastewater by 80 to 85 percent.

Thirty percent of auto laundries recycle wash and/or rinse water with varying degrees of treatment. Washwater can be recycled after settling out solids. Rinse waters have higher purity requirements and can be treated with a germicide and de-emulsifier, then clarified and screened prior to recycling.

Table 4: Auto and Other Laundries–Raw Wastewater Characteristics

Wastewater parameters[b]	Subcategory[a]: Industrial laundries	Linen supply only	Auto washes: Tunnel type	Auto washes: Bay type	Coin operated laundromats only
Conventional					
BOD	650 - 1,300	100 - 800	30 - 80	15 - 170	120 - 250
COD		2,100 - 5,100	150 - 275		65 - 1,400
Oil and Grease	400 - 3,700	100 - 1,200	0 - 0.3	40 - 200	
pH (units)	11 - 13	10.3 - 11.2	8.7 - 9.1		5.1 - 10
TDS	1,500 - 6,500	1,700 - 2,000	570 - 1,700	630 - 2,500	100 - 2,000
TSS	650 - 5,000	500 - 1,500	160 - 230	95 - 850	15 - 800
Toxic					
Cadmium	0 - 0.6	0.04	0 - 0.04		
Chromium	1 - 4	0.06	0 - 1		
Copper	0.2 - 9	0.3	0 - 0.3		
Iron	3 - 125		3 - 4		
Lead	3 - 36	0.7	0 - 1		
Mercury	0.001 - 0.007				
Nickel	1 - 2.5	2	0 - 0.7		
Zinc	0.5 - 9	0.5	0.3 - 0.4		

[a] There is no data at this time for the subcategory "Carpet and Upholstery Cleaning" and there is no wastewater flow for the subcategory "Dry Cleaning, Except Rug Cleaning."

[b] In mg/l, unless otherwise stated.

End-of-Pipe Treatment Technology: Clarification is used by a large number of car washes for solids (approximately 50 percent TSS removal) and oil and grease (approximately 25 percent) removal. Addition of sand filtration improves removals into the 60 to 70 percent range. Other laundries have achieved 70 to 80 percent removal of conventional pollutants and heavy metals using chemical coagulation followed by dissolved air flotation separation. Particulate removal is increased to over 90 percent if the discharge from the separator passes through a diatomaceous earth filter. Ultrafiltration using tubular modules (spiral-wound units were not found feasible) has also shown 60 to 80 percent removal with chemical pretreatment. Other technologies which have been tried but not found economically feasible or not applicable to the pollutants involved are reverse osmosis, foam separation, evaporation, and carbon adsorption.

Type of Residue Generated

Information on the types and characteristics of residues generated by this industry is insufficient and cannot be discussed at this time.

Residual Management Options

Lack of sufficient information on types and characteristics of residues generated does not permit a discussion of specific management options at this time.

BATTERY MANUFACTURING

Sufficient data is not available on any aspect of this categorical industry.

COAL MINING

General Industry Description

Coals are classified into several ranks or types related to chemical composition and physical characteristics. Standard classes are anthracite, bituminous, subbituminous, and lignite. All ranks of coal are mined by the coal industry, which can be divided into the following two segments: (1) the production of anthracite; and (2) the production of bituminous coal, subbituminous coal, and lignite. The industry can also be divided by production processes into coal mining and coal services (coal cleaning and coal preparation).

Once historically significant in the economic and industrial growth of the United States, the importance of anthracite coal as an energy source in this nation has been declining in recent years. Consequently, the anthracite industry is currently under consideration by the USEPA for exemption from BATEA regulations.

Mining of bituminous coal and lignite constitutes the major portion of the coal mining industry. U.S. Geological Survey estimates indicate that bituminous coal and lignite currently comprise over 99 percent of the nation's total coal reserves.

According to the Bureau of Mines, there were 6,168 active bituminous coal and lignite mines in the industry in 1975. The majority of the mines were small operations, with individual production of less than 100,000 tons per year. Although these small mines comprised over 80 percent of the active facilities in 1975, they accounted for less than 20 percent of the bituminous coal production. Large mines producing greater than 100,000 tons per year represented less than 20 percent of the facilities, but produced almost 81 percent of the coal. The current trend has been toward larger mines and consolidation of mining companies.

The coal mining industry currently operates in 25 states located in Appalachia, the Midwest, and the Mountain and Pacific regions. The six leading coal producing states in 1975 were, in order of output: Kentucky, West Virgnia, Pennsylvania, Illinois, Ohio, and Virginia. Production in these states accounted for 74 percent of the total United States output.

Industrial Categorization

For the purpose of establishing pretreatment guidelines, the coal mining industry has been divided into three subcategories: acid or ferruginous mines, alkaline mines, and coal preparation plants and associated areas.

Process Description

Acid or Ferruginous Mines: Acid mine drainage is generated under natural conditions when pyritic coal seams are mined. Pyrites (iron sulfides) contained in coal and associated strata are exposed to the atmosphere during the mining process. In the presence of oxygen, water, and certain species of oxidizing bacteria *(Thiobacillus ferroxidans and Ferrobacillus ferroxidans)*, these sulfides oxidize to ferrous sulfate, forming an acidic, ferruginous leachate.

Mines that are potentially acid make up approximately 50 percent of all surface mines in the United States. The majority of the potential acid mines are located in Maryland, Ohio, Pennsylvania, and northern West Virginia.

Drainage from acid mines presents the most serious threat to the environment from the coal mining category. The acidic nature of this wastewater creates a strong solvent for many metals and minerals, and the data base confirms the elevated levels of many classical pollutants and heavy metals derived from both the coal seams and the associated strata.

Alkaline Mines: Alkaline mine drainage can be generated under natural conditions similar to those found in mines with acid drainage. Iron sulfides, however, are transformed into ferrous bicarbonates, and an alkaline iron-bearing water is produced. Additionally, there are large areas of coal reserves where the naturally occurring associated groundwaters are alkaline. Coal in these areas is usually lower in pyritic sulfur, and the resulting mine drainages do not develop the low pH characteristic of acid mine drainage.

Most bituminous and lignite coal mines are located in areas where the potential for the formation of acid mine drainage does not exist. Approximately 50 percent of the surface mines and 70 percent of the underground mines are in this category. Heavy metals and other toxic pollutants are seldom present in elevated concentrations in alkaline mine drainage, but alkaline wastewaters may be high in suspended solids.

Coal Preparation Plants and Associated: Physical coal cleaning processes used today are oriented toward product standardization and reduction ash, with increasing attention being placed on sulfur reduction. Coal preparation in commercial practice is currently limited to physical processes. In a modern coal cleaning plant, the coal is typically subjected to: size reduction and screening, gravity separation of coal from its impurities, dewatering, and drying.

The commercial practice of coal cleaning is currently limited to separation of the impurities based on differences in the specific gravity of coal constituents (i.e., gravity separation process) and on the differences in surface properties of the coal and its mineral matter (i.e., froth flotation).

Coal preparation can be classified into five general categories. The first three, crushing and drying, coarse size coal beneficiation, and coarse and medium size coal beneficiation are generally used in the preparation of steam coal. Coarse, medium, and fine size coal beneficiation is used for metallurgical grade coal. The last category, "deep cleaning" coal beneficiation, has not yet been commercially demonstrated in the United States. The five categories are described below.

Crushing and Drying — These plants use rotary breakers, crushers, and screens for size control and for removal of coarse refuse. No washing is done and the entire process is dry. Since most removal of pyritic sulfur is accomplished by hydraulic separation, this level of cleaning is inefficient for reducing sulfur levels.

Coarse Size Coal Beneficiation — These coal cleaning plants, in addition to crushing and screening raw coal, also perform wet beneficiation of the coarse material. The fine material is mixed with the coarse product without washing. A finer sizing of the coal is accomplished than in the crushing and drying process. This system provides removal of only coarse pyritic sulfur material and is therefore recommended for a moderate pyritic sulfur content coal.

Coarse and Medium Size Coal Beneficiation — These coal cleaning plants are basically an extension of the coarse size coal beneficiation plants. Coal is crushed and separated into three size fractions by wet screening. The coarse material is cleaned in a coarse coal circuit. Medium fractions are beneficiated by hydrocyclones, concentrating tables, or dense medium cyclones. Fine coal is dewatered and shipped with the clean coal or discarded as refuse. However, the level of beneficiation is not substantially greater than that of coarse size coal beneficiation with respect to sulfur removal, and this system is recommended for use on low and medium sulfur coals which are relatively easy to wash. This process provides rejection of free pyrite and ash, as well as enhancement of energy content.

Coarse, Medium, and Fine Size Coal Beneficiation — In these plants, coal is crushed and separated into three or more size fractions by wet screening. All size fractions are beneficiated. Heavy media processes are used for cleaning coarse and medium size fractions. Froth flotation processes or hydrocyclone processes are used for cleaning fine particles. These coal preparation systems provide high efficiency cleaning of coarse and fine coal fractions with lower efficiency cleaning of the ultrafines. This method accomplishes free pyrite rejection and improvement of BTU content.

"Deep Cleaning" Coal Beneficiation — These plants would be essentially the same as the previous ones in which one size fraction is rigorously cleaned to meet a low sulfur-low ash product specification. Two or three coal products are produced to various market specifications. This level also uses a fine coal recovery circuit to increase total plant recovery.

Wastewater Characteristics

Coal Mining: No wastewater is purposely generated in the extractive portion of coal mining because water is almost always a hindrance and an extra expense to pump and treat. A minor exception is the use of water for dust suppression and equipment cooling. Water enters coal mines via precipitation, groundwater infiltration, and surface runoff, and it can become polluted by contact with materials in the coal, overburden, or mine bottom. Most water entering underground mines passes through the mine roof from overlying strata (rock units). These rock units generally have well developed joint systems, which tend to cause vertical flow. Chemicals used in mining and repair of mining machinery may also be wastewater pollutants. Mine water is therefore considered a wastewater for the mining segment of the coal industry. Table 5 shows raw wastewater characteristic data for acid and alkaline mine wastewaters.

Table 5: Coal Mining–Raw Wastewater Characteristics[a]

	Subcategories		
Parameter	Acid or Ferruginous Mine A	Alkaline Mine B	Coal Prep Plants C
Conventional - mg/l			
COD	62.7	14.3 - 90.7	20,700 - 48,800
pH - units	4.4	6.6 - 8.2	6.6 - 7.3
SS	1.4	< 0.1 - 0.33	56.3 - 247
TSS	134	44.8 - 110.6	9,100 - 34,400
Inorganics - ug/l			
Antimony	0.034	< 0.002 - 0.006	0.002 - 0.034
Arsenic	0.028	< 0.002	0.051 - 0.253
Beryllium	0.009	< 0.001	< 0.010 - 0.057
Cadmium	0.010	0.002	<0.020
Copper	0.043	< 0.006	0.233 - 1.33
Cyanide	< 0.005	< 0.005	<0.005
Lead	0.100	0.020	0.400 - 0.967
Mercury	Not Tested	Not Tested	Not Tested
Nickel	1.000	< 0.005	0.217 - 1.23
Selenium	0.002	< 0.002 -< 0.005	< 0.003 - 0.034
Silver	0.003	< 0.005 - 0.01	< 0.002 - 0.005
Thallium	0.008	< 0.005	< 0.005 - 0.015
Zinc	2.00	< 0.060	< 0.600 - 5.33
Organics - ug/l			
Aromatics			
Benzene			ND - 15
1,2-Dichlorobenzene		ND[b]-< 10	
1,4-Dichlorobenzene		ND -< 10	
2,4-Dinitrotoluene			ND - 6
Ethylbenzene			ND -< 6.7
Nitrobenzene			ND -< 7
Toluene			ND -< 8

(continued)

Table 5: (continued)

Parameter	Subcategories		
	Acid or Ferruginous Mine A	Alkaline Mine B	Coal Prep Plants C
Hydrogenated alphatics			
Carbon tetrachloride		ND - < 10	
Chloroform			ND -< 6.7
1,2-Dichloroethane		ND - < 10	
Methylene chloride		ND - < 10	ND - 82
1,1,1-Trichloroethane		ND - < 10	ND - 7.67
Trichloroethylene			ND - < 10
Nitrogen compounds			
1,2-Diphenylhydrazine			ND - < 3.3
N-nitrosodiphenylamine			ND - 30
Pesticides and metabolites			
Gamma - BHC		ND - <10	
Alpha - endosulfan			ND - < 6.7
Beta - endosulfan			ND - < 6.7
Isophorone			ND - 307
Phthalates			
Bis(2-ethylhexyl) phthalate		ND -< 10	< 10 - 50
Butyl benzyl phthalate		ND -< 10	< 3.3 -< 10
Di-n-butyl phthalate	< 10	<10	< 3.3 -< 10
Diethyl phthalate	< 10	ND - < 10	< 3.3 -< 10
Dimethyl phthalate			ND -< 3.3
Di-n-octyl phthalate			ND -< 3.3
Phenols			
2-Chlorophenol			ND - 36
4-Chlorophenyl phenyl ether			ND - 3.3
2,4-Dimethylphenol			ND - 22
2-Nitrophenol			ND - 19
Phenol		ND - <10	ND - < 10
4,6-Dinitro-o-cresol			ND - 194
Polycyclic aromatic hydrocarbons			
Acenaphthene			ND - < 10
Acenaphthylene			ND - 8
Anthracene		ND - < 10	<10 - 132
Benzo(a)anthracene			6 - 10.3
Benzo(ghi)perylene		ND - < 10	<3.3 - 12
Benzo(k)fluoranthene			<3.3 - 12
2-Chloronaphthalene			ND -< 3.3
Chrysene			ND - 29
Dibenz(ah)anthracene		ND - < 10	ND -< 3.3
Fluoranthene			< 6.7 - 16
Fluorene			< 6.7 - 42
Indeno(1,2,3-cd)pyrene		ND - < 10	ND - < 6.7
Naphthalene			ND - 402
Pyrene			< 3.3 - 19

[a] Majority of numbers represent data from one location.

[b] ND - not detectable.

Coal Preparation: Wastewater from coal preparation emanates from two different sources: process-generated wastewater, and wastewater from associated areas which include coal preparation plant yards, immediate access roads, slurry ponds, drainage ponds, coal refuse piles, and coal storage piles and facilities.

Since cleaning techniques generally require an alkaline medium for efficient and economic operation, process water does not dissolve appreciable quantities of the metallic minerals present in raw coal. On the other hand, some minerals and salts such as chlorides and sulfates of the alkalies and the alkaline earth metals found in raw coal dissolve easily in water.

The liquid discharges from coal preparation plants are often combined with discharges of the associated storage piles, refuse areas, and plant areas prior to final effluent treatment. Wastewater from these areas is characterized as being similar to the raw mine drainage at the mine being served by the preparation plant. Consequently, some refuse piles produce an acid leachate and others produce an alkaline leachate. The origin of the acid leachate is the same as that for acid mine drainage. Table 3 shows raw wastewater characteristic data for coal preparation wastewaters.

Control and Treatment Technology

In-Plant Control Technology: Wastewater from this industry is normally the result of precipitation, groundwater infiltration, and surface runoff. Engineering design (planning) to prevent water contamination is the most beneficial technology. Popular planning controls include chemical grouting within mines to prevent leakage, compaction of coal refuse piles to prevent water from flowing through material, reestablishing cover growth on the refuse piles, and sealing off old abandoned mine entrances.

End-of-Pipe Treatment Technology: Simple clarification of slurries from coal preparation plants will remove over 99 percent of the conventional pollutants and over 80 percent of the important toxic pollutants. Acid mine drainage or runoff can be successfully treated by neutralizing with lime, aeration to oxidize the iron, and clarification at low overflow rates (under 300 gallons per day per square foot). Pilot plant work has also been done on using membrane technologies, carbon adsorption, and starch xanthate.

Type of Residue Generated

Residues generated from coal mining operations are closely related to the type of material found in the mine, both the coal itself, the waste material and overburden, and the groundwater. Wastewaters from mines are essentially rainwater, surface water, or groundwater that has flowed or percolated into the mine and which must be pumped out in order to work the mine. In flowing to, into, and through the mine it picks up sediment and dissolved material by physical, chemical, and biological processes such as erosion, dissolution, and precipitation. It also picks up materials deposited during the mining operation, such as chemicals, grouts, lubricants, and dust.

Residues from the coal processing operations that follow mining are physically and chemically similar to those from the mines they serve.

Treatment of wastewaters from "acid" mines normally involves neutralization with lime followed by clarification. The sludge developed is gelatinous, less than one percent solids and difficult to dewater. The gelatinous sludge has a great capacity for adsorbing other suspended, colloidal and dissolved contaminants. Newer treatment schemes utilizing recirculation of settled sludge can produce sludges with a solids content of over 20 percent.

Treatment of wastewaters from "alkaline" mines normally involves only clarification. The sludge developed is composed mostly of fines from the coal and overburden. Again, there is a substantial tendency for heavy metals and organics to adsorb on the sludge particles.

Treatment of wastewaters from the coal processing operations following mining, such as washing, screening, and storage pile runoff, tends to produce residues similar to those from treatment of the wastewaters of the mine served.

Residual Management Options

The most difficult residue to manage in coal mining wastewater treatment is the thin, gelatinous sludge from acid mine effluent treatment. Several newer techniques are available for dewatering this thin sludge directly using vacuum filters, centrifuges, filter presses, or sand beds, but their usefulness seems very dependent on the specific characteristics of the particular sludge.

Ferromagnetic densification, freezing, and carbon dioxide addition as a pretreatment step also have some potential, but high cost and energy input may preclude their use. Also, as mentioned earlier, changes in the wastewater treatment sequence itself can materially improve the dewaterability of the acid mine treatment sludge.

Historically, the prevalent method for managing acid mine wastewater treatment sludge has been lagooning. Because little release of water occurs during storage in such lagoons, decanting of excess water is not generally practiced. However, some loss of water does occur by evaporation and percolation. Where sufficient space is available to accommodate all the sludge generated during the entire life of the mine, the lagoon may be simply filled and abandoned. More commonly, the sludge is periodically removed from the lagoons. Reinjection into unused parts of the mine is a common method of ultimate disposal. For surface mines, coverage of the deposited sludge with waste overburden to shield it from rainwater is a frequent practice.

Residues from alkaline mine effluent treatment are more easily dewatered, and because of the lesser volume, placement in worked out portions of the mine is a frequent method of ultimate disposal.

It should be noted that wherever the mine wastewater sludges are placed for ultimate disposal, long-term release of the contaminants is the matter of prime concern. The suitability of the site in terms of surface drainage and erosion potential, access of groundwater to the site, its chemical characteristics that might induce leaching of toxic materials and heavy metals, and the potential entrance of leachate into streams or drinking water aquifers should be thoroughly studied beforehand. Where lagoons are to be used for long-term or permanent storage of sludge, the enclosing dikes must be rigorously designed for stability, water tightness, and minimization of erosion.

There appears to be no practical use for coal mine wastewater treatment sludge, nor any practical possibility for by-product recovery.

COIL COATING

General Industry Description

Coil coating is the process of applying protective coatings and paints to strips of sheet metal which are rolled into "coils" for ease of handling. The coated coils are used to produce window frames, door frames, cans for food processing, exterior siding, and many other products.

An estimated 43 billion square feet of painted coil is produced annually by approximately 70 plants in the United States, with a wastewater production of approximately 9 MGD. Over 50 percent is discharged to POTWs.

Industrial Categorization

For the purpose of establishing pretreatment guidelines, the coil coating industry has been divided into three subcategories: cold rolled steel, galvanized steel, and aluminum or aluminized steel.

Process Description

A coil is a long thin strip of metal rolled up in a manner similar to a roll of household aluminum foil. The metal strips range in thickness from 0.25 to 1.25 mm. Rollers uncoil the metal and convey the strips through cleaners, sprayers, and paint rollers. Figure 3 shows a typical process sequence diagram.

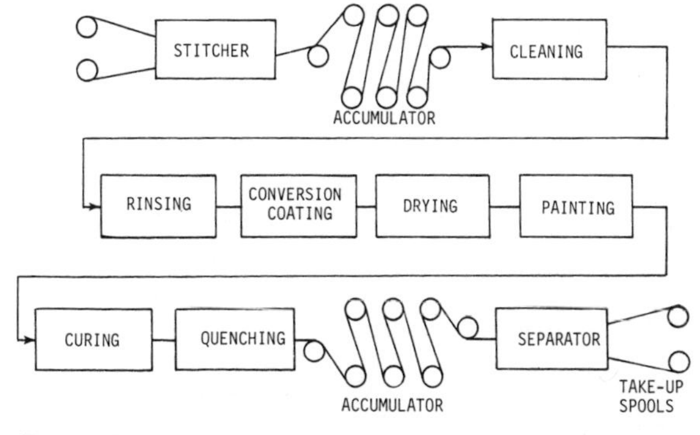

Figure 3: General process sequence for coil coating.

There are three main pollutant producing functions in the coil coating process: cleaning, conversion coating, and painting. A description of each process follows.

Cleaning: This process removes rust and oxide, as well as dirt, grease, and oil before application of coatings. Coil cleaning utilizes a wide variety of acid and alkaline cleaning agents which may appear in the wastewaters. Among these are sodium carbonate, trisodium phosphate, tetrasodium phosphate, sodium tripolyphosphate, sodium hydroxide, sodium metasilicate, and synthetic detergents such as alkyl aryl sodium sulfonate.

Conversion Coating: Conversion coating provides a corrosion resistant film that is chemically and physically bonded to the base metal. It also provides a smooth and chemically inert surface for subsequent application of a variety of paint films. Conversion coatings normally used include: phosphate coatings, chromate coatings, complex oxide coatings, and non-rinse coatings.

Phosphate conversion coatings employ phosphates of iron and zinc, plus a sealing rinse of phosphoric acid.

Chromate conversion coatings can be applied to aluminum or galvanized surfaces although the most common use is with aluminum coils. The process utilizes a chromium salt such as sodium chromate, plus a strong acid such as hydrofluoric or nitric acid to form a film of hexavalent chromium on the surface of the metal.

Complex oxide conversion coatings are principally applied to galvanized surfaces. The complex oxide forms in a basic solution and employs a sealing rinse with a much greater concentration of trivalent and hexavalent chromium ions than sealing rinses for other processes.

No-rinse conversion coating is a recent development in the industry which eliminates the seal rinse step. This process has the potential to minimize the discharge of chromium ions in the wastewater.

Painting: Roll coating with paint is the final step in coil coating. The process uses paints characterized by high pigmentation, adhesion, and flexibility. A prime consideration in the painting stage is the solvent employed to control the viscosity of the paint which may appear in the waste stream.

Wastewater Characteristics

Table 6 shows raw wastewater characteristic data. Concentration of pollutants for the industry can vary widely within a subcategory depending upon the product and in-house process controls.

Control and Treatment Technology

In-Plant Control Technology: Process controls which have been effectively utilized in the coil coating industry include:

Table 6: Coil Coating–Raw Wastewater Characteristics

Wastewater parameter[a]	Subcategory					
	Cold rolled steel		Galvanized steel		Aluminum or aluminized steel	
	mg/l	mg/m²[b]	mg/l	mg/m²	mg/l	mg/m²
Conventional						
Fluorides	3.58	8.34	2.12	9.10	21.00	29.93
Oil and Grease	341	655.17	2.97	150	57.56	33.11
pH (units)	6.15 - 11.45		3.45 - 11.10		2.5 - 11.15	
Phosphorus	42.87	66	14.76	34.16	7.00	8.59
Temperature °C	32.19		34.29		31.0	
Total Dissolved Solids	1,665	10,240	428	1,306	1,130	1,373
Total Suspended Solids	152	669	114.05	404	84.88	120
Metals and Inorganics						
Aluminum	0.61	0.896	1.74	2.94	112	160
Antimony	3.00	16.13				
Arsenic	0.075	0.114				
Barium	0.53	0.798				
Boron	0.25	0.373			9.0	7.92
Cadmium			0.045	0.039		
Calcium	43.1	96				
Chromium (Hexavalent)	4.36	32.55	9.35	29.77	13.1	36.91
Chromium (Total)	6.87	31.13	57.60	60.87	43.5	43.37
Cobalt	0.31	0.491				
Copper	0.051	0.100			0.04	0.03
Cyanide, Amenable to Chlorination	0.016	0.112	0.032	0.154	0.17	0.56
Cyanide (Total)	0.012	0.106	0.082	0.302	0.57	2.30
Iron	10.15	23.31	2.83	4.77	3.45	3.39
Lead	0.14	0.310	0.42	0.91	0.12	0.13
Magnesium	12.9	28.03			20.0	17.60
Manganese	0.53	1.387	0.12	0.19	0.37	0.465
Molybdenum	0.038	0.077			0.05	0.04
Nickel	0.39	0.324	0.39	0.33	0.03	0.06
Silver	0.02	0.030				
Sodium	403.5	1577			170	149
Strontium	0.33	0.495				
Tin	0.32	0.495			0.06	0.05
Titanium	0.042	0.064				
Vanadium	0.031	0.047				
Yttrium	0.021	0.032				
Zinc	7.56	28	25.48	81.82	0.20	0.17

(continued)

Table 6: (continued)

Wastewater parameter[a]	Subcategory					
	Cold rolled steel		Galvanized steel		Aluminum or aluminized steel	
	mg/l	mg/m²[b]	mg/l	mg/m²	mg/l	mg/m²
Organics						
Anthracene	0.064	0.097				
1,2-Benzanthracene	0.056	0.044				
3,4 Benzofluoranthene	0.035	0.023				
Benzo (k) Fluoranthene	0.035	0.023				
Bis (2-Theylhexyl) Phthalate	0.035	0.050	0.30	0.177	0.014	0.047
Butylbenzyl Phthalate	0.15	0.300				
Chrysene	0.023	0.040				
1,1-Dichloroethane	0.018	0.034				
1,1-Dichloroethylene			0.015	0.016		
1,2,3-Dichloroethylene					0.042	0.037
Diethyl Phthalate	0.056	0.158	0.048	0.174	0.055	0.006
2,4-Dimethylphenol	0.021	0.032				
Di-N-Octylphthalate	0.029	0.031				
Fluoranthene	0.040	0.036				
Fluorene	0.03	0.101				
Isophorone	0.600	0.909				
Phenanthrene	0.05	0.097				
Phenols (Total)	0.02	0.117			0.03	0.06
Pyrene	0.016	0.024				
1,1,1-Trichloroethane			0.014	0.067		

[a] In mg/l, unless otherwise stated.

[b] mg/m^2 = milligrams per square meter of coated area.

- Use of optimum alkaline cleaning compound for the basic material and maintenance of optimum solution and rinse temperatures to increase efficiency and minimize wastewater generation.
- Use of more efficient rinse techniques to reduce wastewater volumes.
- Provision of sufficient squeegee action to prevent excessive dragout of alkaline cleaning solution.
- Installation of automatic process controls to decrease water usage and wastewater generation.
- Recycle of rinse water as makeup for the alkaline cleaning tank for up to three stages.
- Recovery of chromating conversion coating chemicals.
- Recycle of quench water through cooling towers.
- Substitution of chromating solutions which do not contain cyanide, thus eliminating need for separate treatment for cyanide removal.
- Use of sealing rinses which contain no zinc.
- Use of the "no-rinse" process for conversion coating, thus reducing wastewater discharge.

End-of-Pipe Treatment Technology: End-of-pipe treatment is practiced by over 80 percent of the coil coating industry. Treatment technologies commonly used are neutralization, clarification, equalization, oil skimming, biological treatment (lagooning is the most common with over ten percent of the industry utilizing) and chemical reduction (over fifty percent of the industry utilize chemical reduction). Other less commonly used technologies are coagulation-precipitation, vacuum and pressure filtration, oxidation, and ion exchange.

Table 7 shows removal efficiencies for several treatment technologies.

Type of Residue Generated

Information on the types and characteristics of residues generated by this industry is insufficient and cannot be discussed at this time.

Residual Management Options

Lack of sufficient information on types and characteristics of residues generated does not permit discussion of specific management options at this time.

COPPER FORMING

Sufficient data is not available on any aspect of this categorical industry.

Table 7: Coil Coating–Percent Removal Efficiencies of Selected End-of-Pipe Treatment Technologies for Conventional and Toxic Pollutants

	Treatment Technologies					
Wastewater parameter	Dissolved air flotation	Electrochemical chromium reduction	Chemical precipitation	Peat adsorption	Sedimentation	Membrane filtration
Conventional						
Oil and grease	72-100				5-99	
Phosphorus			90			
TSS					51-97	38-100
Toxic						
Aluminum					23-100	
Antimony				64		
Arsenic			>99			
Cadmium			99			
Chromium (Hexavalent)		>99		>99		98-100
Chromium (Trivalent)			99			
Chromium (Total)					0-100	>99
Copper			99	>99	0-86	97-100
Cyanide			50-99			0
Fluoride			97			
Iron			99		49-100	>99
Lead			96	>99	73-100	96-98
Mercury			>99	98		
Nickel			>99	97	20	>99
Silver			99	95		
Zinc		97-100	>99	83	0-100	98-100

ELECTRICAL PRODUCTS

Sufficient data is not available on any aspect of this categorical industry.

ELECTROPLATING

General Industry Description

The electroplating category consists of facilities engaged in the electroplating of common and precious metals, anodizing, coating, chemical etching and milling, electroless plating, and printed board manufacture. In electroplating operations, a surface coating is applied, generally by electro-deposition, in order to provide corrosion resistance, improved wear and erosion resistance, anti-friction characteristics, or improved appearance.

Electroless plating and anodizing accomplish similar end results by somewhat different processes. Chemical etching and milling remove specific amounts of metal by chemical dissolution of the basis metal. Printed board manufacture produces solid state printed circuits.

Industrial Categorization

For purposes of wastewater characterization and organization of pretreatment information, the electroplating industry has been divided into seven subcategories: electroplating of common metals, electroplating of precious metals, anodizing, coatings, chemical etching and milling, electroless plating, and printed circuit board manufacture.

Process Description

Electroplating of Common and Precious Metals: In both subcategories, metal ions in acid, alkaline, or neutral solutions are reduced on the workpiece being plated, which is made to act as a cathodic surface by the application of electric current.

Surface preparation is done to remove oil, grease, and dirt which may interfere with the plating processes, and includes cleaning, descaling, and degreasing. A number of different cleaning schemes are commonly utilized by plating facilities. Solvents, alkaline solutions, emulsions, ultrasonic energy, and acid baths are commonly used. A typical surface preparation scheme is shown in Figure 4.

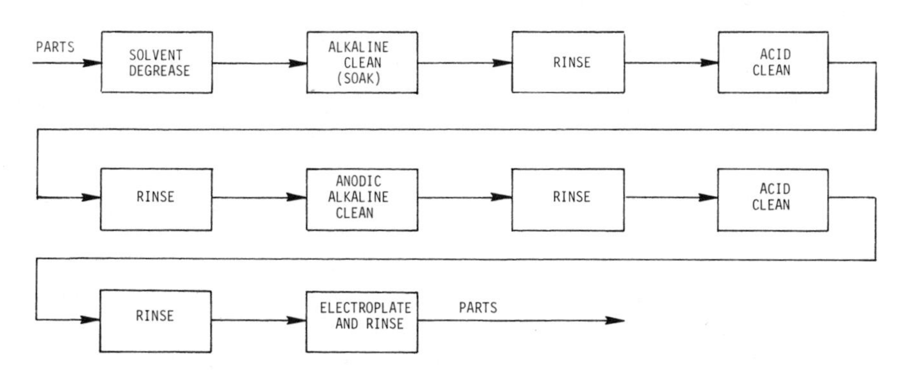

Figure 4: Typical surface preparation scheme for electroplating common and precious metals.

The plating operation is very similar for both subcategories, differing only in the metals and chemical bath used. The two most frequently used plating methods are barrel plating and rack plating. In both cases, a current passing through both solution and workpiece causes plating to occur.

Although the differences in plating schemes are not extremely significant in their impact on process operation, different metal combinations and baths can impact strongly on the types of wastewaters generated. Brief discussions of the more important plating metals and baths used are given below.

Aluminum — Aluminum is mostly used in the plating of uranium and strip steel. A hydride bath must be used, usually containing diethyl ether, aluminum chloride, and lithium aluminum.

Cadmium — Cadmium is generally used to plate iron and steel for corrosion protection; cadmium cyanide is the most widely used bath.

Chromium — Chromium is used to plate steel, nickel-plated steel, and nickel-plated zinc. Chromic acid bath containing silicate or fluoride is used. Chromium plating is unusual in that insoluble lead anodes are used.

Copper — This material is plated onto steel, plastic, and zinc alloy using cyanide, acid sulfate, pyrophosphate, and fluoborate baths. Cyanide bath is the most common, with acid sulfate having increased usage.

Gold — Gold is used for decorative and electrical applications. It is usually plated using a cyanide bath, with the pH of the bath varying with the intended application.

Indium — Indium is utilized for the hardening of bearing surfaces, particularly in the aircraft industry. It is often alloyed before plating; baths used include sulfate, alkalines, fluoborate, sulfamate, chloride, perchlorate, and tartrate baths.

Iron — Plating applications include electroforming and wearing surface rebuilding, as well as soldering equipment plating. Iron is usually plated from low pH ferrous salt baths containing sulfate, chloride, fluoborate, and sulfamate.

Lead — Because of its resistance to acids and corrosives, lead is frequently used to plate the inside of vessels, battery parts, and nuts and bolts. It is plated from fluosilicate and fluoborate baths, with the former generally producing superior results.

Nickel — Plated from Watts (a sulfate-chloride-boric acid bath), sulfamate chloride, and fluoborate baths. The appropriate nickel salt is used, along with a buffer and a wetting agent.

Platinum — Platinum is used primarily in decorative jewelry applications, although limited use occurs in electronics and optical products. Baths used include phosphate, sulfate, chloride, bromide, and molten cyanide.

Silver — Silver is widely used in both decorative and engineering fields. The bath used depends on application, with potassium formulations the most widely used.

Tin and Zinc — Tin and zinc are used to prevent corrosion, and in the case of tin, for its ability to solder. Tin is usually plated from baths of alkaline stannate, acid fluoborate, and/or acid sulfate. Zinc is used with cyanide solutions, alkaline baths using chelating agents such as tetrasodium pyrophosphate, acid or neutral chloride baths, or acid sulfate solutions.

After electroplating, an additional coating is sometimes required. These coatings serve to improve the metal surface for painting, increase lubricity, corrosion protection, or bring about desired color changes. Electroplating baths have extremely long useful lives, hence most wastewater generated by electroplating results from surface preparation, rinses, and post treatments.

Anodizing: Anodizing is an electrolytic oxidation process by which the surface of the metal is converted to an insoluble oxide. Anodizing provides corrosion protection, decorative surfaces, a base for painting and other coating operations, and special electrical and engineering properties. Aluminum, zinc, magnesium, and titanium may be anodized, but aluminum is the major material treated by this process.

The metal is prepared by cleaning as shown in Figure 5. Cleaning etches the metal slightly, which insures an active surface for anodizing. The most common baths used are fluoride-phosphate-chromic acid and potassium hydroxide-aluminum hydroxide-potassium chloride solutions.

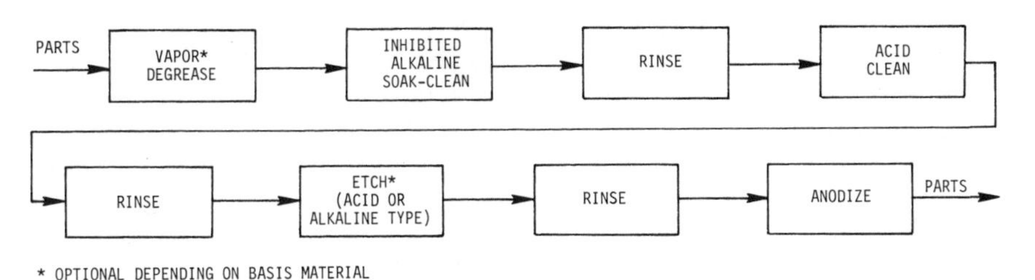

Figure 5: Typical metal cleaning scheme for anodizing of aluminum

Post treatment in anodizing operations generally consists of hot deionized water rinsing. Organic dyes may be added to produce a colored anodizing. With the exception of organic dyes, waste loads are generally low.

Chemical Conversion Coating: Coatings or films are produced on metal surfaces by chromating, phosphating, coloring, or immersion plating. Surface preparation required by these processes is similar to that of electroless plating.

Chromating — A portion of the base metal is converted to one of the components of the film by reaction with aqueous solutions containing hexavalent chromium and other active organic or inorganic compounds. Chromate coatings are most frequently applied to zinc, cadmium, aluminum, magnesium, copper, brass, bronze, and silver. These coatings are used for protective or decorative purposes or as a base for paint when the original material does not have good adhering properties. Chromate conversion coatings are frequently applied to zinc or cadmium-plated parts immediately following electrodeposition. The wastewaters are similar to those for electroplating processes.

Phosphating — Phosphating is the treatment of iron, steel, zinc-plated steel, and other metals by immersion in a dilute solution of phosphoric acid plus other reagents to produce an integral conversion coating on the surface. The process is similar to chromating, and the wastes are similar to electroplating waste streams.

Coloring — Metal coloring by chemical conversion is used to produce a wide range of decorative finishes. Coloring is achieved by brief immersion in a dilute aqueous solution which results in the formation of an oxide or other insoluble film. Copper and brass are frequently colored, as are most ferrous metals. Cadmium, tin, silver and aluminum may also be colored.

Immersion Plating — This is a chemical plating process in which a thin metal deposit is obtained by chemical displacement of the basis material. The less active metal will be deposited from solution while the more active metal (the item being plated) will be dissolved. This process is used to insure corrosion protection or as a preparation for painting or rubber bonding. Preparation for immersion plating consists of an alkaline cleaning step and a pickling step, which produce wastewaters similar to the pretreatment steps described earlier.

Post treatment processes for coating operations may include rinses, dyeing, bleaching, and oil or wax dipping. With the exception of wax or oil treatment, wastewater generated by these processes will be similar to that of electroplating post treatment.

Chemical Milling and Etching: Chemical milling is the process of shaping, machining, fabricating, or blanking metal parts to specific design configurations and tolerances by controlled dissolution with chemical reagents or etchants. Chemical etching is the process of removing relatively small amounts of metal from the surface to improve the surface condition of the basis metal or to produce a pattern. After surface preparation, areas where no metal removal is desired are masked off by dipping, spraying or roll or flow-coating. After the masking step, the part is given an acid dip to activate the surface for etching. Etching solutions include ferric chloride, nitric acid, chromic acid, sodium and ammonium persulfate, and cupric chloride. Wastewaters contain the etching solutions plus concentrations of the particular metal being etched.

Electroless Plating: Electroless plating is a chemical process in which a metallic ion in aqueous solution is catalytically reduced and subsequently deposited on the workpiece. This occurs without the application of electric potential, as required in electroplating. Electroless plating is widely used because of its ability to provide a uniform and dense plating thickness regardless of shape, the capability to plate plastics as well as metals, and the superior hardness and corrosion resistance of certain electroless platings. Iron, gold, cobalt, palladium, and arsenic have been used as electroless plates; however, most operations utilize either copper or nickel.

As in electroplating, surface preparation is necessary before plating. Surface preparation for plating is similar to that used in electroplating operations, except that certain metals and plastics require a surface activation step. Figures 6 and 7 show typical metal and plastic surface preparation operations.

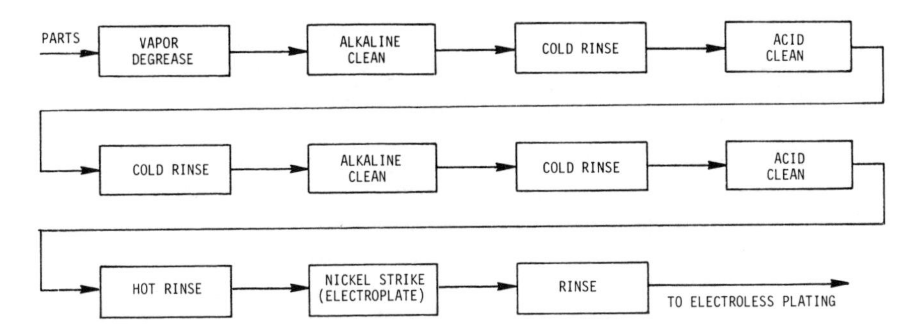

Figure 6: Typical surface preparation operation for electroless plating on metals.

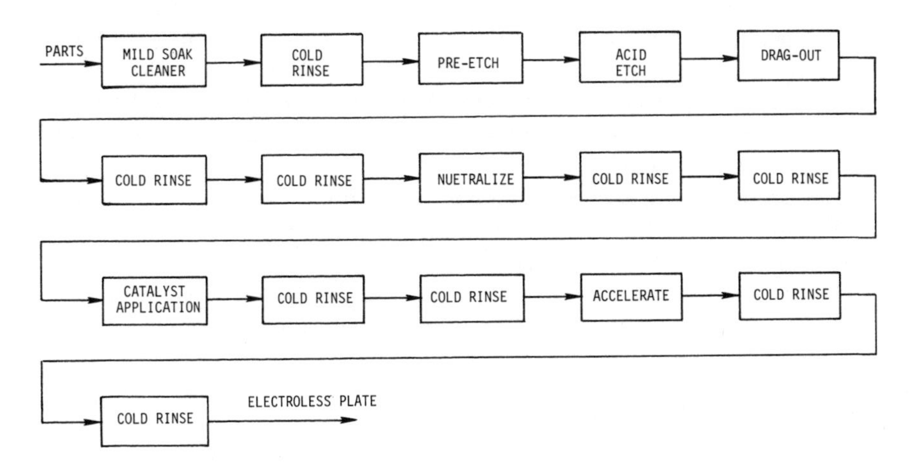

Figure 7: Typical surface preparation operation for electroless plating on plastics.

The solution used in electroless plating consists of a metal source, a reducing agent which reduces the metal to base state, chelating agents to prevent premature reduction, and buffer chemicals to stabilize the bath and extend its life.

Nickel plating baths utilize a nickel salt, such as nickel chloride, and sodium hypophosphite as the reducing agent. Citric or glycolic acid are common chelating agents. Copper plating operations also utilize a metallic salt, such as cupric chloride. Reducing agents used include acetaldehyde, trioxane, hydrazine, and hypophosphite; however, formaldehyde is the most frequently used. The chelating agent used is usually a tartrate or an amine.

Following electroless plating, a variety of finishing processes may be carried out. By far the most common finishing operation is electroplating on the electroless deposited surface. Metal coating and coloring are also done. Wastewater generated differs from electroplating wastes due to increased bath disposal and the addition of chelating agents.

Printed Circuit Board Manufacture: Printed circuit board manufacture consists of forming a circuit pattern of conductive metal, usually copper, on a nonconductive plastic or glass board. Printed circuit boards are utilized in a wide range of electronic applications.

There are three methods used in the production of printed circuit board: the additive, semi-additive, and subtractive processes.

In the additive process, conductive metal is deposited on the board in the desired circuit pattern. After surface preparation, a resist, or nonplatable coating, is applied to prevent plating except on the circuit pattern. Copper is applied by electroless plating, forming the actual circuit. Final finishing includes the addition of contacts, electroplating or coating.

The subtractive process works by removing metal to form the circuit pattern. The entire surface is prepared, catalyzed, and electroless plated. A resistance is then applied in the pattern of the desired circuit and unwanted metal is removed by etching. Finishing operations are the same as in the additive process.

A semi-additive process, which is a compromise between the two basic processes, has also been used to a limited extent.

Individual operations, such as cleaning, electroplating, and electroless plating are essentially the same as in other subcategories. Figures 8, 9, and 10 show process flow diagrams for additive, semi-additive, and subtractive printed board manufacture, respectively.

Wastewater Characteristics

Table 8 shows raw wastewater characteristic data. Sources of wastewater are: rinsing at each operation to remove films of processing solution; spills and leaks; air pollution scrubbers; dumps of cleaning, plating or coating solutions; wash water; and cooling water discharges.

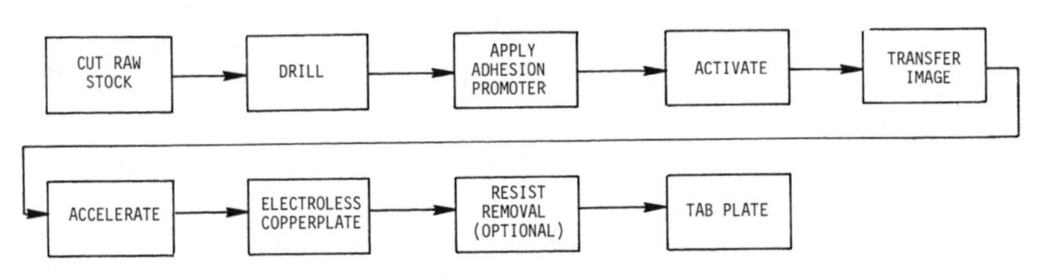

Figure 8: Printed board manufacture—additive process flow diagram.

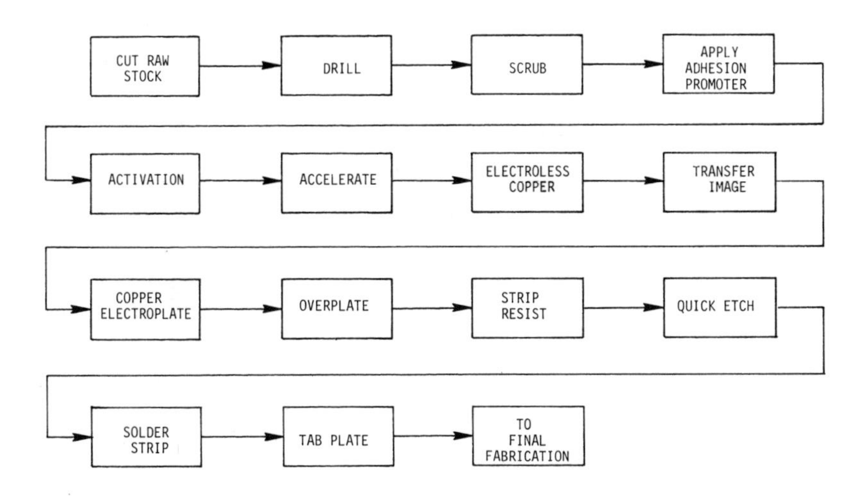

Figure 9: Printed board manufacture–semiadditive process flow diagram.

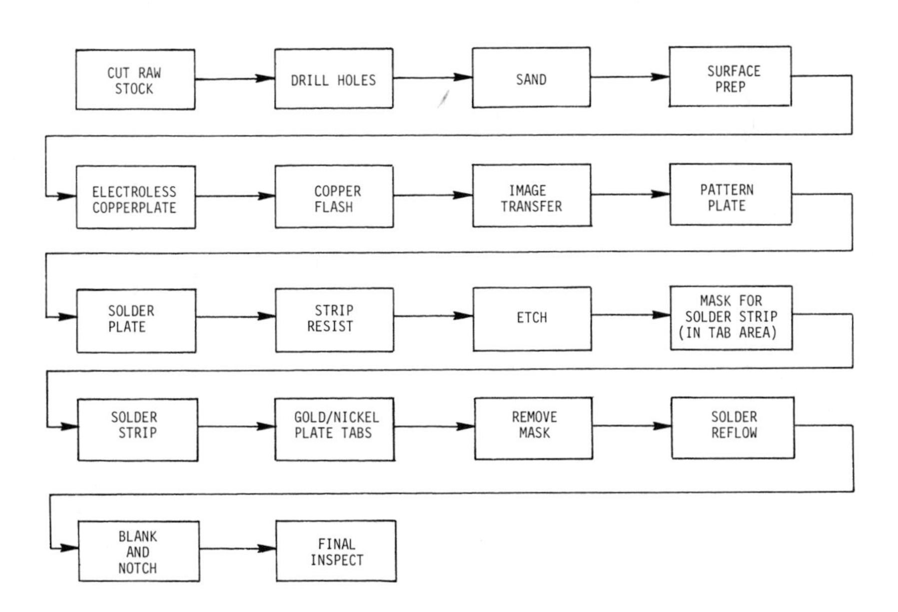

Figure 10: Printed board manufacture–subtractive process flow diagram.

Table 8: Electroplating–Raw Wastewater Characteristics[a]

Wastewater parameter (mg/l)	Subcategory: Common metals plating	Precious metals plating	Anodizing	Coatings	Chemicals etchings and milling	Electroless plating	Printed circuit board
Cadmium	0.007 - 21.60						
Chromium (Hexavalent)	0.005 - 334.5		0.005 - 5	0.005 - 5	0.005 - 334.5		0.005 - 4.4
Chromium (Total)	0.008 - 525.9	0.268 - 79.20	0.190 - 79.20	0.088 - 525.9		0.005 - 47.8	
Copper	0.032 - 272.5				0.206 - 272.5	0.002 - 47.90	0.203 - 535.7
Cyanide A[b]	0.003 - 130	0.003 - 8.420	0.004 - 67.56	0.004 - 67.56	0.005 - 101.3	0.005 - 1.00	0.005 - 9.38
Cyanide (Total)	0.005 - 150	0.005 - 9.970	0.005 - 78	0.005 - 126	0.005 - 126	0.005 - 12.00	0.005 - 10.80
Fluoride	0.022 - 141.7				0.022 - 141.7	0.110 - 18.00	0.280 - 680
Gold		0.013 - 24.89					0.006 - 0.107
Iron	0.252 - 1,482			0.410 - 168	0.075 - 263		
Lead	0.663 - 25.39						0.010 - 10.2
Nickel	0.019 - 2,954					0.028 - 46.80	0.207 - 13.30
Palladium		0.027 - 0.625					0.005 - 0.234
Phosphorous	0.020 - 144.0	0.020 - 144	0.176 - 33	0.060 - 53.30	0.060 - 144.0	0.030 - 109.0	0.051 - 53.6
Platinum		0.112 - 6.457					
Rhodium[c]		0.034					
Silver		0.050 - 176.4					0.001 - 0.478
Tin	0.060 - 103.4			0.102 - 6.569	0.338 - 6.569	0.060 - 90.0	0.060 - 54.0
TSS	0.1 - 9,970	0.1 - 9970	36.1 - 924	19.1 - 5,275	0.1 - 4,340	0.1 - 39.0	1.0 - 611.0
Zinc	0.112 - 252			0.138 - 200	0.112 - 200		

a Reported values are based on historical data from 196 electroplating facilities. No production based data presently available.

b Amenable to chlorination.

c Only one plant had a measurable level of this pollutant.

Significant pollutants in electroplating subcategories include, in addition to the expected metals, cyanide, fluorides, phosphorus, suspended solids, and pH. Organic loads tend to be extremely low. Of particular significance in the electroless plating subcategory is the impact of frequent discharges of bath solution, even though normal electroless wastes are more dilute than electroplating wastes. Also of significance are the chelating agents used both in the electroless plating subcategory and in electroless operations within printed circuit board manufacturing. These chemicals hinder removal of metals by conventional precipitation methods. Low pH flows are characteristic of printed board manufacture due to intensive acid cleaning.

Control and Treatment Technology

In-Plant Control Technology: Wastewater generation may be controlled through the use of the following in-plant controls:

- *Modification of Rinse Schemes* - Rinsing contributes approximately 90 percent of the wastewater generated in electroplating facilities. Most plants utilize single running rinses, which are the least effective type available. Countercurrent, series, spray, and still rinses are more efficient schemes. Which rinse techniques to utilize is determined by both geometry of the part and properties of the bath solution. Rinse water may also be recycled as bath makeup, depending on rinse and process type.
- *Good Housekeeping* - Frequent inspection of equipment, particularly rack coatings and tank liners can reduce bath carry out in rack dip operations. Consideration of equipment and anode materials can reduce formation of untreatable toxics.
- *Chemical Recovery* - Reverse osmosis and ion exchange are two methods of recovering metals and process chemicals. Etchants and spent plating baths are particularly amenable to recovery operations.
- *Process Modification* - Elimination of redundant rinses and concentration of baths, as well as substitution of process chemicals can all reduce waste loadings. Subcategory H, printed circuit board production, can show particular benefit from modification of etching, masking and rinsing operations.
- *Integrated Waste Treatment* - Utilization of a rinse tank as a caustic soda precipitator is an effective means of pollutant reduction. Used with a clarifier to remove soluble pollutants, it may be used in conjunction with other treatment technologies.

End-of-Pipe Treatment Technology: As toxic metals are the major pollutant generated by electroplating facilities, individual physical/chemical treatment processes are typically combined to effect the desired removals. These include:

- Chemical reduction of hexavalent chromium.
- pH adjustment and precipitation.
- Filtration - reverse osmosis, diatomaceous earth filters, ultrafiltration, membrane filtration.
- Oxidation - ozonation, chlorine oxidation, electrolytic oxidation, peroxide oxidation.
- Air flotation using various physical configurations and chemical additives.
- Various electrochemical processes.

Applicability of individual treatment technologies to the wastewaters of various electroplating processes is summarized in Table 9. Removal efficiencies of some of the treatment technologies are shown in Table 10. These technologies may be utilized as either in-line systems, with treatment segregated streams, or combined as end-of-pipe treatment systems. A comparison of toxic pollutant removal capabilities of end-of-pipe treatment, in-plant control, and in-line treatment systems is shown in Table 11.

Type of Residue Generated

Except for the cyanide ion, which may be changed to nitrogen and carbon dioxide gases, all of the metals, wetting and cleaning agents, chelates, and particulate material removed from a plating wastewater stream during wastewater treatment are concentrated into a residue. Regardless of the specific anion and metal cation used in a particular plating process, the residuals generated from pretreatment will be very similar in their physical and chemical nature. This concentrated residue is most commonly a sludge, although some newer treatment methods produce a brine (brines are generated in schemes involving recycling or recovery). Table 12 summarizes the residues produced by the various treatment unit processes.

The most common type of sludge produced is a metal hydroxide sludge. The hydroxide sludges are precipitated and are stable at high pH. A low (acid) pH is likely to result in release of the metal cations. These sludges can also have small amounts of chelated metal complexes and free cyanide ions.

Various newer or alternate processes generate sludges containing metal sulfide, carbonate, xanthate, fluoride, or phosphate precipitates.

Because sulfides and xanthates are somewhat biodegradable, the stability of such sludges during long-term storage or "permanent" disposal is more questionable than for hydroxide sludges. Fluoride and phosphate sludges have sufficient solubility that leaching is possible. Acid pH's can also affect the stability of all the sludges mentioned.

Quantities of electroplating sludges, such as the above, produced in "job" (that is independent) shops in the United States have been estimated as 22,000 tons, dry weight, in 1975; 56,500 tons in 1977; and 74,000 tons

Table 9: Electroplating Industry—Applicability of Individual Treatment Technologies

Application	Treatment Technologies										
	Cyanide oxidation	Chromium reduction	Clarification	Segregated clarification for chelated metals	Sludge drying	Reverse osmosis	Evaporation	Ion exchange	Electrolytic recovery	Membrane filtration	Advanced rinsing techniques
Acid metal plating recovery						X	X	X	X		X
Anodizing		X	X		X						
Catalyst deposition			X		X						
Chromating		X	X		X			X			
Common metals plating	X	X	X		X						
Cyanide metal plating recovery							X		X		X
Electroless plating	X	X	X	X	X						
Etching and milling		X	X		X	X	X		X		
Immersion plating	X		X		X						
Mixed plating waste treatment						X	X	X		X	
Phosphating			X		X			X		X	
Precious metals plating	X	X	X		X		X	X	X		X
Stripping	X	X	X		X						

in 1983. Quantities from "captive" shops (that is, units of larger, integrated manufacturing operations) have not been estimated, but are believed to be much greater.

Table 10: Electroplating Industry–Percent Pollutant Removal by Individual Treatment Technologies

Wastewater parameter	Treatment Technology Removal Efficiencies - Percent: Diatomaceous earth filter	Chlorine oxidation	Ozonation	Ion exchange	Reverse osmosis	Membrane filtration	Activated Carbon[a] A	Activated Carbon[a] B
Conventional								
Phosphate				79	86-90			
Phosphorus				79				
Sulfate				90	88-92			
TSS	98					99		
Toxic								
Cadmium				100	90-92		99.6	98.6
Chromium (Hexavalent)						98		
Chromium (Total)				99		99	98.2	99.3
Copper	94			98-100	90-92	99	90.0	96.0
Cyanide (Total)		99.6	100		88-96			
Iron	96			97				
Lead						98		
Nickel	98			100	88-90	99	99.5	37.0
Silver				100			98.0	99.0
Tin				100		40	92.0	98.5
Zinc	99			99	88-90	98	76.0	94.0

[a] Activated carbon "A" is with lime precipitation and Activated Carbon "B" with ferric chloride addition.

Table 11: Electroplating Industry–Toxic Pollutant Removal by Combined Treatment Systems[a]

Wastewater parameter	End-of-pipe system mg/l	End-of-pipe system mg/op-m² [c]	In-plant control[b] mg/l	In-plant control[b] mg/op-m²	In-line treatment system mg/l	In-line treatment system mg/op-m²
Chromium	0.54	24	0.61	23	0.02	0.8
Copper	0.49	21	0.56	21	0.02	0.8
Cyanide	0.01	0.4	0.01	0.4	ND[d]	ND
Fluoride	2.0	87	2.0	76	2.0	75
Nickel	0.99	43	1.13	43	0.02	7.5
Total suspended solids	15.0	655	16.5	622	16.7	628

[a] Based on a comparative study at a sample electroplating facility. Values are for final effluent.

[b] Counter current rinsing only.

[c] Mg/op-m² - milligrams pollutant per square meter plated for each plating operation.

[d] ND - not detectable.

Table 12: Summary of Residues Generated in Treating Electroplating Wastewaters

Wastewater treatment process	Principle use	Residual produced	Comments
Chromium reduction	Conversion to less toxic and more easily removed form	Chromic ion	Preliminary waste treatment step.
Clarification	Removal of toxic and objectionable pollutants from the effluent	Liquid sludge	The main waste treatment step. Usually done at high pH using lime as a coagulant.
Cyanide oxidation	Detoxification by conversion to innocuous products	N_2 and CO_2 gas	Frequently done concurrently with clarification at high pH.
Membrane processes, ion exchange, freezing, electrolytic recovery, solvent rinsing, ion flotation	Concentration of rinse waters or spent plating baths	Brine	Main purpose is recycle of valuable components of plating baths, which indirectly reduces the amount of residue needing disposal.
Sludge clarification for chelated metals	(Same as Clarification)	(Same as Clarification)	(Same as Clarification). Chelates hinder precipitation of the complexed metals.
Sludge drying (inclusive of thickening and dewatering	Dewatering of sludge	Semi-solid sludge	A main waste treatment step that usually follows clarification.

Residual Management Options

Sludges produced by treatment of electroplating wastes almost always contain toxic components, and, if not treated for recovery of valuable components, are landfilled. Alternates used in the past, such as ocean dumping or deep-well injection, are no longer acceptable. The emphasis in landfilling of electroplating sludges is on ensuring long-term stability of the sludge and guarding against leaching or migration of toxic materials into aquifers or surface waters. The trend today, however, is to plating process changes and waste treatment techniques that allow recycling of plating bath components and rinse waters, both of which tend to obviate the production and disposal of waste sludges. Some sludge reclaiming processes are also under development for recovering valuable materials from the sludge, but none are widely applied at present. These are described in abbreviated form in Table 13.

Processing of the sludge generated by precipitation of the metal ions, usually with lime, typically involves gravity thickening, and dewatering. The capabilities of a particular dewatering technique are very sensitive to the composition of the wastewater flow, including the presence of extraneous material such as wetting agents and chelates, to the coagulants used, to the pH, and to the operating parameters used for the mechanical dewatering equipment. Lime-precipitated metal hydroxide sludges are typically more easy to dewater than gelatinous ferric or aluminium hydroxide sludges. Solids concentrations of two to five percent after gravity thickening are common. Cakes of 15 to 20 percent solids are achievable from

Table 13: Summary of Electroplating Sludge Reclamation Processes

Name	Relationship to plating process	Starting material	Product	Sludge reclamation process description
Electrochemical and roasting process for metal recovery	"End of pipe" treatment of mixed wastes.	Mixed hydroxide sludges	Metal, exact form is unclear.	The sludge is re-dissolved, by acidification, and the liquor subjected to controlled potential electrolysis. Roasting is necessary only to separate metals with nearly equal electro-potentials, such as chromium and zinc.
Integrated treatment	Inline - between plating bath and first rinse tank.	Carryover from plating bath	Dual function rinse rinse and treatment solution, recycled to inline tank.	Plated part immersed in treatment/rinse solution which is treated continuously in an off-line tank and recycled.
Single metal sludge recovery	Offline, similar to clarification but for a segregated stream from one or more identical plating lines.	Rinse water	Single component metal hydroxide sludge.	Chemical treatment to reconvert the sludge directly back to a high-quality plating solution, which is then recycled.
Solvent extraction	"End of pipe" treatment of mixed wastes.	Mixed hydroxide sludges	Separated metal salt, exact form is unclear.	The sludge is re-dissolved. by acidification, and the liquor is then extracted with organic solvents that are selective for specific metal ions.
Spent bath recovery	(Similar to single metal sludge recovery.)	Plating bath	(Similar to single metal sludge recovery.)	(Similar to single metal sludge recovery.)

centrifuges, 20 to 25 percent from vacuum filters, and 40 to 60 percent from pressure filters.

Measures taken to ensure the long-term stability of the dewatered and landfilled sludge may include:

1. Minimizing contact with rainwater, surface water, and groundwater by covering with impervious backfill after each placement of sludge; lining the site with an impervious layer such as compacted clay or silt; grading the surface of the backfill so as to divert surface water away from the site and prevent erosion; and choosing a landfill site that is well drained and substantially above the groundwater table.
2. Guarding against low pH by covering and/or bedding the deposited sludge on a layer of lime; and not placing electroplating sludges in contact with organic materials such as garbage or sewage sludge, as these materials tend to form organic acids during anaerobic decomposition.
3. Encapsulation of the sludge in impervious containers; chemical fixation of the sludge with a stable binding material; or conversion of the hydroxide sludges to essentially insoluble oxides by baking at high temperatures.

General points to be remembered in disposing of plating sludges are:

1. Such sludges are relatively inert if covered with relatively impervious fill, and kept dry, but they should never be placed on cropland or vegetated land.
2. Most fine-grained soils have a high exchange capacity for the toxic metals in these sludges, so that, even if unfavorable conditions occur and release of the toxic metal cations takes place, migration any great distance is possible, but unlikely. The above does not apply to toxic anions such as fluoride or chromate.
3. Electroplating sludges should not be incinerated, as certain toxic metals such as zinc, lead, and cadmium can be volatilized at the temperatures commonly reached in incinerators.

EXPLOSIVES MANUFACTURING

General Industry Description

The explosives manufacturing industry includes facilities engaged in the manufacturing operations which: produce explosives, blasting agents, solid propellants, pyrotechnics, and initiating explosive compounds; load, assemble and pack ammunition and military ordnance; and demilitarize or dispose of obsolete, off grade, contaminated, or unsafe explosives and propellants.

Industrial Categorization

This industry has been divided into the following subcategories: manufacture of explosives; manufacture of propellants; load, assemble, and pack explosives; manufacture and load, assemble, and pack initiating compounds; formulation and packaging of blasting agents, dynamite, and pyrotechnics.

Process Description

The general production process for the manufacturing of explosives involves the nitration of an organic molecule. Raw materials used in this process are nitric acid, acting as the nitrate source, and sulfuric or acetic acid, acting as a dehydrating agent. Examples of the organic molecules used are glycerin, toluene, resorcinol, hexamine, and cellulose. After nitration, these organic molecules produce the following products: nitroglycerin and dinitroglycerin; trinitrotoluene and dinitrotoluene; trinitroresorscinol; nitromanite; and nitrocellulose, respectively. Additional production processes involve the formation of highly sensitive initiators with nitrogen salts as a nitrogen source, such as lead azide.

Figure 11 is a flow diagram for the production of nitroglycerin.

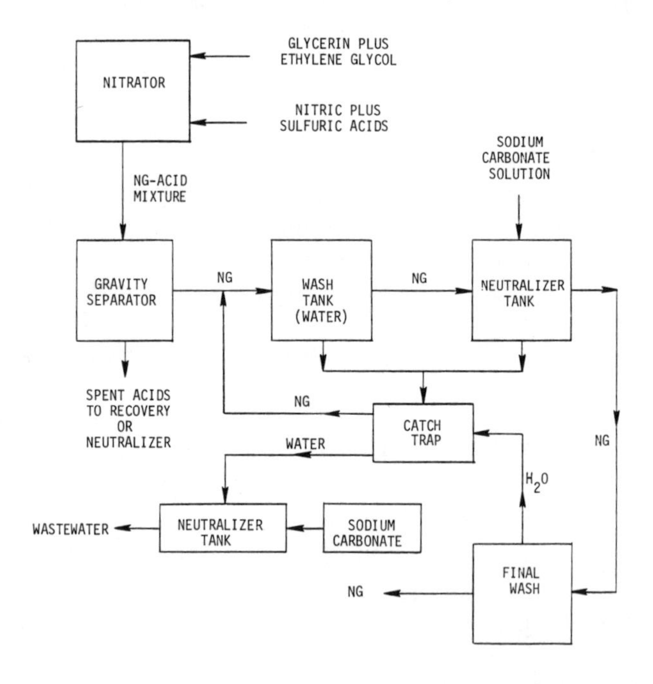

Figure 11: Typical process flow for the manufacture of nitroglycerin.

Production is generally either batch or semi-continuous. Semi-continuous production schemes involve certain individual production steps being run continuously, while the remainder of the production method is on a batch basis. Specific products for each of the subcategories and sources of wastewater are discussed below.

Manufacture of Explosives: This subcategory includes manufacturing operations that produce explosive compounds by mixed acid nitration of organic material. Products manufactured include nitroglycerine, PETN, RDX, HMX, TNT, and nitrostarch. Initiating compounds are excluded from this subcategory.

Manufacture of products in this subcategory results in the production of relatively large volumes of wastewaters from the neutralization and washing of the final product. Wastewater is usually low in suspended solids and contains soluble nitrate and nitrite salts.

Manufacture of Propellants: This subcategory includes the manufacture of nitrocellulose-based propellants and gas generators. Propellants differ from explosives in the designed rate of reaction upon ignition. Propellants are designed to burn in a controlled manner, as opposed to explosives, which detonate instantly. Relatively large volumes of wastewater are generated as the result of water used to transport propellant between unit operations, to remove solvents from the final product, to cool and lubricate in the cutting and machining of the final product, and to clean and wash down process equipment. Suspended solids are present, and the presence of organic solvents often makes organic loads higher than those that result from explosives manufacture.

Load, Assemble, and Pack Explosives: In the load, assemble, and pack (LAP) operations the necessary explosives and propellants are obtained from outside sources and then mixed and packed as a final product. Examples of this type of operation in the commercial sector would be small-arms manufacturers involved in the loading and assembling of small-to-intermediate-caliber ammunition. Also included would be manufacturers of explosive devices such as jet piercers and shaped charges. In the military sector, large-caliber shells, bombs, grenades, and other munitions are filled with blends of TNT and other ingredients. Propellants and small explosive devices are usually loaded dry, whereas larger explosives are usually melted in kettles and poured as a liquid. Wastewater results from the clean up of process equipment and wash down of work areas. Wastewater volumes and waste load are relatively small when related to production.

Manufacture and Load, Assemble, and Pack Initiating Compounds: This subcategory includes plants which manufacture "sensitive" explosive compounds, such as trinitroresorcinol, hexanitromannite, isosorbide dinitrate, tetryl, tetracene, lead azide, lead styphnate, and mercury fulminate. The extremely sensitive nature of these compounds make it almost impossible for them to be safely transported in bulk quantities. Following manufacture they are usually conveyed to a load, assemble, and pack operation at the same site where they are used to manufacture primers, detonators, and caps. The small quantities of explosive materials used make this type of LAP operation fundamentally different from that included in the "Load, Assemble, and Pack Explosives" subcategory.

Wastewater is generated in the neutralization and purification of these compounds during manufacture, and in the cleaning of equipment and wash down of work areas. Additional water may be used for safety purposes, such as keeping floors, work areas, and drain lines wet. Water used per unit of production is higher in this subcategory than in the other explosives subcategories. Pollutant loads result from the manufacturing process, as well as from chemical treatment of catch basins and sumps to desensitize any initiating compounds trapped therein.

Formulation and Packing of Blasting Agents, Dynamite, and Pyrotechnics: Included in this subcategory are operations that manufacture blasting agents commonly referred to as ANFO, slurry, water gels, and nitrocarbonitrates. All ANFO (Ammonium Nitrate/Fuel Oil mixture), ANFO-based dry mixes, and dry nitrocarbonitrates are designated ANFO compositions. Slurry also refers to water gels, which are high energy, waste resistant blasting agents. Also included in this subcategory, because of the similarity in process and wastewater characteristics, is the manufacture of dynamite, black powder, and pyrotechnics. Pyrotechnics include fireworks, illuminating flares, smoke generators, and distress or signal flares. Products in this category are produced by batch dry blending operations. Often, dry ingredients are blended with nitroglycerin to produce an explosive. Manufacturing operations that would be expected to fall into this subcategory, but that are excluded are those which formulate pyrophoric materials and require liberal use of water to prevent spontaneous ignition.

Generally, very small volumes of wastewater are generated per unit of production. The primary source of contaminated wastewater is the clean up of mixing equipment and the wash down of work areas. Wastewater composition reflects the ingredients of the products being manufactured. Some additional wastewater may be generated in the manufacture of dynamite through the use of wet scrubbers to control dust while the ingredients are being blended. Although the volume of wastewaters emitted from these scrubbers may be significant, particularly if no water recirculation is employed, the contaminant load is small.

This subcategory comprises 95 percent of the total explosives industry. Because of this, future study efforts will be restricted to this segment of the industry. It is the intention of the USEPA to initially promulgate regulations for only this subcategory and to defer regulations for the rest of the industry until a later date.

Wastewater Characteristics

Wastewaters are generally very low in pH and can be high in BOD, COD, and nitrates. Wastes can also contain concentrations of the explosives produced. Tables 14 and 15 show raw wastewater characteristics for conventional and toxic pollutants, respectively. Table 16 shows production based data for the "Formulation and Packing of Blasting Agents, Dynamite, and Pyrotechnics" subcategory. The availability of more data for this subcategory than for the other subcategories is due to its dominance in the industry.

Control and Treatment Technology

In-Plant Control Technology: Many products are manufactured by a dry process, so that waste streams come from only the cleanup of spills and leaks. Some manufacturing processes use water to transport the material throughout the facility and to purify the product. When this water need not be of high quality, recycle can reduce the waste produced from this operation. Separation of contact and noncontact waters can also reduce the size of treatment systems and the volume of flow discharged.

End-of-Pipe Treatment Technology: Of the 24 military plants and 280 commercial plants comprising this industry, all of the former and 180 of the latter are direct dischargers. The remaining 100 commercial plants have zero discharges. Thus, discussion of pretreatment and "indirect" discharge through POTWs is somewhat academic. Nevertheless, for the hypothetical situation where indirect discharge might be proposed for a commercial explosives plant, the most suitable pretreatment methods would be those used by some direct dischargers. These would include: ponding, with percolation and evaporation to balance liquid input; slow-rate land application using spray application — dilution may be necessary if the nitrogen concentration in the wastewater is too high; biological treatment using activated sludge; flotation; or one of several advanced waste treatment (AWT) processes.

Table 14: Explosives Manufacturing–Raw Wastewater Conventional Pollutant Characteristics

	Subcategory				
Wastewater parameters	Manufacture of explosives, mg/l	Manufacture of propellants, mg/l	Load, assemble, and pack explosives, mg/l	Initiating compounds, mg/l	Formulation of blasting agents, etc., mg/l
BOD	20-1,000	200	1,000	1,000-12,000	9-3,400
COD	60-3,400	200-1,200	8-8,500	11,000-50,000	130-6,000
Explosives	present	present	present	present	present
NH_3-N					45-215
Nitrogen (nitrate)	25-7,000	1-4,000	0.4-12	0.5-5,000	317-934
pH	variable	variable	variable	variable	variable
TKN					69-245
TOC	12-1,500	30-130	5-500	5,700	
TSS	10-300	100-1,000	1-700	1-60,000	

Table 15: Explosives Manufacturing–Raw Wastewater Toxic Pollutant Characteristics

		Priority Pollutant Level - ug/l												
		Metals						Organics						
Product Manufactured	Plant Identification[a]	Sb	Cu	Pb	Ni	Ag	Zn	Bis (2-ethylhexyl) pthalate	Cyanide	Chloro- form	Di-n-octyl pthalate	Isopher- one	Methylene chloride	Phenol
ANFO[b]	1	ND		ND			ND							
Slurry	1	350		80			980							
Ammonium Nitrate	1	ND		20			116							
ANFO	2	ND	5	ND		ND	140							70
Slurry	2		10	20			35		6					
Ammonium Nitrate	2		30	ND		2	24							
ANFO	7	ND	940	110	100	ND	1,300		14					65
Dynamite	8								10					20
Slurry	8								2,600			ND		
Dynamite	9			10			1,450							
ANFO	9													
Nitrocar- bonitrate	9			20			2,740							
Pyrotechnics	10	ND	37	ND	ND		45	72	154	544	ND		3.4×10^6	15

a Signifies identification number for plants sampled as part of EPA Verification Study for the Explosives Industry: Subcategory E.

b ANFO - ammonium nitrate/fuel oil mixture. Sample prepared by mixing 2,000 mg of ANFO in nitrocarbonitrate product per liter of distilled, deionized water.

Table 16: Explosives Manufacturing–Summary of Raw Waste Loads from Formulation and Packing of Blasting Agents, Dynamite, and Pyrotechnics Subcategory

Product produced	Plant identification[a]	Wastewater flow in cubic meters per million gallons of product produced	Associated pollutant levels in grams per million gallons of product						
			BOD_5	COD	TKN	NH_3-N	NO_3-N	TDS	TSS
Slurries	1	0.135	305	543	779	753	1,050	5,130	267
Slurries	2	0.048	1	1	2	2	29	34	1
Slurries	3	0.017				95	104		
Slurries[b]	4	0.216	425			1,205	1,490		
Slurries	5	0.046				51			
Slurries	6	0.011			42	36	32		
ANFO[c]	7	0.0026	8	7	35	32	52	210	3
Dynamite[d]	8	1.04	4	20	38	34	23		
Slurries	8	0.307	3,303	4,170	2,530	1,520	2,820	5,530	252
Dynamite	9	3.60	29	54	162	119	86	115	29
Pyrotechnics	10	0.435	4,430	5,520		2	478		135

[a] Signifies identification numbers for plants sampled as part of EPA Verification Study of the Explosives Industry: Subcategory E.

[b] Data from industry estimates; represents only one of two waste streams from this plant.

[c] ANFO - ammonium nitrate/fuel oil mixture.

[d] Represents scrubber effluent only, no washdown data available.

Type of Residue Generated

Of the treatment methods described, ponding, land application (pre-application treatment step), and biological treatment would all produce biological sludges. Nonbiological sludges might be produced by sedimentation, flotation, or various AWT processes. Most importantly, all the above sludges would be "contaminated" with heavy metals from the reactants or products of the manufacturing process, or from corrosion of the production equipment. Various toxic organic compounds would also be present.

Residual Management Options

Specific information is not available on residues from this industry.

FOUNDRIES

General Industry Description

The foundry industry comprises facilities that pour or inject molten metal into a mold to produce intricate metal shapes that cannot be readily formed by other methods.

The foundry industry in the United States employs over 400,000 workers in 4,400 foundries that produce over 19 million tons of product annually. This production includes cast pieces made of iron, steel, aluminum, brass, and copper, as well as other metals.

The basic foundry process is essentially the same regardless of the method of melting, molding, or finishing. A raw material charge is melted in a furnace, from which the molten metal is withdrawn as needed. The mold for the product is a sand cast or a set of metal die blocks which are locked together to make a complete cavity. The molten metal is ladled into the mold, and then cooled until the metal solidifies into the desired shape. The rough product is further processed by removing excess metal, quenching, cleaning, and chemical treatment.

Industrial Categorization

The foundry industry includes a number of foundry types as well as processes within these foundry types. Plants are capable of casting one or more metals on a site and each site may utilize one or more processes that can generate wastewater.

Nine subcategories have been selected to describe the foundry industry based on the type of metal cast, the type of process, plant size, location and age, wastewater characteristics and treatability, and method of effluent disposal. These subcategories are: iron and steel foundries, aluminum casting, zinc casting, copper casting, magnesium casting, lead casting, tin casting, titanium casting, and nickel casting.

Process Description

Iron and Steel Foundries: Iron is the world's most widely used metal. When alloyed with carbon, it has a wide range of useful engineering properties. Four general classes of iron are produced in foundries: gray, ductile, malleable, and steel. The same general processes are used with all four classes of metal in the production of products ranging from cooking utensils and pipe fittings to steel railroad car wheels.

Ferrous foundries (iron and steel) have five operations that may produce wastewater in some form. Dust collection, melting furnace scrubber, slag quenching, casting quench and mold cooling, and sand washing operations are all similar to those described for other foundries. Sand is washed by three methods (dry, thermal, and wet), of which only wet washing produces wastewater. The sand used in the molding operation is slurried and agitated to remove particles of burnt clay or impurities from its surface. The slurry is then sent to a classifier for separation and the sand is removed and dried. The water may be reused until the solids content is high.

Aluminum Casting: Aluminum is a light metal with good tensile strength. It is easily cast, extruded, or pressed, and it weighs half as much as a similar product made from steel.

The aluminum foundry industry can be subdivided into five operations which represent different processes within the foundry. Investment (also known as precision or lost wax) casting operations use molds that are produced by surrounding an expendable pattern with a ceramic slurry that hardens at room temperature. The pattern, normally a wax, is then melted or burnt out of the hard mold. These molds provide very close tolerances. After the molten metal is poured into the mold and solidifies, the mold is broken away. Thus, a new mold is needed for each casting. Final cleaning to remove mold residues is by high-pressure water jets, which creates a wastewater source. Wastewater is also produced by wet scrubbing the fumes, particulates, and smoke from the melting furnaces. Common control methods include stack gas quenching, washing coolers, and wet scrubbers. The quenching operation changes the cast metal to a desired grain structure. It may also reduce the machine cycle time. Quench water may enter the wastewater stream.

In theory, die casting should produce no wastewater; however, because of machine and mold cooling water leakage, and mold spray for cleaning and lubrication, it is a potential major source of wastewater. Die casting molds are lubricated to prevent the casting from sticking to the die. The lubricants used are dependent on the temperature of the metal, the operating temperature of the die, and the alloy to be cast. The lubricants are generally organic compounds and may enter the wastewater stream by leaks in the mold machinery.

Zinc Casting: Zinc, with a lower melting point than most metals, is generally die cast, making its process different from other foundry subcategories. Because it is not as strong as most metals, it is usually alloyed with metals such as copper, aluminum, or magnesium.

Copper Casting: Copper is second only to aluminum in importance among the nonferrous metals. It is often alloyed with tin, lead, and zinc to produce brass and bronze. Copper castings are produced by several methods which include centrifugal molds, green sand molds, and die casting. Products include bushings and bearings, propellers, and other cast products.

Operations in copper and copper alloy foundries that produce wastewater include dust collection, mold cooling and casting quench, and continuous casting.

Magnesium Casting: Most magnesium is cast in sand molds. This is to prevent metal-mold reactions, which may occur because of the reactive nature of molten magnesium. Often inhibitors such as sulfur, boric acid, or ammonium fluorosilicate are mixed with the sand to prevent these reactions.

Magnesium foundries generate wastewater in the grinding, scrubber, and dust collection operations. Because of the violent reaction of fine magnesium particles with air, wet scrubbers are used to control the dust from the grinding operations.

Lead Casting: Lead foundries produce lead castings such as lead wheel balancing weights and sash balances, as well as white metal castings.

Tin, Titanium, and Nickel Casting: The remaining three subcategories, Tin Casting, Titanium Casting, and Nickel Casting, have been recommended for Paragraph 8 exclusion under the NRDC Consent Decree. This recommendation is based on the low number of plants and low production by these subcategories.

Wastewater Characteristics

Table 17 shows raw wastewater characteristic data. Because each subcategory operation often involves different processes, raw wastewater pollutants from each subcategory vary significantly.

Control and Treatment Technology

In-Plant Control Technology: Wastewater recycle is an extensively used control method with many facilities (20 to 25 percent) recycling 100 percent and becoming zero dischargers.

Table 17: Foundries–Raw Wastewater Characteristics

Parameter	Subcategories[a,b,c] Iron and steel	Aluminum casting	Zinc casting	Copper casting	Magnesium casting
Conventional - mg/l					
Oil and grease	29	1,200	5,700	23	6
ph - units	7.7	7.3	6.8	7.7	8.7
Sulfides	2.1				
TSS	2,700	540	1,300	230	8.3
Metals and Inorganics - ug/l					
Antimony	270				
Arsenic	29				
Cadmium	180			100	
Chromium	180	130			
Copper	620	450	150	37,000	40
Cyanide	23	5	9	49	
Lead	24,000	330		9,300	55
Mercury	1.4	1.5	0.3	0.3	
Nickel	74	16	30	720	
Selenium		< 40			
Silver	30				
Zinc	37,000	1,700	140,000	67,000	1,200
Organics - ug/l					
Esters					
Bis (2-chloroethyl) ether		9			
Bis (2-chloroethxy) methane	20				
Bis (2-chloroisopropyl) ether	3				
Phthalates					
Bis (2-ethylhexyl) phthalate	140	100,000	1,500	11	35
Butyl benzyl phthalate	32	49		62	
Di-n-butyl phthalate	20	710	79	1	10
Diethyl phthalate	6	240		6	40
Dimethyl phthalate	320	19		22	
Di-n-octyl phthalate	41	4	2,800		
Nitrogen compounds					
N-nitrosodiphenylamine	250				
N-nitroso-di-n-propylamine		120			
Phenols					
2-Chlorophenol	39	53	120		
2,4-Dichlorophenol	330	1,400	25		
2,4-Dimethylphenol	190	37	61	36	
2-Nitrophenol	ND[d]	95			
4-Nitrophenol	18		1,600		
Pentachlorophenol	39	530		14	11
Phenol	2,800	4,500	170	25	6
2,4,6-Trichlorophenol	80	130	58		
p-Chloro-m-cresol	55	90	22		
4,6-Dinitro-o-cresol	17	32			
Aromatics					
Benzene	20	840	150		
2,4-Dinitrotoluene	29				
2,6-Dinitrotoluene	29			4	
Ethylbenzene		78			
Nitrobenzene	94				
Toluene	15	540	27		18
1,2,4-Triclorobenzene	7				
Polycyclic aromatic hydrocarbons					
Acenaphthene	17	70			11
Acenaphthylene	19	16			32
Anthracene	81	130			30
Benzo (a) anthracene	13	13,000			
Benzo (a) pyrene	< 30	28			
Benzo (b) fluoranthene	14				
Benzo (k) fluoranthene	6				
Chrysene	11	4,600			10
Fluoranthene	54	100	14		
Fluorene	110	150			3
Naphthalene	27	48	50		
Phenanthrene	81	120			< 30
Pyrene	140	65			7

(continued)

Table 17: (continued)

Parameter	Subcategories[a,b,c] Iron and steel	Aluminum casting	Zinc casting	Copper casting	Magnesium casting
Polychlorinated biphenyls and related compounds					
Aroclor 1016, 1232, 1248, 1260	41	210	29		< 5
Aroclor 1221, 1254	45	320	24		< 5
Halogenated aliphatics					
Carbon tetrachloride	20	480	29		
Chloroform	19	80	57		
Dichlorobromomethane	37	2			
1,1-Dichloroethane		55			
1,2-Dichloroethane		73			
1,2-Trans-dichloroethylene	< 11		43		
Methylene chloride	50	310	100		44
1,1,2,2-Tetrachloroethane		18			
Tetrachloroethylene	53	46	130		
1,1,1-Trichloroethane	20	4,000	140		
1,1,2-Trichloroethane	20				
Trichloroethylene	39	58	230		
Pesticides and metabolites					
α-BHC		26			< 5
β-BHC	20	70			
δ-BHC	20	7			
γ-BHC	20				< 5
Chlordane		38			
4,4'-DDE	20	10			
4,4'-DDD					< 5
4,4'-DDT	20				< 5
Dieldrin	20				
α-Endosulfan					< 5
Endrine aldehyde	11				< 5
Heptachlor	20				

[a] At the time of writing this manual, no specific data concerning the lead casting industry was available. The tin, titanium, and nickel casting subcategories are not included as they have been recommended for Paragraph 8 exclusion under the NRDC Consent Decree.

[b] Blanks indicate that data is not currently available.

[c] Values indicated are average values.

[d] ND - not detectable.

End-of-Pipe Treatment Technology: A common pretreatment scheme in this industry is to have oil and grease removal (emulsion breakers — sulfuric acid or alum — are normally added) followed by some form of solids-liquid separation such as lagooning, clarifiers, or cyclone separators. Recently ultrafiltration has been utilized at some plants to remove organics before discharging to the POTW.

Type of Residue Generated

Oil and grease are present in many of the waste streams generated from this industry, most of it being generated from housekeeping and machinery leaks. In many instances, emulsion breakers — sulfuric acid or alum — must be added to allow effective removal from the wastewater stream.

Other liquid sludges are generated from wet scrubber dust collection systems, cooling waters, and wastewater treatment systems. These sludges normally have high concentrations of heavy metals (chrome, lead, and zinc).

Residual Management Options

Several options are available for final disposal of the oil and grease residuals. In some cases the collected oil and grease can be sufficiently cleaned on site to be used in the foundry. Solids removed during the cleaning process would be added to the other liquid sludges produced at the site. In other applications it becomes cost effective to modify existing boilers to use the material as a supplementary fuel source not only reducing the disposal problem but also reducing energy cost. Removal by waste haulers is a third alternative. If land is available, land farming may be practicable. In this technique, the residue is spread on cleared land and disked into the ground, both with regular farming equipment. The residue is degraded over a period of time, principally by bacterial action. Active cultivation of crops on the land is not part of the technique, although the soil structure is said to be improved by the organic degradation products. The attitude of regulatory authorities toward the technique is not yet clear, as the potential presence of toxic "contaminants" in the residue is a source of concern. Surface runoff and possible groundwater contamination would have to be monitored and controlled in any application of this technique.

The most common management technique in the past for the other residuals generated has been to thicken in gravity thickeners, dewater, and dispose of in a local landfill. Because of high levels of metals and other priority pollutants, this sludge will probably be classified as hazardous and disposal in a local sanitary landfill may no longer be a viable option.

Of the several ultimate disposal options available (lagooning, deep well injection, salt deposit disposal, cementation, pyrolysis) only cementation and pyrolysis seem to have merit for the foundry industry although at this time research is only in the beginning stages.

GUM AND WOOD CHEMICALS

General Industry Description

The Gum and Wood Chemicals manufacturing industry includes establishments primarily engaged in manufacturing hardwood and softwood distillation products, wood and gum naval stores, charcoal, natural dyestuffs, and natural tanning materials. More specifically, the materials produced by this industrial category are char and charcoal briquets, pyroligneous acids and other by-products; gum rosin and turpentine; wood rosin, turpentine and pine oil; tall oil rosin, fatty acids and pitch; essential oils, terpenes, hydrocarbons, alcohols or ketones; esters, modified esters, and alkyds; and sulfate turpentine.

The volume of wastewater produced by a given plant in the Gum and Wood Chemicals industry ranges from 2,300 to 2,000,000 gallons per day.

Raw wastewaters contain floating and emulsified oils; the organic components of the wastewater include terpenes, natural components of wood, and various solvents. Metals are used as catalysts in some processes of the industry. The type of manufacturing process determines the type of metal found in the waste stream.

Industrial Categorization

This industry has been divided into the following subcategories: char and charcoal briquets; gum rosin and turpentine; wood rosin, turpentine, and pine oil; tall oil rosin, fatty acids, and pitch; essential oils; rosin derivatives; and sulfate turpentine. USEPA has recommended exclusion of char and charcoal briquets, gum rosin and turpentine, and essential oils under Paragraph 8 of the NRDC Settlement Agreement. These subcategories have also been excluded from further consideration herein.

Process Description

The Gum and Wood Chemicals industry began in the United States when early colonists harvested pine oleoresin for use in construction of naval vessels. Since that time, the industry has grown and expanded as new uses have been found for pine products. One of the more significant innovations has been the development of by-products from the Kraft paper process — tall oil and sulfate turpentine — as raw materials for the Gum and Wood Chemicals industry. A description of the manufacturing processes in each of the pertinent subcategories is given below.

Wood Rosin, Turpentine, and Pine Oil: Figure 12 shows a typical process flow diagram. Pine stumps are washed and the wash water flows to a settling pond from which water recycles back to the washing operation. Wood hogs, chippers, and shredders mechanically reduce the wood stumps to chips approximately 2 inches in length and 1/16 inch thick. Chips are fed to a battery of retort extractors, which employ the following steps: water is removed from the chips by azeotropic distillation with a water-immiscible solvent; the resinous material is extracted from the wood chips with a water-immiscible solvent; and residual solvent is removed from the spent wood chips by steaming. After the steaming step, spent chips are removed from the retort and sent to the boilers as fuel. Vapors from the entrainment separator are condensed and proceed to one or more separators where the solvent-water mixture separates. The solvent is recycled for use in the retorts.

Extract liquor is sent to a distillation column to separate the solvent from the products. Overhead from the column is condensed and enters a separator where condensed solvent is removed and recycled to the retorts. The vapor phase from the separator condenses in a shell-and-tube ex-

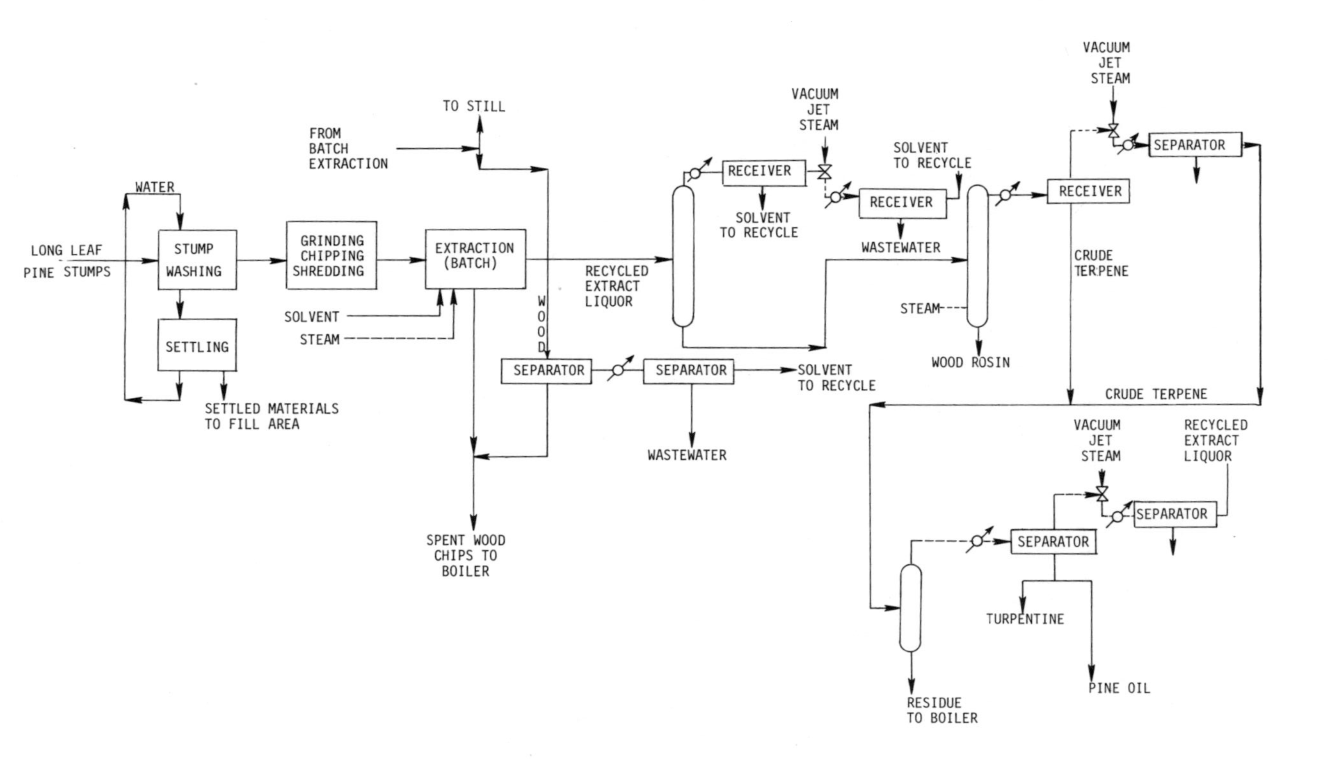

Figure 12: Wood rosin, turpentine, and pine oil process flow diagram via solvent extraction.

changer and enters a separator in which the remaining solvent is separated. Solvent is sent to recycle and wastewater to treatment.

Bottom stream from the first distillation column enters a second distillation column. Steam introduced into the bottom of the tower strips off the volatile compounds. This overhead steam enters a condenser and separator. A portion of the condensed liquor phase is refluxed back to the distillation column, but a larger portion is stored as crude terpene for further processing. The nonaqueous phase from the separator is stored as crude terpene while the aqueous phase is removed as wastewater. Bottom stream from the second distillation column is the finished wood rosin product. Crude terpene removed in the second distillation column is stored until a sufficient quantity accumulates for processing in a batch distillation column.

Tall Oil Rosin, Fatty Acids, and Pitch: Figure 13 shows a typical process flow diagram. Crude tall oil is treated with dilute sulfuric acid to remove some residual lignins as well as mercaptans, disulfides, and color materials. Acid wash water is discharged to the process sewer. The stock then proceeds to the fractionation process. In the first fractionation column, pitch is removed from the bottoms and is either sold, saponified for production of paper size, or burned in boilers as fuel. The remaining fraction of the tall oil (rosin and fatty acid) proceeds to the pale plant, which improves the quality of the raw materials by removing unwanted materials such as color bodies. The second column separates low-boiling point fatty acid material, while the third column completes the separation of fatty and rosin acids.

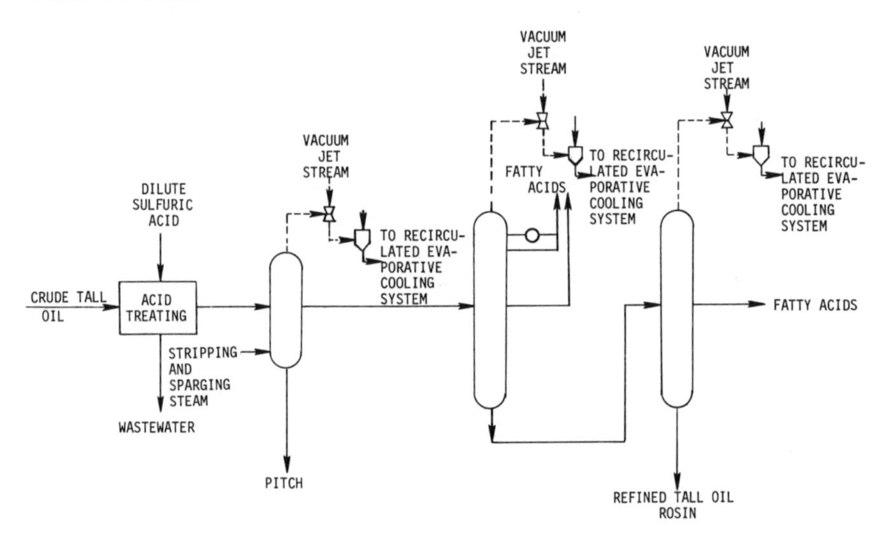

Figure 13: Typical crude tall oil fractionation and refining–process flow diagram.

Wastewater generated in this subcategory results from vacuum jet ejectors on the distillation towers. This water generally is recycled, with excess water discharged to the plant sewer.

Rosin Derivatives: Figure 14 illustrates a typical process flow diagram. Process operating conditions in the reaction kettle depend on product specifications, raw materials, and other variables. A simple ester is produced from stump wood rosin and U.S.P. glycerin under high-temperature vacuum conditions. A steam sparge (lasting approximately two to three hours) removes excess water of esterification. This allows completion of the reaction and removes fatty acid impurities for compliance with product specifications. Condensable impurities are condensed in a noncontact condenser on the vacuum leg and stored in a receiver. Noncondensables escape to the atmosphere through the reflux vent and steam vacuum jets. Production of phenol and maleic anhydride modified tall oil resin ester is similar to simple rosin ester production except that steam sparging is seldom used and other polyhydric alcohols may be used in the product formulation.

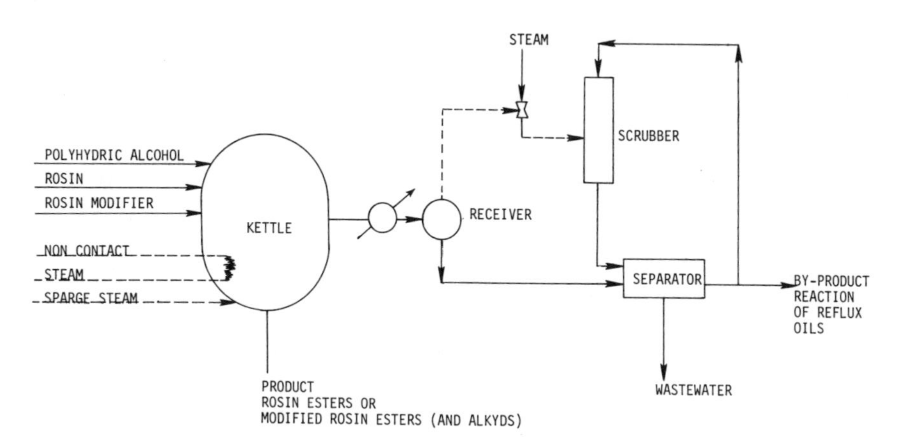

Figure 14: Typical rosin derivative process flow diagram.

Wastewater comes from the chemical reaction, separation of product, and wash down of reaction vessels.

Sulfate turpentine is condensed from the relief gas from the digestor of the Kraft pulping process. During distillation, the first tower usually strips odor-causing mercaptans from the turpentine. Subsequent fractionation breaks the turpentine into its major components: alpha-pinene, beta-pinene, dipentene, and sulfated pine oil. Minor components include, camphene, and anethole.

Distillation of sulfate turpentine is an intermediate production step. Some of these turpentine components are marketed after distillation, but the majority of them remain in the plant for further processing.

Operations are usually batch reactions that take place in reaction kettles in the presence of some organic solvent and metal catalyst. Selection of catalysts and solvents depends on the desired products, of which there are approximately 200.

Wastewater is usually generated from the condensation in the distillation tower and from washdown of reactors.

Wastewater Characteristics

Table 18 shows conventional and toxic raw wastewater characteristic data for the nonexcluded subcategories. Oil and grease, volatile organics and metals are significant pollutants, as is phenol, an acid extractable.

Table 18: Gum and Wood Chemicals—Raw Wastewater Characteristics

	Subcategory[a]			
Wastewater parameter	Wood rosin, turpentine, and pine oil	Tall oil rosin, fatty acids, and pitch	Rosin derivatives	Sulfate turpentine
Conventional (mg/l)				
BOD	1,500	42	1,260	1,200 - 4,800
COD	1,200	1,100	31,000	7,000 - 19,000
Oil and Grease	10	48	92	14 - 1,800
Suspended Solids	240	44	71	170 - 300
Toxic (ug/l)				
Acid Extractables				
Phenols			2,300 - 10,620	10 - 150
Phenols (total)	460	550	41,000 - 53,000	530 - 6,000
Metals				
Arsenic	10		41 - 53	10 - 110
Cadmium			95 - 120	
Copper	33	150	180 - 300	33 - 2,700
Chromium	1,500	83	34 - 62	49 - 510
Lead	15	14	49 - 72	10 - 57
Nickel		19	34 - 100	66 - 220
Selenium		11		10
Zinc	160	50	38,000	
Volatile Organics				
Benzene			170 - 710	90 - 180
Bis (2-ethylhexyl) phthalate				10
Carbon Tetrachloride		120		
Chloroethane			520	
Chloroform	10	10		80 - 1,000
Ethylbenzene	50	20	2,200 - 28,000	6,600 - 67,000
Methylenechloride	190	7.0	2,700 - 7,300	510 - 16,000
Toluene		20	4,000 - 17,000	10 - 40,000
1,1,1-Trichloroethane			830	10 - 640

[a] Char and Charcoal Briquets, Gum Rosin and Turpentine, and Essential Oil categories are recommended for exclusion under Paragraph 8 of the NRDC Settlement Agreement; therefore, no data are presented for these subcategories. Production based data is not available.

Control and Treatment Technology

Approximately 40 percent of the Gum and Wood Chemical industry facilities are indirect dischargers. At present, most of these utilize primary treatment. While little data is available to the treatability of gum and wood chemical waste, considerable data exist on treatability of electroplating wastes, which gum and wood chemical waste somewhat resembles. This should aid in future application of more advanced waste treatment technologies.

In-Plant Control Technology: Two in-plant controls have particularly wide usage in Gum and Wood Chemicals facilities. In the Wood Rosin, Turpentine and Pine Oil subcategory, stump wash water is frequently recycled. After settling, wastewater is returned to the stump wash cycle. Recycle is also practiced in the Tall Oil Rosin, Pitch, and Fatty Acids subcategory. Oily water from tall oil barometric condensers is circulated through "oily cooling towers" which skim excess oil and allow subsequent reuse in the facility's main cooling towers. Both of these methods reduce total wastewater flow and therefore allow more efficient treatment.

End-of-Pipe Treatment Technology: Oil-water separation is the most widely used treatment technology in the Gum and Wood Chemicals Manufacturing category. Baffle separators are commonly used. Clarification, often in conjunction with the addition of chemical aids, is also utilized by much of the industry.

Equalization, air flotation, neutralization, metals precipitation, biological treatment, carbon adsorption, and fly ash slurry adsorption have all seen varying degrees of utilization in the industry. Table 19 shows removal efficiencies obtained from sampling conducted at three plants for the following treatment technologies: biological treatment, granular activated carbon, and fly ash slurry adsorption.

Type of Residue Generated

Information on types and characteristics of residues generated by this industry is insufficient and cannot be discussed at this time.

Residual Management Options

Lack of sufficient information on types and characteristics of residues generated does not permit a discussion of specific management options at this time.

INORGANIC CHEMICALS

General Industry Description

Inorganic Chemicals manufacturing plants tend to be large, producing multiple products by application of basic chemical reactions and/or physical separation techniques from ores or natural brines. Plants tend to be located near raw materials or sources of needed chemicals.

Table 19: Gum and Wood Chemicals—Removal Efficiencies of Selected End-of-Pipe Treatment Technologies—Conventional and Toxic Pollutants[a]

Wastewater parameter	Treatment technology removal efficiencies - percent		
	Biological treatment[b]	Granular activated carbon[c]	Fly ash adsorption[d]
Conventional			
BOD	90-95	80	
Chlorides		48	
COD	53-85		34
Dissolved Solids		16	
Oil and Grease	33-44	92	
Suspended Solids	56-70	84	
Total Solids		20	
Toxic			
Acid Extractables			
Phenols		88	
Phenols (Total)	0		
Metals			
Arsenic	0-30		
Cadmium		76	
Chromium	79-85	77	
Copper		72	
Lead	23		
Nickel		68	
Zinc	66-81	74	
Volatile Organics			
Benzene			90
Chloroform	66		
Ethylbenzene			100
Methylene Chloride	0		
Toluene	98-100		
Trichloroethylene			100

[a] Reported values are based on the performance of one sample plant for each treatment technology.

[b] Treatment system consists of equalization, ash settling, and aerated lagoon. Removals shown are for aerated lagoon only.

[c] Treatment system consists of oil-water separation, neutralization, dissolved air flotation, filtration, and granular activated carbon. Removals shown are for granular activated carbon only.

[d] Ash is from wood chip fired boiler.

Industrial Categorization

The inorganic products segment of this industry has been divided into a total of 62 subcategories as shown in Table 20. For the purpose of effluent limitation guidelines, only the nine most significant subcategories (Table 20, subcategories 1, 3-7, 11, 12, and 14) are discussed herein. Of these nine, the hydrogen peroxide (3) subcategory is also being considered for possible exclusion based on wastewater characteristics. Table 20 also summarizes the rationale for the exclusion of the remaining subcategories.

Process Description

Chlor-Alkali-Mercury Cell Process (1A): Caustic and chlorine are produced from either NaCl or KCl as shown in Figure 15. Raw material is

Table 20: Inorganic Chemicals Industrial Categorization[a]

Designation	Subcategory	Comments
1	Chlor-Alkali	Significant Subcategory
1A	Mercury Cell Process	Significant Subcategory
1B	Diaphragm Cell Process	Significant Subcategory
2	Hydrofluoric Acid	No Discharge to POTWs
3	Hydrogen Peroxide	Significant Subcategory
4	Titanium Dioxide	Significant Subcategory
4A	Sulfate Process	Significant Subcategory
4B	Chloride Process	No Discharge to POTWs
5	Aluminum Fluoride	Significant Subcategory
6	Chrome Pigments	Significant Subcategory
7	Hydrogen Cyanide	Significant Subcategory
7A	Andrussow Process	Significant Subcategory
7B	Acrylonitrile By-Product	Zero Discharge
8	Sodium Dichromate	No Discharge to POTWs
9	Carbon Dioxide	No Process-Related Priority Pollutants Generated
10	Carbon Monoxide/Hydrogen	No Process-Related Priority Pollutants Generated
11	Copper Sulfate	Significant Subcategory
12	Nickel Sulfate	Significant Subcategory
13	Silver Nitrate	Included in Inorganic Chemicals-Phase II
14	Sodium Bisulfite	Significant Subcategory
15	Sodium Hydrosulfite	No Discharge to POTWs
16	Hydrochloric Acid	Characterized by small flows and waste loads
17	Nitric Acid	No Process-Related Priority Pollutants Generated
18	Sodium Carbonate	Only one plant with Wet Process Discharge
19	Sodium Metal	Insignificant Priority Pollutant Discharge
20	Sodium Silicate	Characterized by small flows and waste loads
21	Sulfuric Acid	Characterized by small flows and waste loads
22	Ammonium Chloride	Characterized by small flows and waste loads
23	Ammonium Hydroxide	Zero Discharge
24	Barium Carbonate	Insignificant Priority Pollutant Discharge
25	Boric Acid	Characterized by small flows and waste loads
26	Calcium Carbonate	Insignificant Priority Pollutant Discharge
27	Copper Oxide	Only one plant with Wet Process Discharge
28	Manganese Sulfate	Only one plant with Wet Process Discharge
29	Strong Nitric Acid	No Process-Related Priority Pollutants Generated
30	Oxygen and Nitrogen	Characterized by small flows and waste loads
31	Potassium Iodide	Characterized by small flows and waste loads
32	Sodium Hydrosulfite	Characterized by small flows and waste laods
33	Sodium Silicofluoride	Included in Inorganic Chemicals-Phase II
34	Sodium Thiosulfate	Insignificant Priority Pollutant Discharge
35	Sulfur Dioxide	Insignificant Priority Pollutant Discharge
36	Bromine	Zero Discharge
37	Calcium Hydroxide	Zero Discharge
38	Chromic Acid	Zero Discharge
39	Fluorine	Zero Discharge
40	Hydrogen	Zero Discharge
41	Iodine	Zero Discharge
42	Potassium Chloride	Zero Discharge
43	Aluminum Chloride	Zero Discharge
44	Zinc Sulfate	Zero Discharge
45	Calcium Carbide	Zero Discharge
46	Calcium Oxide	Zero Discharge
47	Potassium	Zero Discharge
48	Potassium Sulfate	Zero Discharge
49	Sodium Bicarbonate	Zero Discharge
50	Borax	Zero Discharge
51	Ferric Chloride	Zero Discharge
52	Lead Monoxide	Zero Discharge
53	Sodium Fluoride	Zero Discharge
54	Aluminum Sulfate	Zero Discharge
55	Potassium Dichromate	Zero Discharge
56	Calcium Chloride	Included in Inorganic Chemicals-Phase II
57	Sodium Chloride	Included in Inorganic Chemicals-Phase II
58	Sodium Sulfite	Included in Inorganic Chemicals-Phase II
59	Potassium Permanganate	Only one plant with Wet Process Discharge
60	Zinc Oxide	Only one plant with Wet Process Discharge
61	Lithium Carbonate	Only one plant with Wet Process Discharge
62	Ferrous Sulfate	Covered under Subcategory 4A

[a] Phase II subcategories have not been addressed by USEPA to date.

purified by dissolution in water followed by barium carbonate treatment to precipitate magnesium and calcium ions. Brine is then fed to the mercury cell, where chlorine is liberated at one electrode and sodium-mercury amalgam is formed at the other. The chlorine is cooled, dried in a sulfuric acid stream, purified, compressed, and sold. The mercury-sodium amalgam is decomposed by water treatment in a "denuder" to form NaOH (or KOH) and hydrogen, as products. Wastewaters consist of purification muds [$CaCO_3$, $Mg(OH)_2$, and $BaSO_4$] from brine purification, some spent brine solutions, and condensates from chlorine and hydrogen compressions. Wastes also contain mercury.

Chlor-Alkali-Diaphragm Cell Process (1B): Products and wastes of the diaphragm cell process are similar to those from the mercury cell process except that the cell is manufactured differently and mercury is not usually present in the effluent.

Hydrogen Peroxide (3): Hydrogen peroxide is manufactured by three different processes: (1) an organic process involving the oxidation and reduction of anthraquinone; (2) an electrolytic process; and (3) a by-product of acetone manufacture from isopropyl alcohol.

In the organic process, anthraquinone (or an alkylanthraquinone) in an organic solvent is catalytically hydrogenated to yield a hydroanthraquinone. This material is then oxidized with oxygen or air back to anthraquinone, with hydrogen peroxide being produced as a by-product. The peroxide is water-extracted from the reaction medium, and the organic solvent and anthraquinone are recycled. Recovered peroxide is then purified and shipped.

In the electrolytic process, a solution of ammonium bisulfate is electrolyzed. Hydrogen is liberated at the cathodes of the cells, and ammonium persulfate is formed at the anode. Persulfate is then hydrolyzed to yield ammonium bisulfate and hydrogen peroxide which is separated from the solution by fractionation. The ammonium bisulfate solution is then recycled, and peroxide is recovered for sale. Raw wastes consist of ammonium bisulfate losses, ion exchange losses, boiler blowdowns and some cyanide wastes from the special batteries used in electrolysis. Wastewaters contain alkalinity, dissolved solids, and some metals (e.g. iron).

Titanium Dioxide-Sulfate Process (4A): The sulfate process, shown in Figure 16, uses ilmenite ore (containing 40 to 70 percent titanium dioxide) and sulfuric acid as raw materials. Titanium dioxide-bearing ores are dissolved in sulfuric acid at high temperatures to produce titanium sulfate as an intermediate product. In some cases, small amounts of antimony trioxide are also added. The acid solution is clarified, a portion of the iron sulfates is removed by crystallization, and the titanium sulfate is hydrolyzed to form a white, nonpigmentary hydrate. The hydrate is calcined to form crystalline titanium dioxide, which is milled, surface treated, and packaged for sale. The process generates large amounts of sulfuric acid

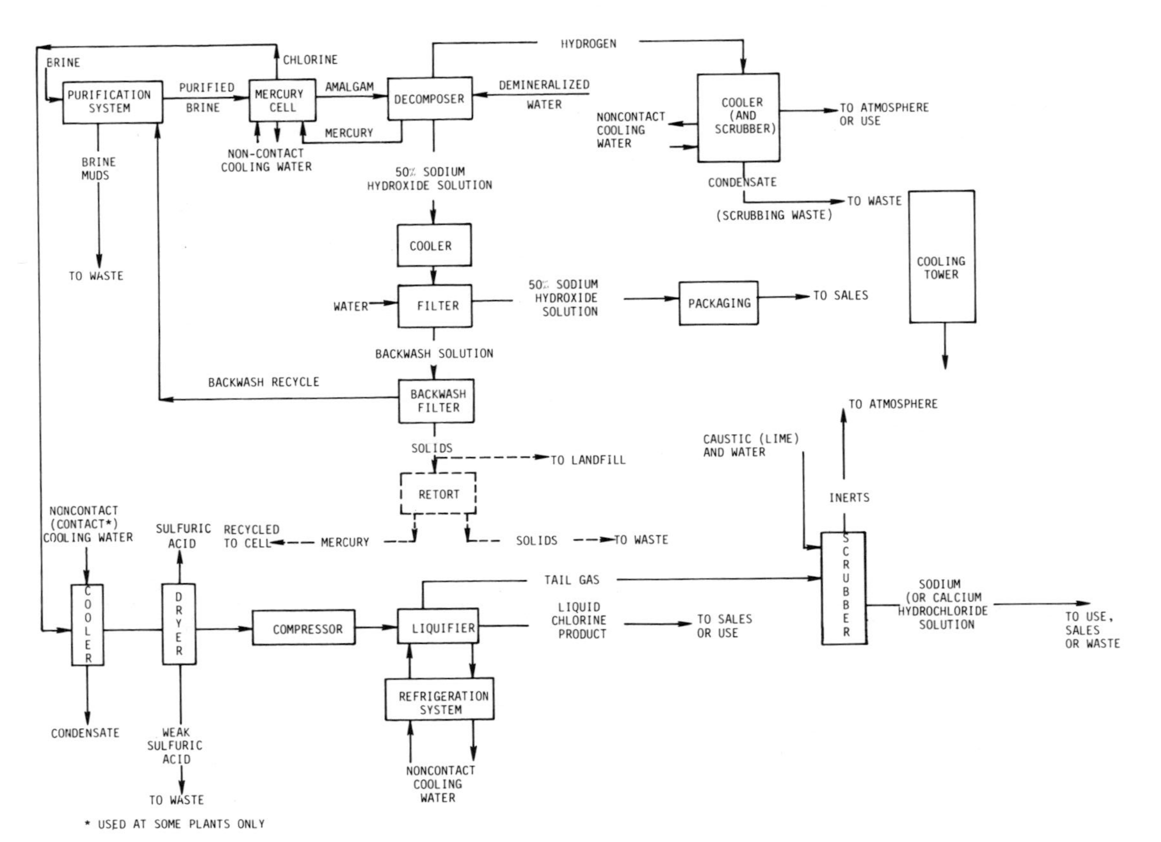

Figure 15: General process flow diagram for production of chlorine/caustic by mercury cells.

waterborne wastes which contain dissolved solids, suspended solids, and metal salts. Solid wastes include ferrous sulfate and a hydrated by-product.

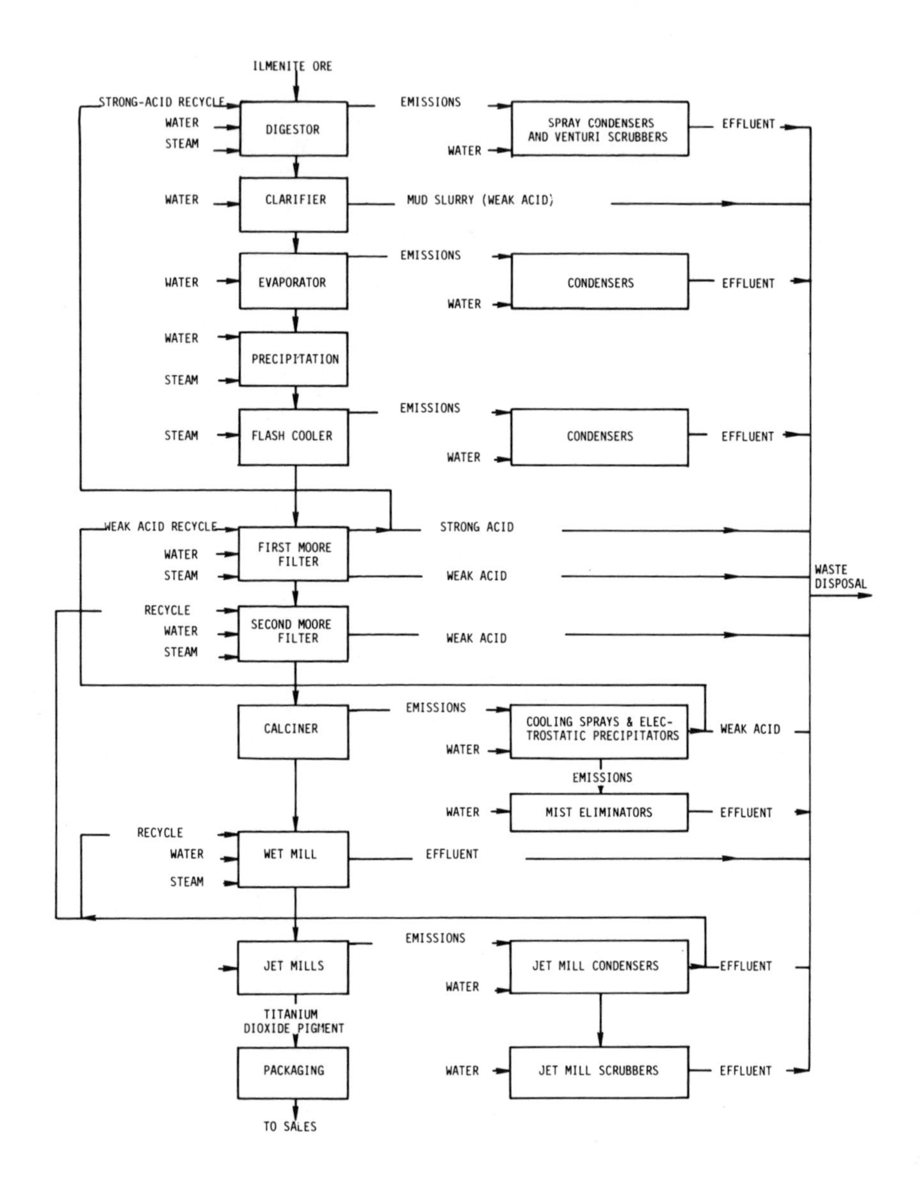

Figure 16: General process flow diagram for production of titanium dioxide by sulfate process.

Aluminum Fluoride (5): Figure 17 shows a general process flow diagram for production of aluminum fluoride. Partially dehydrated alumina hydrate (Al_2O_3) is reacted with hydrofluoric acid (HF) gas. The product, AlF_3, is formed as a solid, and is cooled with noncontact cooling water before being sent for milling and shipping. Gases from the reactor are scrubbed with water to remove unreacted HF before being vented to the atmosphere. Sources of wastewater include the noncontact cooling water, floor and equipment washings, and scrubber wastewater. The major source of wastewater is the scrubber wastewater contaminated with hydrofluoric acid, aluminum fluoride, aluminum oxide, sulfuric acid, and silicon tetrafluoride.

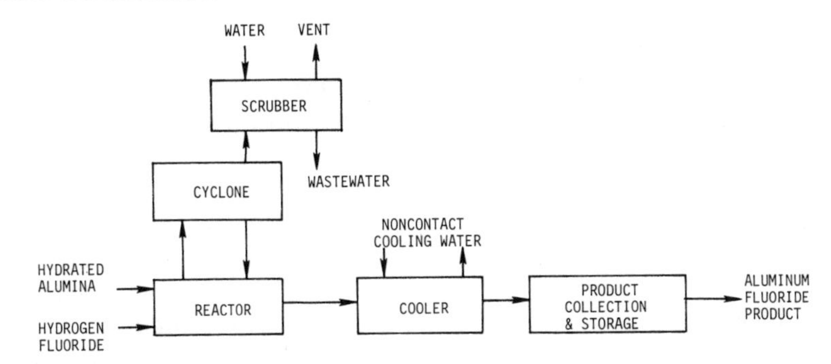

Figure 17: General process flow diagram for production of aluminum fluoride.

Chrome Pigments (6): Chrome pigments are a family of inorganics, including anhydrous chromium oxide, hydrated chromium oxide (Guigets Green), chrome yellow, chrome orange, molybdenum orange, chrome green, and zinc yellow (zinc chromate). Figure 18 shows a generalized process diagram applicable to all chrome pigment plants. All wastewaters generated in the chrome pigments subcategory contain dissolved chromium and pigment particulates. Therefore, wastewater sources are similar for all pigment products except that at chrome oxide plants an additional scrubber waste is generated. Additional pollutants that can be present are given below for each major pigment group.

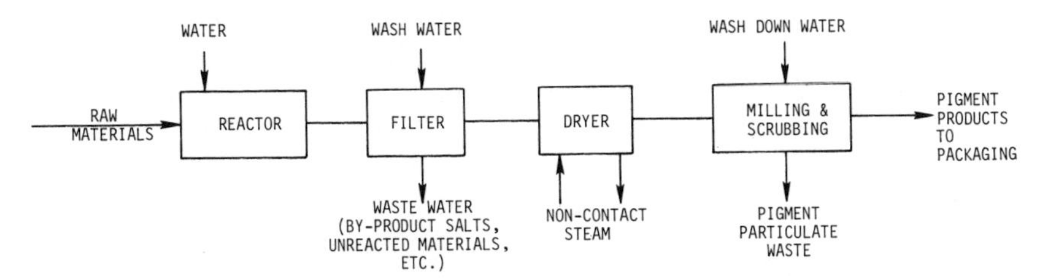

Figure 18: General process diagram for production of chrome pigment complexes.

Anhydrous Chrome Oxide — Raw wastewater contains sodium sulfate if sodium dichromate and sulfur are used as raw materials, and contains sodium carbonate if sodium dichromate and carbon are used as raw materials instead.

Hydrated Chromium Oxide (Guigets Green) — Boric acid is used in the preparation of hydrated chromium oxide; the wastewater contains sodium borate and boric acid.

Chrome Yellow and Chrome Orange — Raw wastewater contains sodium acetate, sodium chloride, sodium nitrate, sodium sulfate, and lead salts.

Molybdenum Orange — Raw wastewater contains sodium chloride, sodium nitrate, sodium sulfate, chromium hydroxide, lead salts, and silica.

Chrome Green — Raw wastewater contains sodium nitrate, sodium chloride, ammonium sulfate, ferrous sulfate, sulfuric acid, and iron blue pigment particulates.

Zinc Yellow — Raw wastewater contains hydrochloric acid, sodium chloride, potassium chloride, and soluble zinc salts.

Hydrogen Cyanide-Andrussow Process (7A): Air, ammonia, and methane are reacted at elevated temperatures (900 to 1000°C) over a platinum catalyst to produce the dominant product hydrogen cyanide and water. Waterborne wastes from the process, including distillation bottoms, scrubber streams, cooling water blowdowns, and wash waters, consist principally of ammonia, sulfates, cyanide, and nitriles.

Copper Sulfate (11): Copper sulfate is produced by reacting copper metal with sulfuric acid, air, and water. The resulting copper sulfate solution is either sold or fed to crystallizers and driers producing copper sulfate crystals. Some plants do not start with copper metal but use a by-product stream from copper refineries which consists of copper, sulfuric acid, and a small amount of nickel. The solution needs to be strengthened by the addition of more copper but the same chemical reaction applies. Waste sources may include noncontact cooling waters, wash downs, spills, steam condensates from evaporators, and copper sulfides as solid waste. Plants that produce copper sulfate in the liquid form have no contact waste streams from the process.

Nickel Sulfate (12): Nickel sulfate is produced by reacting pure nickel (or pure nickel oxide) with pure sulfuric acid. The nickel sulfate solution produced is filtered and sold, or processed further using a crystallizer, a drier, and a screen to produce a solid nickel sulfate product. Figure 19 shows a general process flow diagram for production of nickel sulfate. The use of impure raw materials produces a nickel sulfate solution which must be purified in sequence with oxidizers, lime, and sulfides to precipitate impurities which are then removed by filtration. Purified nickel sulfate solution can be sold, or processed further for production of solid nickel sulfate product. Noncontact cooling water used for nickel sulfate production in the reactor and in crystallizers is the main source of wastewater. Direct process contact water comes from the preliminary preparation of spent plating solutions used in the process. Plants which use impure nickel raw materials generate a filter backwash waste stream with high impurity levels. Wash downs, spills, leaks, and maintenance uses account for the remaining wastes.

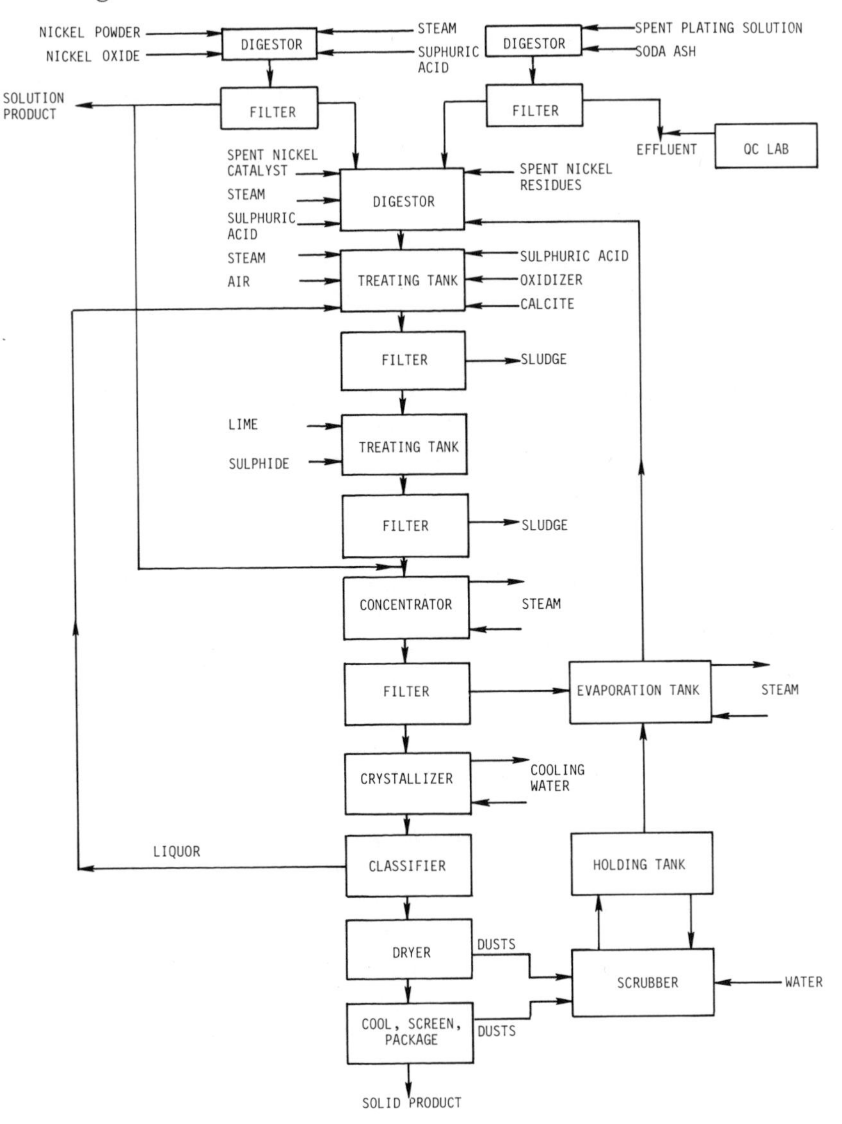

Figure 19: General process flow diagram for production of nickel sulfate.

Sodium Bisulfite (14): A slurry of sodium bisulfite crystals is produced by reacting sodium carbonate with sulfur dioxide and water. It can be sold or processed further to form anhydrous sodium metabisulfite. This requires thickening, centrifuging, drying, and packaging operations. Direct process contact water is the main source of wastewater which must be treated, together with miscellaneous wastes such as water used for maintenance purposes, wash downs, and spill cleanup.

Wastewater Characteristics

Table 21 shows raw wastewater characteristic data. Table 22 shows raw wastewater characteristics as a function of manufacturing production.

Control and Treatment Technology

In-Plant Control: The following in-plant control measures can reduce wastewater effluents from the inorganic chemicals manufacturing industry:

- Muds produced in various processes can be clarified and disposed of in landfills. For example, mercury from subcategory 1A (Chlor-Alkali - Mercury Cell Process) can be precipitated with sulfides. Asbestos in Subcategory 1B (ChlorAlkali - Diaphragm Cell Process) can be filtered or clarified. Some salts can be recycled back into the process.
- Clarification with skimming can reduce the organic solvents and suspended solids in subcategory 3 (Hyrdogen Peroxide Organic Process). Scrap iron decomposes the hydrogen peroxide in the waste stream. The Electrolytic Process produces a very small process waste stream, and total evaporation has been used to eliminate this stream.
- Chemical precipitation followed by clarification and land disposal of sludges is practiced for titanium dioxide production in subcategory 4A.

End-of-Pipe Treatment Technology: Wastewater treatment technologies commonly used in the Inorganic Chemicals Manufacturing industry include hydroxide precipitation, ferrite co-precipitation, sulfide precipitation, settling, filtration, flotation, centrifugation, magnetization, the Xanthate Process, ion exchange, reduction processes, oxidation processes, membrane processes, adsorption, and fluoride removal processes. The type, degree, and cost of treatment depend upon the particular chemical. The important technologies are briefly described below:

- Hydroxide precipitation is a widely used technology for removing trace metals from wastewaters, with lime or caustic soda commonly used to supply the hydroxide ions. Under suitable conditions, the metals form insoluble metal hydroxides which can be separated

Table 21: Inorganic Chemicals Industry—Raw Wastewater Characteristics

Wastewater parameter[b]	Subcategories[a]										
	1A	1B[c]	1B[d]	3	4A	5	6	7A	11	12	14
Conventional Pollutants											
Ammonia								1,381			
BOD_5								1,549			
Chemical Oxygen Demand											27,970
Cyanide							812	31			
pH (units)	6.7 - 8.5										
Total Suspended Solids	5,400		150		8,000			160	392	8,000	3,730
Toxic Pollutants											
Metals and Inorganics											
Antimony	0.95	1.91	0.04	X	1.40	X	7.7	X	0.33	X	0.65
Arsenic	0.4	0.68	0.66	X	0.34	0.48	X	X	3.50	X	X
Beryllium	X	X	X	X	X	X	X	X	X	X	X
Cadmium	0.79	0.05	0.06	X	0.34	0.03	1.25	X	0.87	0.16	0.04
Calcium									X	X	X
Chromium	0.24	0.3	18.75	X	123.6	1.14	349	X	0.14	1.3	3.36
Copper	1.48	7.45	16.65	X	1.48	0.24	7.5	X	1,850	73.3	0.93
Iron									X	X	X
Lead	1.9	1,631	2.0	X	5.19	X	69	X	0.18	0.12	1.05
Mercury	27.6	0.07	0.35	X	X	0.01	X	X	X	0.01	0.02
Nickel	2.54	0.64	54.4	X	6.37	0.29	0.74	X	112	1,115	0.46
Selenium	X	X	0.09	X	X	0.11	0.03	X	0.01	0.14	X
Silver	1.46	X		X	0.06	X	0.02	X	X	X	0.03
Thallium	0.65	X	0.01	X	0.04	X	X	0.025	X	0.02	X
Organic Compounds											
Bis (2-ethylhexyl) Phthalate	X	0.120	X	X	X	X	X	X	X	X	X
Benzene	X	0.015	X	X	X	X	X	X	X	X	X
Carbon Tetrachloride	X	0.197	X	X	X	X	X	X	X	X	X
Chloroform	X	0.691	X	X	X	X	X	X	X	X	X
Dichlorobromomethane	X	0.309	X	X	X	X	X	X	X	X	X
1,2-Dichloroethane	X	0.621	X	X	X	X	X	X	X	X	X
2-4 Dinitrophenol	X	X	X	X	X	X	X	X	X	X	X
Hexachloroethane	X	0.090	X	X	X	X	X	X	X	X	X
Napthalene	X	X	X	0.011	X	X	X	X	X	X	X
Pentachlorophenol	X	X	X	4.850	X	X	X	X	X	X	X
Phenol	X	X	X	0.029	0.060	X	0.073[e]	X	X	X	X
Tetrachloroethylene	X	0.196	X	X	X	X	X	X	X	X	X
1,1,1-Trichloroethane	X	X	X	X	X	X	X	0.244	X	X	X

[a] See Table 20 for identification of subcategories.

[b] Maximum concentration in mg/l except as noted; X denotes priority pollutants in insignificant concentrations; a blank space indicates that the pollutant was not found in wastewater samples.

[c] Graphite anode.

[d] Metal anode.

[e] Organic pigment process.

Table 22: Inorganic Chemicals–Raw Wastewater Characteristics–Production Based Data[a]

Pollutant parameter	Subcategory[b]									
	1A	1B[c]	1B[d]	4A	5	6	7A	11	12	14
Conventional										
Aluminum					4.4					
Ammonia nitrogen						0.16				
BOD							20.2			
COD										2.94
Cyanide (total)	0.41	1.20								
Fluoride					11.9					
Iron				602						
TSS	2.19		0.069	211	53.7	70.4	1.94	0.097		0.27
Inorganics										
Antimony	0.00045	0.00026		0.21		0.87		0.00069		0.000052
Arsenic	0.0003	0.0028		0.014	0.0016			0.0078		
Cadmium	0.00005			0.027	0.0002	0.16		0.0019	0.00017	0.00001
Chromium	0.00009	0.00026	0.00095	2.11	0.0035	21.5			0.00025	0.011
Copper	0.00033	0.0019	0.00041	0.12	0.0033	0.86		4.11	0.01	0.00046
Lead	0.00032	0.273	0.00004	0.089		6.49		0.00039	0.0001	0.00009
Mercury	0.016	0.00002			0.00005				0.00003	
Nickel	0.00026	0.00054	0.00064	0.12	0.003	0.0325		0.25	1.20	0.00031
Selenium					0.0015				0.00003	
Silver	0.00022						0.0009			
Thallium	0.0003			0.0078						
Zinc	0.0023	0.00054	0.00024	0.57		8.63		0.024		0.0053
Toxic organics										
Bis(2-ethylhexyl) phthalate			0.00001							
Carbon tetrachloride		0.0003								
Chloroform			0.0011							
Dichlorobromomethane			0.0005							
1,2-dichloroethane			0.001							
Hexachlorobutadiene			0.00005							
Hexachloroethane			0.00014							
Phenol				0.002		0.015				
Tetrachloroethylene			0.00046							

[a] Units are in pounds of pollutant per 1,000 pounds of product produced, unless otherwise noted.

[b] See Table 20 for identification of subcategory.

[c] Graphite anode.

[d] Metal anode.

from solution by settling, filtration, flotation, centrifugation, and/ or membrane processes.

- Ferrite co-precipitation is a process for the removal of heavy metals from acidic wastewater. Treatment is by adding a ferrous salt to the metal bearing wastewater, then neutralizing and oxidizing the complex heavy metal-ferrous hydroxide precipitate by aeration to form the stable insoluble ferrite (ferromagnetic oxide) co-precipitates with large particle sizes. Ferrites, which are marketable residuals, can be easily separated by magnetization or by filtration. Very high removal efficiencies can be achieved for most of the common heavy metals, including mercury and hexavalent chromium.
- Sulfide precipitation involves the addition of sulfide to the metal-bearing wastewater to precipitate the metals as metal sulfides, and subsequent physical separation of insoluble metal sulfides by gravity settling or filtration. Sodium sulfide and sodium bisulfide are two chemicals commonly used. Toxicity of sulfides warrants both care in application and post treatment systems to remove excess sulfide.
- The Xanthate Process uses xanthate for the removal of trace metals from waste streams. Xanthates contain functional groups capable of forming insoluble complexes with metals, and the sludge so formed can be separated by settling, filtration, or centrifugation. Xanthates are effective in removing cadmium, chromium, copper, iron, lead, mercury, nickel, silver, and zinc from wastewaters.
- Ion exchange is a chemical reaction between the ions in solution and the ionic sites on an exchange resin, and is a proven technology for reducing metals down to low concentration levels. The technology is used only in limited industrial pollution abatement applications because of the high cost associated with the process.
- Reduction processes involve the addition of a reducing agent to convert the soluble metals from a higher valency state to a lower one in order to facilitate a desired chemical reaction. Sodium bisulfite, sodium metabisulfite, sulfur dioxide, and ferrous salts can convert soluble, toxic, hexavalent chromium to soluble, much less toxic, trivalent chromium. The latter can form an insoluble chromium hydroxide by hydroxide precipitation, and be subsequently removed by settling or filtration. Another common reduction process is the application of sodium borohydride to reduce metals in waste streams. Soluble divalent mercury ion can be reduced by sodium borohydride to insoluble metallic mercury which is then removed from solution by carbon adsorption.
- Oxidation processes include wet oxidation, incineration, biological oxidation processes, and chemical oxidation. Wet oxidation is a process in which an aqueous waste, such as acrylonitrile liquor, can be oxidized in the liquid phase in a closed, high temperature, high

pressure vessel. Incineration can be a combination of oxidation and pyrolysis, in which oxidation involves actual reaction of organic pollutants with oxygen, while pyrolysis refers to rearrangement or breakdown of organic molecules. Biological oxidation includes activated sludge, trickling filter, rotating biological contactor, aerated lagoon, etc. Chemical oxidation involves the use of oxidizing agents, such as ozone, hydrogen peroxide, chlorine, hypochlorite, potassium permanganate, chlorine dioxide, oxygen, etc. for the removal of cyanides, sulfite, ammonia, and other harmful species in dilute waste streams. For example, in the presence of an oxidizing agent, cyanide can be oxidized to cyanates which can be further decomposed into nitrogen and carbon dioxide by excess chlorination or acid hydrolysis.

- Membrane processing units have semipermeable barriers which allow the transport of some ionic molecules and retain others. The driving force can either be electropotential differences (electrodialysis) or pressure difference (reverse osmosis and ultrafiltration).
- Adsorption is a surface phenomenon in which a substance (pollutant) is accumulated on the surface of an adsorbent, such as activated carbon, or activated alumina.
- Fluoride ions in the wastewater can be removed by lime and/or alum treatment, followed by sedimentation, filtration, and activated alumina adsorption.

Table 23 shows removal efficiencies of various treatment technologies. A number of options for the final disposal of waterborne wastes from inorganic chemical manufacturing are available, depending upon quantity and characteristics of the waste stream. They include discharge to surface water, land disposal, and evaporation ponds.

Type of Residue Generated

Information on types and characteristics of residues generated by this industry is insufficient and cannot be discussed at this time.

Residual Management Options

Lack of sufficient information on types and characteristics of residues generated does not permit a discussion of specific management options at this time.

IRON AND STEEL MANUFACTURING

General Industry Description

Steel mills may range from comparatively small plants to completely

Table 23: Inorganic Chemicals Removal Efficiencies of Various Treatment Technologies

Treatment Technology	Treatment Technology Removal Efficiency - Percent														
	Sb	As	Cd	Be	Cu	Cr^{+3}	Cr^{+6}	Pb	Hg^{+2}	Ni	Ag	Se	Tl	Zn	Fe
Activated Alumina		96-99													
Activated Carbon		63-97					85-98								
Alum/Filter	62				93							48	31		
Chloride Precipitation											97^{+}				
Ferrite Coprecipitation/Filter			99^{+}		99^{+}	99		99.9	99.9					99^{+}	
Ferric Chloride/Filter	65	98									98	75-80	30		
Ferric Sulfate/Filter		90						98.5			72-83	75-82			
Ferrous Sulfide (Sulfex)			99^{+}		99			99.9		99.9				99^{+}	
Lime/Ferric Chloride/Filter		98													
Lime/Filter	28	72-85	92-98	99^{+}	90-98	98		$98\text{-}99^{+}$		94-98	80-93	33-38	60	77-93	
Lime/Sulfide			98^{+}												
Sodium Carbonate/Filter								99^{+}							
Sodium Hydroxide/Filter								99^{+}			72			97	
Sulfide/Filter									$87\text{-}99^{+}$		high			97	
Xanthate Process			98		91-96	93	93	99		96				96-97	88-97

integrated steel complexes where great quantities of raw materials and resources are brought together to ultimately produce steel. Even the smallest of plants will generally represent a fair-sized industrial complex. Because of the wide product range, the operations vary significantly within each facility. Great quantities of water are used, both for processing and for cooling purposes. As a result, the iron and steel industry generates large volumes of wastewater.

Industrial Categorization

The industry can be broadly subdivided into six major operational areas: coke making, burden preparation, iron making, steel making, forming and finishing, and miscellaneous. The number and type of pollutant parameters of significance vary with the operation being conducted and the raw materials used. Waste volumes and waste loads also vary with the operation. For the purposes of raw waste characterization and delineation of pretreatment information, the industry is further subcategorized primarily along operational lines with permutations where necessary, as shown in Table 24. The typical integrated steel mill in the industry will embody several of these subcategories and the discharges may be combined.

Table 24: Iron and Steel Manufacturing Subcategorization

Main category	Subcategory	Designation
1. Coke Making	By-product Coke	A
	Beehive Coke	B
2. Burden Preparation	Sintering	C
3. Iron Making	Blast Furnace - Iron	D
	Blast Furnace - Ferromanganese	E
4. Steel Making	Basic Oxygen Furnace (Semi-wet Air Pollution Control Methods)	F
	Basic Oxygen Furnace (Wet Air Pollution Control Methods)	G
	Open Hearth Furnace	H
	Electric Arc Furnace (Semi-wet Air Pollution Control Methods)	I
	Electric Arc Furnace (Wet Air Pollution Control Methods)	J
	Vacuum Degassing	K
	Continuous Casting	L
5. Forming and Finishing	Hot Forming - Primary	M
	Hot Forming - Section	N
	Hot Forming - Flat	O
	Pipe and Tubes	P
	Pickling-Sulfuric Acid-Batch	Q
	Pickling-Hydrochloric Acid-Batch and Continuous	R
	Cold Rolling	S
	Hot Coatings - Galvanizing	T
	Hot Coatings - Terne	U
6. Miscellaneous	Miscellaneous Runoffs - Storage Piles, Casting & Slagging	V
	Cooling Water Blowdown	W
	Utility Blowdown	X
	Maintenance Department Wastes	Y
	Central Treatment	Z

Process Description

Five basic steps are involved in the production of steel in a modern integrated steel mill:

1. Coal is converted into coke by either the by-product process (A) or the beehive process (B). Coke fines generated in these processes are screened out before the coke is used in the blast furnace.
2. Coke is combined with iron ore and limestone in a blast furnace to produce iron (D, E). Waste materials from the blast furnace include sizeable quantities of fine dust which are high in iron content.
3. Iron is converted into steel in either a basic oxygen furnace (F, G), an open hearth furnace (H), or an electric furnace (I, J). Further refinements include degassing (K) by subjecting the steel to a high vacuum. Steel is cast either by continuous casting (L) or in ingot molds. The slag generated in the steel making processes is transported and subjected to a slagging operation where the steel scrap is reclaimed and the slag crushed into a saleable product. Waste materials from the steel making processes include sizeable quantities of fine dust which are high in iron content.
4. Iron bearing waste fines from the blast furnace and steel making processes are blended with limestone and coke fines in a sintering operation (C) for the purpose of agglomerating and recycling the fines back to the blast furnace. Processing of steel plant wastes (burden preparation) by pelletizing or by briquetting has also been proven on a pilot scale and several such plants are due on line in the near future.
5. The final step includes forming and finishing operations. Ingots are reduced to slabs or billets and ultimately to plates, shapes, strips, etc. through the forming operations. The steel finishing operations do little to alter the size or dimensions, but impart desirable surface or mechanical characteristics to the product.

A flow diagram of a typical steel mill is shown in Figure 20.

By-Product Coke (A): In 1980 by-product process produced about 99 percent of all metallurgical coke. Bituminous coal is heated in ovens out of contact with air to drive off the volatile components. The residue in the ovens is coke; the volatile components are recovered and processed to produce tar, light oils, and other materials of potential value, including coke oven gas. Typical products from the carbonization of coal are gas, tar, ammonia, tar acids, hydrogen sulfide, light oil, coke, and coke breeze.

Significant liquid wastes are excess ammonia liquor, final cooling water overflow, light oil recovery wastes, and indirect (non-contact) cooling water. In addition, wastewaters may result from coke wharf drainage, quench water overflow, and coal pile runoff. Final cooling water is a po-

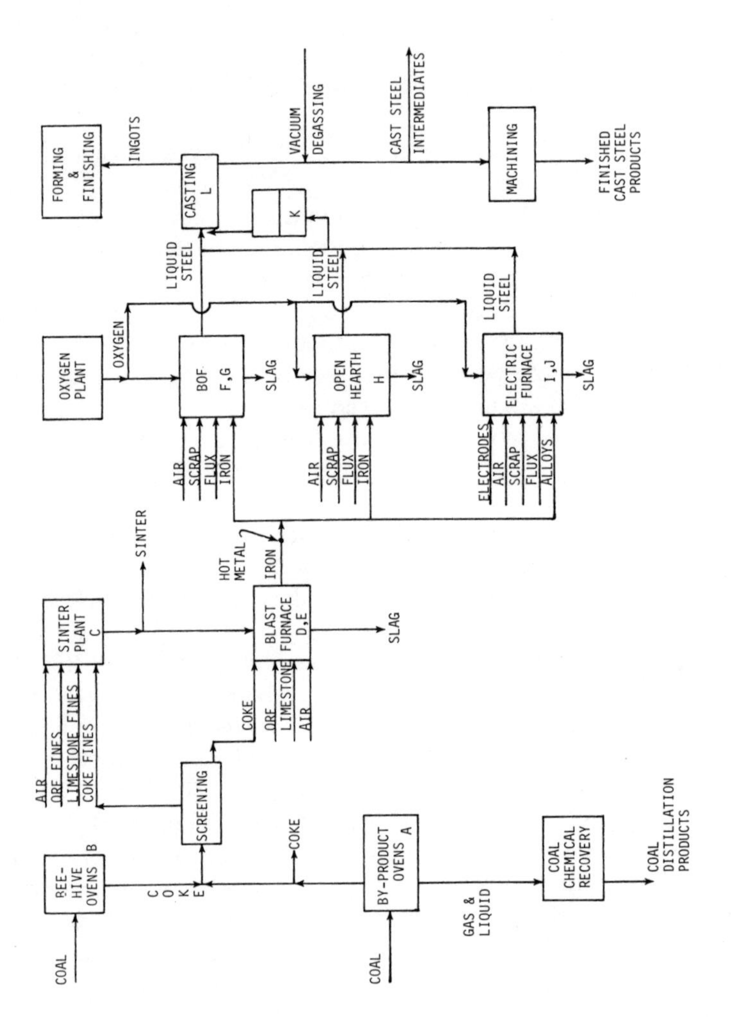

Figure 20: Steel production manufacturing process flow diagram.

tential source of toxic cyanogen compounds. Light oil recovery wastes contain primarily phenol, cyanide, ammonia, and oil. Effluent from the quenching of coke, which is permitted to overflow to the sewer in some plants, contains trace amounts of cyanide, phenol, and solids. Condensed steam and cooling water constitute the bulk of wastewaters discharged to the sewer in this subcategory.

Beehive Coke (B): In this process for manufacturing coke, air is admitted to the coking chamber in controlled amounts for the purpose of burning volatile products distilled from the coal to generate heat for further distillation Beehive produces only coke and no other by-products are recovered. Water is used only for coke quenching.

A properly controlled beehive oven has very little water discharge. In some instances, an impoundment lagoon is provided to collect overflow water and settle out coke fines. Discharges from this pond can contain phenol and cyanide, however recycle with zero discharge is currently practiced in some plants.

Sintering (C): Sintering plants have the primary function of agglomerating and recycling iron bearing waste fines back to the blast furnace. Sintering is achieved by blending iron bearing components and limestone with coke fines which act as a fuel. The mixture is spread on a moving down draft grate and ignited. The down draft keeps the coke burning and the bed is brought to fusion temperature. Hot sinter is crushed, cooled, sized, and formed into pellets and briquets.

Raw wastes from the sintering process emanate from the material handling dust control equipment and the dust and volatized oil in the process gases.

Most modern plants have fabric type dust collectors with no aqueous discharges. However, several plants utilize wet scrubbers and generate wastewaters which contain significant concentrations of suspended matter, oil, sulfide, and fluoride. Aqueous discharges are usually associated with the pelletizing or briquetting operations. However, there is potential for wastewater from wet methods of dust control.

Blast Furnace Iron (D): Virtually all iron made in the world today is produced in blast furnaces which reduce iron ore to metallic iron. Iron ore, limestone, and coke are charged into the furnace. Coke is burned to produce carbon monoxide which reacts with the ore to produce carbon dioxide and metallic iron. The major impurity of most iron ores and coke is silica, which is removed by the limestone which combines with the silica to produce a molten mass called slag. As molten iron leaves the blast furnace, floating slag is skimmed off. Auxiliary operations associated with a blast furnace are raw material storage and handling, air compression and heating, gas cleaning, iron and slag handling, and dust handling.

The blast furnace has two basic water uses — cooling water and gas washer water. Continuous circulation of cooling water is required to prevent the furnace walls from burning through. The principal wastewaters result from the gas cleaning operation. These wastewaters contain significant concentrations of cyanide, phenol, ammonia, sulfide, and suspended solids. Phenol, cyanides, and ammonia originate in the coke and are particularly high if the coke has been quenched with wastewater or has not been completely coked. Suspended solids result from the fines in the burden being carried out in the gas.

Blast Furnace - Ferromanganese (E): The blast furnace charge consists of iron and manganese ores, limestone, and coke. Principal wastewaters are from the gas cleaning operation and contain significant concentrations of cyanide, phenol, ammonia, sulfide, manganese, and suspended solids. Cyanide formation, due to the reaction of carbon from the coke with nitrogen from the blowing air, is particularly high at the higher temperatures of a ferromanganese furnace as compared to an iron furnace.

Basic Oxygen Furnace (F,G): Raw materials for this steel making process are hot metal (iron), scrap steel, limestone, burnt lime, fluorspar, dolomite, and iron ores. Alloying materials such as ferromanganese, ferrosilicon, etc., may be used to finish steel to the required specifications. The basic oxygen furnace uses pure oxygen to refine the hot metal (iron) and all other metallics into steel by oxidizing and removing the elements present such as silicon, phosphorus, manganese, and carbon. Oxides such as silicon dioxide, manganese oxide, phosphorus pentoxide, and iron oxide are fluidized in the slag which floats on the metal surface while oxides of carbon are emitted as gas. Wastes from this process are heat, airborne fluxes, slag, carbon monoxide and dioxide gases, and oxides of iron.

Basic oxygen furnaces are always equipped with gas cleaning systems for containing and cooling huge volumes of hot gases (1650°C) and submicron fumes released. Water is used to quench the off-gases. Two main process types are used for gas cleaning; precipitators and venturi scrubbers. For venturi scrubbers, the gases are quenched and saturated to 80°C, whereas for the precipitators the gases are cooled to about 250°C. If venturi scrubbers are used, the majority of airborne contaminants are mixed with water and discharged as effluents. Generally, water clarification equipment is provided for the treatment of this effluent.

In addition to the fume collection cooling water system, the basic oxygen furnace has three main water systems. The oxygen lance cooling water system is either a "once-through" or a "closed recirculation" system. The furnace trunnion ring cooling is generally a "once-through" system with a discharge differential temperature increase of about 20°C. The hood cooling water system may be a recirculating type using induced draft cooling towers with chemical treatment. If water of good quality and sufficient quantity is available, "once-through" cooling systems are sometimes employed.

Open Hearth Furnace (H): Open hearth furnaces can utilize an all-scrap steel charge but generally a 50-50 charge of hot metal and steel scrap is used. The furnace front wall is provided with water cooled lined doors for charging raw materials into the furnace. A plugged tap hole at the base of the wall opposite to the doors is provided to drain the finished molten steel into ladles. Fuel in the form of oil, coke oven gas, natural gas, pitch, creosote, tar, etc., is burned at one end of the furnace to provide heat for melting of scrap and other process requirements.

The open hearth process has two plant water systems: the furnace cooling water system and the fume collection water system. Furnace cooling is a once-through system with heated aqueous discharges of 17 to 22°C differential temperature. Fume collection systems are either wet, high energy venturi scrubbers or dry precipitators.

Aqueous discharges from precipitators are zero except for any waste heat boiler blowdown. Discharges from the scrubbers are wastewaters from the primary quenchers with concentrations of fluoride, nitrates, suspended solids, and zinc.

Electric Arc Furnace (I,J): The electric arc furnace steel making process produces high quality alloy steel in refractory lined cylindrical furnaces utilizing a cold steel scrap charge and fluxes. Sometimes, a lower grade of steel produced in the basic oxygen furnace or the open hearth furnace is alloyed in the electric arc furnace. Heat for melting the scrap charge, fluxes, etc., is furnished by passing an electric current through the scrap or steel bath by means of consumable cylindrical carbon electrodes inserted through the furnace roof. The heat cycle generally consists of charging, meltdown, molten metal period, oxidizing, refining, and tapping. Pure oxygen is sometimes lanced across the bath to speed up the oxidation cycle. Waste products from the process are smoke, slag, carbon monoxide and dioxide gases, and oxides or iron emitted as submicron fume. Zinc oxides from galvanized scrap may be released depending upon the type and quality of scrap.

The electric arc furnace has two main plant water systems: the furnace cooling water system and the fume collection cooling water system. The former is generally a "once-through" system but may be a "closed recirculation" system; the latter can range from completely dry to wet systems using precipitators, bag houses, or high energy, venturi scrubbers. Semi-wet systems are generally "once-through", with a temperature differential of 17 to 22°C in cooling waters. However, recycle with no discharge is also practiced. The wet high energy venturi scrubber fume collection systems produce aqueous discharges similar to the basic oxygen wastewater.

Vacuum Degassing (K): In the vacuum degassing process, steel is further refined by subjecting the ladle to a high vacuum in an enclosed refractory lined chamber. Steam jet ejectors with barometric condensers are employed to draw the vacuum. Certain alloys are added which may be drawn into the gas stream. The system is purged with nitrogen to eliminate residual carbon monoxide.

Wastewater from this process contains suspended solids, zinc, manganese, lead, and nitrates.

Continuous Casting (L): In the continuous casting process, billets, blooms, slabs, and other shapes are cast directly from the hot metal, thus eliminating the ingots, molds, soaking pits, and stripping facilities. Three water systems serve the casting machine: mold cooling, machine cooling, and spraying. Mold and machine cooling are performed in closed recycle streams. Wastewaters result from washing scale from the steel surface with spray water and contain significant quantities of suspended matter and oil.

Hot Forming - Primary (M): Hot forming defines the initial stages in forming useful products from steel ingots by hot-rolling. The basic operation of a primary mill is the gradual cross-sectional reduction of a hot steel ingot into blooms and slabs between the surfaces of two rotating steel rollers, and the progression of the ingot through the space between the rolls. The hot steel ingots are transferred to the primary mills for rolling from soaking pit furnaces which consist of square rectangular, or circular, fuel-fired refractory lined pits. During the rolling operation, cooling water is sprayed extensively over the table and mill stand rolls. This water is discharged to trenches beneath the rolling mill equipment. It is also necessary to use high pressure (2,000 psi) descaling water for spray over the hot ingot to flush away iron oxide scales that form on the hot ingot. Blooms are passed through hot-scarfing machines after leaving the bloom shares to remove defects from the surface of the bloom. Fume control is required and water sprays carry the iron oxide wastes through a trench under the mills to a collection system.

Hot Forming - Section (N): Blooms from the primary mill are conveyed directly to the billet mill without reheating. Billets are further processed to produce material with small sections, such as tube rounds, bar and rod, and special products. Modern billet mills utilize continuous mills which have alternate horizontal and vertical stands. A continuous mill consists of a series of roll stands, arranged one after the other so that the piece to be rolled enters the first stand and travels through the mill, taking one pass in each stand and emerging from the last set as a finished product. Descaling water and cooling water are sprayed at the stands and rolls with the discharge going to the trenches under the mills. After billet mills, the product is cut to the desired finish piece length. Billets are cooled on cooling beds and pushed into cradles, from which they can be loaded into cars for shipment or transferred for further processing. Smaller quantities of mill scale are generally generated in the hot forming-section subcategory than in the primary rolling operation but the particle size may be smaller and more difficult to settle out.

Hot Forming - Flat (O): This subcategory embodies operations associated with plate mills, hot strip mills, and skelp mills. The basic operation of a plate mill is the reduction of a heated slab to the weight and dimensional limitations of plates. This is accomplished by heating the slabs, descaling, rolling to plates, leveling or flattening, cooling, and shearing to the desired size. Descaling is completed on the delivery side of the mill as the slab is passed through top and bottom high pressure hydraulic sprays operating at 1,000 to 1,500 psi. About four percent of the spray water evaporates and the balance is discharged through a trench under the mills to an iron oxide and water collection system. During the rolling operation, cooling water is sprayed externally over the table and mill stand rolls.

The basic operation of a hot strip mill is the reduction of slab to flat strip steel in thicknesses of 0.04 to 1.25 inches; widths of 24 to 96 inches; and lengths of up to 2,000 feet. Principal water uses include descaling water sprays and cooling water.

Skelp is a hot-rolled strip used to make butt-weld pipe or tube. Skelp is rolled from a heated bloom and has a width which corresponds to the circumference of the pipe and a gauge which corresponds to the thickness of the wall. Descaling water, cooling water, and water-soluble oil sprays accompany the rolling operation.

Pipe and Tubes (P): Typical steel tubular products are standard pipe, conduit pipe, line pipe, pressure pipe, structural pipe, oil-country tubular goods, pressure tubes, mechanical tubes, and stainless steel pipe and tubes. Butt-welded pipe or tube is made from a hot-rolled strip. By heating this skelp to the welding temperature and drawing it through a die or roll pass, it is bent into cylindrical shape and its edges pressed firmly together into a butt-weld, thus forming a pipe. Seamless tubular products are made either by "piercing" or by "cupping". In the former process, a solid round bar or billet is heated, pierced, and then shaped to the desired diameter and wall thickness. In cupping, a circular sheet or plate is forced by successive operations through several pairs of conical dies until the plate takes the form of a tube or cylinder with one end closed. Electric-resistance-welded tubing (ERW) is made from strip sheet or plate. Steps in the manufacture of ERW are: forming, welding, sizing, cutting, and finishing. Plates are converted into pipes by the electric-weld process by shearing, planing, crimping, bending, welding, expanding, and finishing.

Significant pollutants in the wastewaters resulting from this subcategory include suspended solids and oil and grease. Wastewaters originate from contact cooling waters such as roll spray cooling waters and cooling bed or spray quench waters. Suspended solids can be traced to the scale which is flushed off the pipe surface by the roll cooling spray waters. Oil and grease originate in the hydraulic and lubricating systems.

Pickling-Sulfuric Acid-Batch (Q) and Pickling-Hydrochloric Acid-Batch and Continuous (R): Pickling is the chemical removal of surface oxides (scale) from metal by immersion in a heated solution. Carbon steel pickling is almost universally accomplished by using either sulfuric acid (Q) or hydrochloric acid (R). Acid conditions vary with the type of material to be pickled. In addition, bath temperature, use of inhibitors, and source of agitation are also varied depending on the material to be pickled. Pickling is done by either continuous strip or batch type operations. Continuous strip pickling lines use horizontal pickling tanks. Large, open tanks of a wide range of sizes are used for batch type pickling, principally for rod coils, bars, billets, sheet, strip, wire, and tubing. Pickling is also applicable to forgings, castings, structural parts, and other items.

In continuous pickling, fresh acid solution is added to the last tank section and cascades through the tanks to an overflow located in the first section. In batch type pickling, the tanks are generally rubber lined and brick sheathed and hold a large volume of heated acid solution. Sulfuric acid is most often used for this purpose. After a certain iron build-up due to iron scale removal, the batch acid solution is considered spent and dumped. Pickling is followed by the rinse operation which may vary from one-step dunk to more sophisticated multi-stage rinsing. It may be possible to use the contaminated rinse water as input water to the fume scrubber, prior to its final disposition as pickle recycling system makeup water.

Most continuous strip pickling lines employ the traditional approach to rinsing: flood the strip with hundreds of gallons of water per minute to wash away the few gallons of acid that may be dragged out of the pickling tanks. Multi-stage spray rinsing systems can easily be incorporated into new continuous strip pickling lines, and they can be installed in existing lines in place of the present rinsing sections.

Acid fumes are prevalent in the pickling process and must be removed in order to provide a good working environment. To remove the acid from the exhaust stream, washing or filtration methods may be applied. The acid mist filter controls air pollution and simultaneously recovers acid for reuse.

Wastewaters in the pickling-sulfuric acid-batch subcategory (Q) originate in either of two forms: as spent solutions of concentrated waste pickle liquor containing iron and sulfuric acid; or as dilute solutions resulting from dunk or spray rinsing of pickled product. Significant pollutants in the pickling-hydrochloric acid-batch and continuous subcategory (R) include suspended solids, total iron, ferrous iron, dissolved iron, and pH.

Cold Rolling (S): In cold rolling cooled hot strip mill product is passed through a pair of rolls for the purpose of reducing its thickness, producing a smooth dense surface, and developing controlled mechanical properties in the metal. Cold reduction is a special form of cold rolling in which the thickness of the starting material is reduced by relatively large amounts with each pass through the rolls. In tempering, the thickness of the material is reduced only a few percent to impart the desired mechanical properties and surface characteristics to the final product. During rolling, the steel becomes quite hard and unsuitable for most uses. As a result, the strip must undergo an annealing (heating) operation to return its ductility and to effect other changes in mechanical properties suitable for its intended use. This is done in either a batch or continuous annealing operation.

Wastewaters from this subcategory originate when water, oil, oil-in-water emulsions, oil-water-detergent solutions, or combinations of any of these rolling solutions, used for cooling or lubricating the rolls, are dumped. Suspended solids and oil and grease are the important pollutants.

Hot Coatings - Galvanizing (T) and Hot Coatings - Terne (U): Coating is the application of a layer of one substance to completely cover another. In the iron and steel industry, coatings are applied for a variety of reasons. Most often, a relatively thin layer of a metallic element such as zinc, chromium, or aluminum is applied to carbon steel, imparting such desirable qualities as resistance to corrosion, safety from contamination, or decorative appearance. In addition to metallic coatings, non-metals, simple and complex organic compounds, miscellaneous inorganic materials such as vitreous enamel, and metallic powders in silicate paints are also used as coating materials. All methods of applying coatings to steel surfaces require careful surface preparation which is the primary and most important step in the process. Commonly used for this purpose are alkaline or solvent cleaning for grease removal, acid pickling for removing scale or rust, and physical desurfacing using abrasives or brushes. Following surface preparation, metallic coatings may be applied by one of the following processes: hot dip process, electroplating, metal spraying, metal cementation, fuse welding, metal cladding.

Hot dipped coating using steel baths of molten metal is practiced as a batch-dip operation. In hot coatings-galvanizing (T), the coated products are withdrawn from the bath, subjected to drying with a warm air blast, or chemically treated with ammonium chloride, sulfur dioxide, chromate, or phosphate solutions to produce special galvanized finishes and surface characteristics. Terne is an inexpensive, corrosion-resistant, hot-dripped coating (U) consisting of lead and tin. A major portion of all terne coating materials is used in the automobile industry to manufacture gasoline tanks, automotive mufflers, oil pans, air cleaners, and radiator parts. Batch and continuous processes are both used, however continuous process is by far the larger portion of the market.

Wastewaters in the hot coating-galvanizing (T) subcategory result from cleaning operations, chemical treatment, and rinses applied to the product before or after coating as well as batch discharges from the various solutions and baths. Suspended solids, oil and grease, zinc, chromium, and pH are the principal pollutants. Wastewaters in the hot coating-terne subcategory originate from similar sources and contain suspended solids, oil and grease, lead, tin, and pH.

Subcategories V through Z: There are no iron and steel manufacturing processes associated with subcategories V through Z, as this miscellaneous category covers ancillary operations within a mill. Wastewaters resulting from there operations are highly variable in both quality and quantity.

Wastewater Characteristics

Table 25 shows raw wastewater characteristic data. In addition to the pollutants listed in the table, thermal discharges may also be generated.

The steel industry operates throughout the year and generates wastewaters over a 24-hour day. Wastewater volume and characteristics are subject to hourly variations. Process wastewaters are generally treated on site before disposal.

Wastewaters are subject to wide variations in flow within individual subcategories. This is largely due to the diversity in the plant cooling water systems and fume collection and cooling systems.

BOD in wastewaters of the steel industry is mostly due to coke manufacturing processes. However, cold rolling and blast furnace wastewaters will also contribute some BOD. Coking process waters are generally amenable to biological treatment only if they comprise less than 25 percent of the total wastewater.

Control and Treatment Technology

In-Plant Control Technology: Significant in-plant control of both waste quantity and quality is possible for some important subcategories of the iron and steel manufacturing industry. In by-product coke making (A) wastewaters are generated by the coking process and there also is usually a wastewater discharge from the coke quenching operation. Wastewaters from the by -product coke making operation (A) are highly contaminated and require intensive treatment. Wastewater from coke quenching can be reduced by dry coke quenching or simply by routing the wharf drains to the quench tower as make-up water and not allowing any overflow from the quench tower. Zero liquid discharge from modern coke plants can be achieved by evaporation of all liquid since the pollutants are mostly volatile except approximately one percent dissolved solids (chlorides, etc.), but this would be accompanied by potential air pollution problems. Effluent gases from less than optimum incineration of the wastewater can be expected to contain high concentrations of nitrogen oxides, sulfur oxides, and some particulate matter.

Liquid discharges from blast furnace subcategories (D,E) can be significantly reduced by recycling the gas cleaning and cooling water. Modern blast furnace practice has shown that this water could be put through settling chambers to remove the suspended solids and over a cooling tower to remove the heat.

Table 25: Iron and Steel Manufacturing–Raw Wastewater Characteristics[a]

Wastewater parameter	Subcategory Designation[b]								
	A	B	C	D	E	F	G	H	I
Ammonia	39 - 7,330	0 - 0.33		1 - 12	141				
BOD_5	12 - 1,550	0 - 3							
Chlorides[c]									
Cyanide[c]	7.7 - 110			0 - 1	23.6				
Fluoride			0 - 6	0 - 2		0 - 2	0 - 11	16 - 20	
Lead									
Manganese					833				
Nitrate								20 - 23	
Oil and grease	2 - 240		457 - 504						
Phenol	6.1 - 910	0 - 0.1			0.13				
Sulfates									
Sulfide	4.2 - 629		64 - 188	0 - 40					
Suspended solids	23 - 421	29 - 722	4,340 - 19,500	307 - 1,720	5,000	321 - 396	180 - 5,330	388 - 3,880	77 - 863
Total iron									
Zinc								2 - 880	0 - 13

Wastewater parameter	Subcategory Designation[b]								
	J	K	L	M	N	O	P	O	R
Ammonia									
BOD_5									
Chlorides[c]									3 - 200,000
Cyanide[c]									
Fluoride	10 - 15								
Lead		0.4 - 1							
Manganese		5 - 13							
Nitrate		3 - 25							
Oil and grease			20.5 - 22	2 - 14	2 - 14	2 - 10	0 - 61		
Phenol									
Sulfates								105 - 26,000	
Sulfide									
Suspended solids	2,160 - 42,800	23 - 70	7 - 74	4 - 91	12 - 125	6 - 57	27 - 103	21 - 159	
Total iron								42 - 7,900	134 - 11,700
Zinc	405 - 5,637	2 - 8							

(continued)

Table 25: (continued)

Wastewater parameter	Subcategory Designation[b]							
	S	T	U	V	W	X	Y	Z
Ammonia								
BOD_5				15				
Chlorides								
Cyanide[c]				3.2				
Fluoride								
Lead			0.20					
Manganese								
Nitrate								
Oil and grease	54 - 41,140	19	73				Present	
Phenol								
Sulfates				592 - 890				
Sulfide								
Suspended solids	90 - 900	98	8 - 48	412	Present	Present	Present	
Total iron								
Zinc		14.5				Present		

[a] All concentrations represent net raw wastes, and are in mg/l, except as noted.

[b] See Table 24 for names of various subcategories.

[c] With acclimation, higher levels can be tolerated.

Liquid discharge exclusive of noncontact cooling water for all of the steel making processes - basic oxygen (F,G), open hearth (H), and electric furnace (I,J) - results from the gas cleaning operations. Although technology for dry gas cleaning lags behind the requirements for gas cleanliness, reductions in flow or pollutant loads from these subcategories are still feasible by use of recycle systems and closeup of semi-wet systems.

In the hot forming-primary subcategory (M), an important control measure relating to all contact cooling, descaling, and scarfing wastewaters is periodic cleaning of scale pits to remove buildup of mill scale which otherwise will wash through. This measure is also applicable to the hot forming-section (N) and hot forming-flat (O) subcategories. Complete recycle with no blowdown, makeup as needed, and cooling tower or pond cooling for hot mills will result in zero discharge of wastewaters from the pipe and tubes subcategory (P). While this is practiced in some mills, it may not be accomplished under all circumstances. In the pickling-sulfuric acid-batch subcategory (Q), on-site recovery of acid from concentrates, rinses, and fume scrubber effluents, and the recovery of iron as ferrous hepthahydrate crystals can eliminate aqueous discharges. However, high initial capital costs are involved which may be eventually balanced by recovery of usable products. In hydrochloric acid pickling (R), reuse of all acid rinse waters to make up fresh batches of pickle liquor is possible. In cold rolling (S), recycle of rolling solutions and use of treated wastewaters on cold rolling lines can significantly reduce discharges. In hot coatings (T,U), control of wastewater volumes through counter-current rinses and by use of fume hood scrubber recycle systems, and special attention to maintenance of equipment designed to reduce loss of solution, are effective means for reducing discharge loads.

End-of-Pipe Treatment Technology: The iron and steel manufacturing industry utilizes a broad range of treatment technology in control of its effluents. Table 26 presents a brief summary of treatment practices employed in each subcategory, and the pollutant removals achievable with each treatment process.

Type of Residue Generated

There are a multitude of residues generated in iron and steel plants. In physical character this includes sludges, skimmings, volatile solids, and liquids. In chemical character, it includes iron oxides; such coking derivatives as ammonia, phenol, cyanide, light oils and tars; finishing mill derivatives such as waste pickle liquors; rolling oils (i.e., emulsified vegetable oils); and lubricating oil drippings from all parts of the mill. Heavy metals, mostly zinc and chromium, from the large amount of scrap used in the steel-making step of the plant are also present in the residues.

Sedimentation is the most widely applicable waste treatment tech-

Table 26: Iron and Steel Manufacturing—Removal Efficiencies (percent) of Various Wastewater Treatment Processes

Pollutant and Method	Subcategory Designation[a,b]																		
	A	B	C	D	E	F	G	H	I	J	K	L	M	N	O	P	S	T	U
Suspended Solids																			
1. Chemical treatment and clarification			99		99	91	99	98	100	99									
2. Chemical treatment, clarification, and filtration													99	99	99	99	99		
3. Clarification																80		80	
4. Clarification and filtration	74			97							97	97							
BOD																			
1. Activated sludge	98																		
2. Clarification	48	80																	
Ammonia																			
1. Solvent recovery, ammonia stripping and clarification	93																		
2. Clarification		40		25															
Phenol																			
1. Activated sludge or solvent extraction	99			90															
Nitrate																			
1. Bio-denitrification								91			94								
Zinc																			
1. Chemical treatment and clarification								70											
2. Clarification and filtration											99								
Oil and Grease																			
1. Primary and secondary clarification including skimming													85	85		85	80		85
2. Chemical treatment, air flotation and clarification						90													
Fluoride																			
1. Chemical treatment and clarification								42	60	10									

[a] See Table 24 for names of various subcategories.

[b] Data currently not available for subcategory designations Q, R, V, W, X, Y, and Z.

nique in a fully integrated steel mill. Settled material ranges from coarse mill scale generated during hot rolling processes, to fine dusts from wet scrubbers used on off-gases from blast, basic oxygen, open-hearth, and electric arc furnaces. Direct recycling of dewatered coarse mill scale into steel-making furnaces is a common practice in large integrated mills. Sintering of fine dusts and sludges for similar recycling is also widely practiced, but this is dependent to a certain extent on the characteristics of the particular sludge. Details of the residues generated in particular parts of the mill are given in Table 27.

Residual Management Options

As outlined in Table 27, the various operations in Iron and Steel Manufacturing produce residues in several different physical forms and compositions. Recovery of valuable materials from wastes for recycling or sale has long been a practice in this industry. Much development work has been going on to extend this principle to other waste streams in an attempt to eliminate the necessity for the disposal of troublesome wastes and waste residues.

One might categorize the wastewater treatment residues from this industry into three groups: sludges, skimmings, and chemical by-products. The sludges are composed of: carbon fines, (i.e., coke "breeze"); iron oxide, (i.e., mill scale and flue dust); limestone and silica, (also in flue dust); toxic heavy metals (especially in flue dusts); precipitated iron hydroxides and sulfates, (i.e., pickle liquor neutralization); and precipitated inorganic coagulants, (i.e., chemical coagulation and pickle liquor neutralization). Those sludges with substantial contents of heavy metals, such as zinc, chromium, lead, copper, manganese, and nickel, are usually unsuitable for recycling because the heavy metals impair product quality. Such sludges are therefore usually landfilled. With that exception, the large amounts of carbon and/or iron oxide-rich sludges have sufficient value that they are already recycled. In this regard, the limestone and silica contents are largely inert, and do not impair the sludge's recycle. This leaves the pickle liquor treatment sludges as the most troublesome residual. Although process substitution, (i.e., hydrochloric acid pickling) is a popular method of avoiding this residue disposal problem, neutralization of the sulfate pickle liquor and lagooning of the sludge is also a common technique. Methods of ultimate disposal of the sludge from the lagoon are obscure, although scavenger disposal has been practiced. Other practices used include: deep-well injection of partially treated pickle-liquor (severely restricted now in most states), use of the waste pickle liquor as a coagulant in water and wastewater treatment, and sale of the sludge for use in paint manufacturing.

Table 27: Types of Residues Generated in Iron and Steel Manufacturing

Iron or steel production process	Applicable waste treatment process	Residue Characteristics		Recycle potential of residue	Comments
		Physical	Chemical		
By-product coke	Clarification and filtration	Fine particulate sludge	Carbon fines, with some adsorbed organics	Yes - send to unit	
	Ammonia stripping and solvent recovery	Various organic liquids and aqueous-inorganic solutions		None - recover by-products for sale	
	Activated sludge	Colloidal, biological sludge	Biomass	None - send to landfill	
Bee-hive coke	Clarification	Particulate sludge	Carbon fines, with some adsorbed organics	Yes, but not common	
Sintering	Chemical coagulation and thickening of flue gas scrubber wastewater	Colloidal sludge	Iron oxide, with some carbon and limestone	Yes - incorporate in sintering unit feed	
Blast furnace	Chemical coagulation and thickening of flue gas scrubber wastewater	Colloidal sludge	Iron oxide, with some carbon and limestone, silica, and traces of heavy metals	Yes - send to sintering unit	Fe-Mn blast furnaces may have greater amounts of heavy metals in the sludge
Basic oxygen furnace	Chemical coagulation and thickening of flue gas scrubber wastewater	Colloidal sludge	Iron oxide, silica and limestone; small amount of heavy metals	None at present because of heavy metal content	Heavy metals content is related to the proportion of steel scrap used in the furnace charge
Open hearth furnace	Chemical coagulation and thickening of flue gas scrubber wastewater	Colloidal sludge	Iron oxide, silica and limestone; small amount of heavy metals	None at present because of heavy metal content	Heavy metals content is related to the proportion of steel scrap used in the furnace charge
Electric arc furnace	Chemical coagulation and thickening of flue gas scrubber wastewater	Colloidal sludge	Iron oxide, silica and limestone; small amount of heavy metals	None at present because of heavy metal content	Heavy metals content is related to the proportion of steel scrap used in the furnace charge

(continued)

Table 27: (continued)

Iron or steel production process	Applicable waste treatment process	Residue Characteristics: Physical	Residue Characteristics: Chemical	Recycle potential of residue	Comments
Vacuum degassing	Clarification and filtration				
Continuous casting	Clarification and filtration	Particulate sludge	Iron oxide (mill scale)	Yes - send to sintering unit	
Hot rolling, slabs, billets, structural shapes and bars	Clarification	Coarse particulate sludge	Iron oxide (mill scale)	Yes - send to sintering unit	
	Skimming	Oil/water/grease mixture	Hydrocarbons with some entrained mill scale	None - dewater and use as boiler fuel, or incinerate	
Hot rolling, plate, strip, pipe, and tube	Clarification	Coarse particulate sludge	Iron oxide (mill scale)	Yes - dewater and send to sintering unit or directly to blast furnace	Applies to front end (beginning of mill)
	Chemical coagulation clarification, and filtration	Fine particulate sludge	Iron oxide (mill scale) and coagulant	Yes - send to sintering unit	Applies to back end to mill
	Skimming	Oil/water/grease mixture	Hydrocarbons with some entrained mill scale	None - dewater and use as boiler fuel, or incinerate	
Pickling, sulfuric acid - batch	Neutralization and clarification	Gelatinous sludge	Mixture of $FeSO_4$ $Fe(OH)_2$, $CaSO_4$ (if lime is used for neutralization) and some metallic iron debris and adsorbed heavy metals	None - collect in lagoons for ultimate disposal	Disposal of this sludge is one of the major problems in steel mill wastewater treatment
Pickling, hydrochloric acid - batch and continuous	Neutralization and clarification	Gelatinous sludge	Sludge: $Fe(OH)_2$, with some adsorbed heavy metals, and some metallic iron debris. Supernatant: $CaCl_2$ brine	None - collect in lagoons for ultimate disposal	Recycling of the HCl pickling solution in a closed cycle is a basic feature of continuous HCl pickling

(continued)

Table 27: (continued)

Iron or steel production process	Applicable waste treatment process	Residue Characteristics		Recycle potential of residue	Comments
		Physical	Chemical		
Cold rolling	Chemical coagulation clarification, and and skimming	Fine particulate sludge, and oil/water/grease skimmings	Sludge: iron oxide, metallic iron, and entrained oils	None - dewater the skimmings and landfill or incinerate	Quantity of this sludge small relative to that from hot rolling
			Skimmings: mixture of mineral oils and greases (equipment lubricants) and coagulated emulsion of vegetable oils rolling oils)		The rolling oil itself can be reused if partially treated by clarification and skimming without coagulation
Hot coatings, galvanizing	Clarification		Substantial heavy Metal (Zn) content		
Hot coatings, terme	Skimming				
Fugitive (i.e., stormwater) runoffs and maintenance department wastes	Varies	Can generate particulate sludges; and skimmings	Sludges: iron oxides, metallic limestone, with adsorbed heavy metals	Varies	
			Skimmings: mineral oil and grease droppings of equipment lubricants		
Cooling water and utility blowdowns	Varies	Can generate gelantinous sludges	Heavy metal (e.g., Cr^{+3})	None - dewater and landfill hydroxide sludge	
Central treatment	----------any of the above----------				

The skimmings removed after sedimentation are usually mineral oil and grease lubricants or vegetable oils used during cold rolling of sheet and strip. The latter oil is used as an oil-in-water emulsion for cooling and lubricating the product being rolled and the rolls themselves. The emulsion is usually reused many times because it is too expensive not to do so. Nevertheless, dumps would be very objectionable if they were to get into a municipal sewerage system because the emulsified oil forms a cloudy or milky liquid which is visually conspicuous and has a high BOD. Before being removed by flotation, the emulsion must be "cracked," (i.e., destabilized), usually by depressing the pH with waste pickle liquor. Although messy and difficult to handle, the most practical way of disposing of these skimmed oils is to charge them into one of the many furnaces for their fuel value.

Chemical by-products come primarily from the coke oven operations. Ammonia, phenol, light oils, and coal tar have significant market value and can be sold. These by-products, together with cyanide, can be recovered by well-established processes. Wastewater from these recovery processes still contains small concentrations of the listed compounds, which can then be removed by standard biological wastewater treatment methods such as activated sludge. The waste activated sludge is similar in character to that from domestic waste treatment and may be disposed of in like manner. Because there may be small quantities of nonbiodegradable organics, such as heterocyclic compounds known to be carcinogenic, in wastewaters from the coke plant, and because such compounds tend to be adsorbed on the sludge floc, some care should be used in disposing of this waste activated sludge.

LEATHER TANNING AND FINISHING

General Industry Description

Tanning is the process of converting animal skins or hides into leather. Hides are dehaired, tanned by reacting with one or a combination of tanning agents, dyed, and finished to produce a flexible long-lasting leather. Total industry wastewater flow is approximately 52 million gallons daily (MGD), of which 47 MGD are discharged to POTWs and 5 MGD are discharged directly to navigable waters.

Industrial Categorization

For the purpose of establishing wastewater effluent limitations guidelines and standards of performance for new sources, the Leather Tanning and Finishing Industry has been subcategorized as shown in Table 28.

Table 28: Leather Tanning and Finishing Subcategorization

Subcategory	Designation
Hair Pulp, Chrome Tan, Retan-Wet Finish	A
Hair Save, Chrome Tan, Retan-Wet Finish	B
Hair Save, Non-Chrome Tan, Retan-New Finish	C
Retan-Wet Finish	D
No Beamhouse	E
Through-the-Blue	F
Shearling	G

Process Description

The subcategories in Table 28 include various combinations of four basic operations: beamhouse; tanyard; retan, color, and fatliquor; and finishing.

These four basic operations are described below, followed by a definition of the subcategories based on these operations. Figure 21 shows a product and wastewater flow diagram for generalized leather tanning and finishing plants.

Beamhouse: In the beamhouse, hides are processed to prepare them for the tanning operation. Hides that have been cured with salt or brine are received, stored, trimmed, and soaked to restore moisture and to remove salt. Wash waters from this operation contain dirt, salt, blood, manure, and proteins.

Degreasing operations, with either hot water and detergent or solvent, are performed on pig and sheep skins. Quantities of grease and solvent enter the waste stream in the degreasing operations.

Fleshing, the removal of fatty tissue and meat from the hides, is accomplished on a fleshing machine, through the use of rotating blades. Cold water, necessary to keep fat congealed, generates a fatty wastewater. Fleshings are recovered and sold to plants for rendering or for conversion to glue.

Beamhouse opeations are classified according to one of two hair removal practices. Machine removal permits hair recovery, and is practiced in a "save hair" beamhouse. The dissolving process is referred to as "pulping" and is practiced in a "pulp hair" beamhouse.

Prior to unhairing, the hides are slurried with lime and other additives, primarily sulfide sharpeners, to loosen the hair before its removal. Following unhairing, hides are sometimes relimed to make the hide swell for easier splitting and to assure complete hair follic removal.

The liming and unhairing processes are among the principal contributors to the waste effluent. In a save hair operation with good recovery of hair, the contribution to the effluent is substantially lower than in the pulp hair operation.

Tanyard: The purpose of the tanning process is to produce a durable

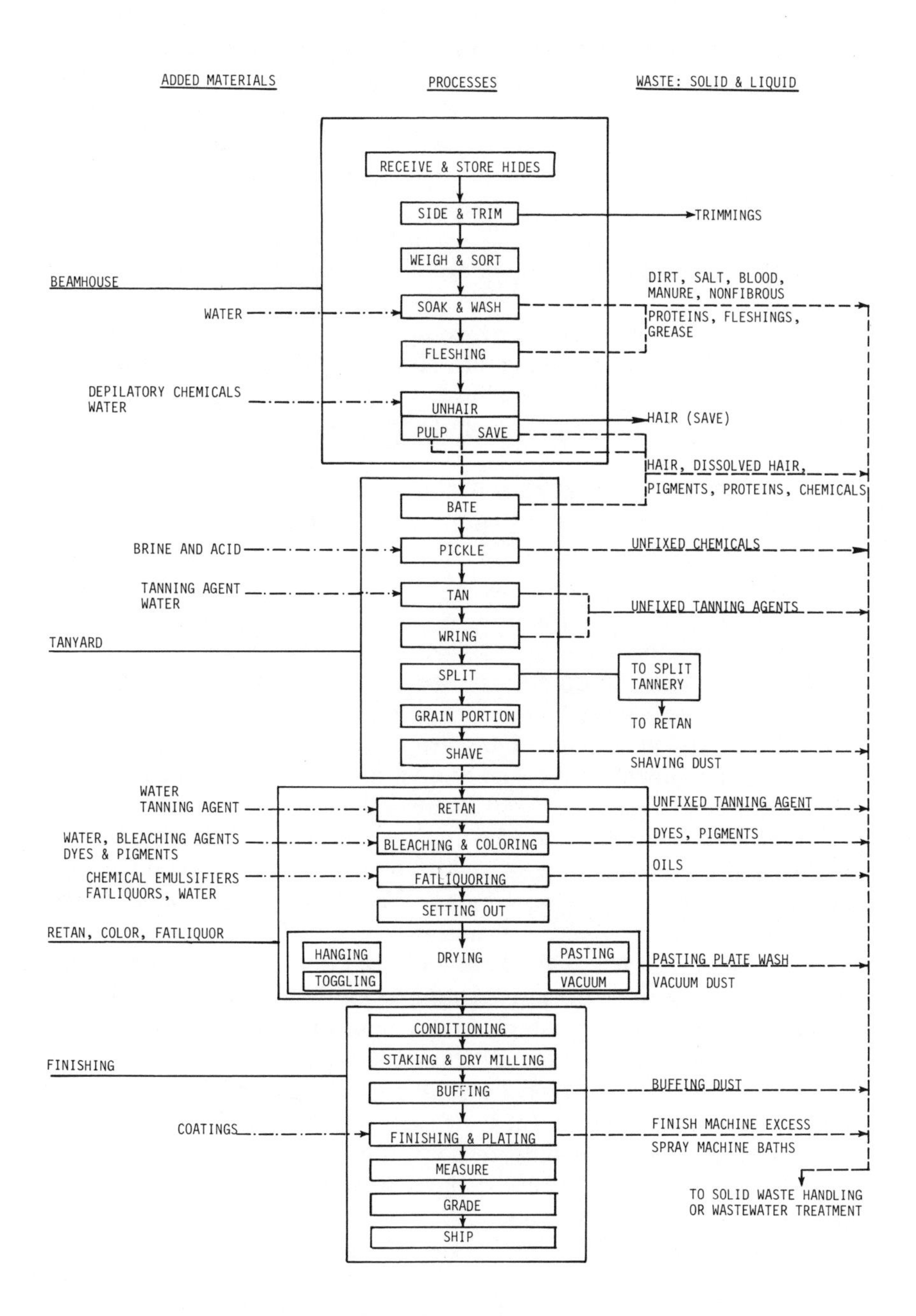

Figure 21: Product and wastewater flow diagram for generalized leather tanning and finishing plants.

material from the animal skin or hide which is not subject to degradation by physical or biological mechanisms such as rain, cold, or sweat.

Bating is the first step in preparing the hide for the tanning process. Hides are placed in vats or drums which contain a solution of ammonium salts and enzymes, which delime the skins, reduce the swelling, peptize the fibers, and remove protein degradation products.

Bating is followed by pickling in a brine and acid solution to condition the hide for receiving the tanning agent. Principal waste constituents are acid and salt.

Tanning is accomplished by reacting the hides with a tanning agent, usually chrome or vegetable tannins, although alum, metal salts, and formaldehyde can be used. Waste effluents from the tanning process are substantial. Recycle and recovery of tanning agents are becoming more common.

Retan, Color, and Fatliquor: These three operations are usually performed in one drum. Tanning solution is added to provide additional penetration into the hides (retan); synthetic or vegetable dyes are added to color the hides (color); oils are added to replace the natural oils of the skin that were lost in the tanning process (fatliquor). High strength, low volume discharges containing oil and color are generated.

Finishing: Many finishing processes, including drying, coating, staking, and sanding, are principally dry processes. Other processes (i.e. pasting and wash up) generate a high strength, low volume wastewater.

The seven subcategories and the various combinations of processes which comprise these subcategories are defined as follows:

Hair Pulp, Chrome Tan, Retan-Wet Finish (A) — Facilities which primarily process raw or cured cattle or cattle-like hides into finished leather by chemically dissolving the hair (hair pulp); tanning with chrome; and retanning and wet finishing.

Hair Save, Chrome Tan, Retan-Wet Finish (B) — Facilities which primarily process raw or cured cattle or cattle-like hides into finished leather by chemically loosening and mechanically removing the hair; tanning with chrome; and retanning and wet finishing.

Hair Save, Non-chrome Tan, Retan-Wet Finish (C) — Facilities which process raw or cured cattle or cattle-like hides into finished leather by chemically loosening and mechanically removing the hair; tanning with primarily vegetable tannins, alum, syntans, oils, or other chemicals; and retanning and wet finishing.

Retan-Wet Finish (D) — Facilities which process previously unhaired and tanned hides or splits into finished leather through retanning and wet finishing processes including coloring, fat-liquoring, and mechanical conditioning.

No Beamhouse (E) — Facilities which process previously unhaired and pickled cattlehides, sheepskins, or pigskins into finished leather by tanning with chrome or other agents, followed by retanning and wet finishing.

Through-the-Blue (F) — Facilities which process raw or cured cattle or cattle-like hides into the blue tanned state only, by chemically dissolving or loosening the hair and tanning with chrome, with no retanning or wet finishing.

Shearling (G) — Facilities which process raw or cured sheep or sheep-like skins into finished leather by retaining the hair on the skin; tanning with chrome or other agents; and retanning and wet finishing.

Wastewater Characteristics

Most processes are batch operated, generating large fluctuations in wastewater strength and flow. Table 29 shows raw wastewater characteristics. Table 30 shows raw wastewater characteristics as a function of production-based data.

Characteristics of the wastewater discharged by tanneries vary depending upon the mix of production processes at a given plant. General wastewater constituents, which contribute to numerous problems for POTWs and industrial treatment facilities, include large pieces of scrap hide and leather and excessive quantities of hair and other solids that clog or foul operating equipment and cause fluctuations in wastewater flow and pH. The wastewater contains high levels of suspended and settleable solids, biodegradable organic matter, and significant quantities of toxic pollutants.

Control and Treatment Technology

In-Plant Control Technology: Wastewater flows and masses of waste constituents can be reduced by the following in-plant control measures:

- *Process Changes.* While tanning operations traditionally employed the batch system, it is possible that more of the chemical applications as well as washing and rinsing could be handled more efficiently on a counter-current continuous flow basis. This would achieve maximum utilization of all active ingredients, leaving only concentrated wastes of small volumes for treatment and disposal. Use of effluents from one process as makeup water in another is also generally feasible at some points within a tannery.
- *Substitution of Process Ingredients.* Difficulties caused by high concentrations of contaminants in spent tan liquors from vegetable tanning processes can be lessened through recovery and reuse of these spent liquors in segregated, concentrated waste streams, and through the use of synthetic tanning agents (syntan).
- *Water Conservation and Reuse.* Use of hide processors and other specially designed vessels to permit use of lower float volumes; use of

Table 29: Leather Tanning and Finishing—Raw Wastewater Characteristics

Wastewater parameters	Subcategory[a]						
	Hair pulp, chrome tan, retan-wet finish	Hair save, chrome tan, retan-wet finish	Hair save, non-chrome tan, retan-new finish	Retan-wet finish	No beamhouse	Through-the-blue	Shearling
Conventional - mg/l							
Ammonia	17 - 380	0.4 - 660	23 - 680	58 - 160	6.2 - 99	382 - 613	8.7 - 35
BOD$_5$	213 - 4,300	140 - 2,790	1 - 7.77	201 - 1,600	20 - 29,800	1,310 - 11,000	100 - 3,920
COD	182 - 27,200	704 - 5,700	1080 - 75000	1,200 - 4,800	140 - 37,900	10,500 - 32,900	370 - 31,500
Oil and Grease	15.4 - 10,000	49.3 - 620	2 - 1,340	57.5 - 854	85.2 - 1,160	67 - 6,170	56 - 1,210
Sulfide	0.80 - 198	0.030 - 300	0.10 - 328	0.16 - 2.40	0.09 - 6.40	137 - 680	0.08 - 68
TKN	90 - 626	63 - 3,650	130 - 1,200	110 - 480	22 - 160	960 - 1,780	39 - 750
TSS	24.8 - 36,000	94 - 8,580	28 - 8,210	96 - 7,440	124 - 37,400	1,220 - 14,600	118 - 7,680
Inorganics - ug/l							
Chromium	79,667	90,500	5,132	89,000	74,000	550,000	365,000
Copper	173	56	380	250	187	100	78
Cyanide	40	35	80	30			10
Lead	1,667	700	138	1,300	787	28	75
Nickel	40	22	61	45	15	160	24
Zinc	427	315	490	198	1,045	980	345
Organics - ug/l							
Acenaphthene	32			P			
Acenaphtylene	16					P	
Anthracene	94	56	8	120	122	P	36
Benzene	15	10	10	10	80		8
Benzidine	27						
Bis (2-ethylhexyl) phthalate	51	32					93
Bromodichloromethane			10		10		
Chlordane						P	
Chlorobenzene		10					
Chloroform	20	26	24	10	10	P	16
2-Chloronaphthalene							P
1,2-Dichlorobenzene	255		126		36		61
1,3-Dichlorobenzene						P	
1,4-Dichlorobenzene	54		20		13	P	20
1,1-Dichloroethane	20						

(continued)

Table 29: (continued)

Wastewater parameters	Subcategory[a]						
	Hair pulp, chrome tan, retan-wet finish	Hair save, chrome tan, retan-wet finish	Hair save, non-chrome tan, retan-new finish	Retan-wet finish	No beamhouse	Through-the-blue	Shearling
Dichloromethane	10		138	10	10	P	177
2,4-Dichlorophenol		114					
Diethyl phthalate			P	P			
2,4-Dimethylphenol	P					P	
Dimethyl phthalate	118						
Di-n-butyl phthalate			P	P			
Ethylbenzene	88	150	58	80	80	P	
Fluoranthene		2					
Fluorene						P	
N-nitrosodiphenylamine				247			
Naphthalene	46	49	32	P	27	P	26
Nitrobenzene	425						
Pentachlorophenol		6,200	1,455		3,550		400
Phenol	3,700	2,876	9,050	3,200	6,200	P	91
Phenols (total)	9,010	1,920	2,435	3,038	4,630	8,480	515
Pyrene		1					
Tetrachloroethene	150	10	23		40		
1,1,2,2-Tetrachloro-ethane	10		10				18
Toluene	275	80	12	10	80	P	10
1,2-Transdichloro-ethene	30						
1,1,1-Trichloroethane	P	10				P	
1,1,2-Trichloroethane	10						
Trichloroethene	20				10		
2,4,6-Trichloro-phenol	3,390	4,800	915	573	3,270	P	

[a] A blank space indicates pollutant is not of concern.

[b] "P" means pollutant was found, but no concentration was recorded.

Table 30: Leather Tanning and Finishing—Raw Wastewater Characteristics—Production-Based Data[a,b]

Wastewater parameter	Hair pulp, chrome tan, retan-wet finish (A)	Hair save, chrome tan, retan-wet finish (B)	Hair save, non-chrome tan, retan-wet finish (C)	Retan-wet finish (D)	No beamhouse (E)	Through the-blue (F)	Shearling (G)
Ammonia	0.417 - 20.6	0.012 - 38.9	0.433 - 10.5	0.868 - 4.50	0.060 - 5.02	8.12 - 13.02	0.146 - 3.59
BOD_5	2.10 - 275	5.45 - 620	0.03 - 203	1.90 - 24.2	9.57 - 924	21.9 - 234	6.90 - 445
Chromium (total)	0.062 - 20.5	0.001 - 6.87	0.008 - 3.939	0.045 - 5.50	0.072 - 17.2	4.95 - 8.43	0.002 - 14.4
COD	6.01 - 612	62.6 - 230	28.6 - 2220	32.3 - 76.0	12.3 - 29,200	223 - 669	22.6 - 3,580
Oil and grease	0.411 - 261	2.90 - 44.6	0.010 - 40.5	0.544 - 12.4	1.26 - 894	1.42 - 177	2.52 - 137
Phenol	0.007 - 2.87	0.016 - 0.238	0.004 - 0.786	0.003 - 0.289	0.003 - 0.705	0.204	0.002 - 11.3
Sulfide	0.021 - 9.94	0.004 - 16.1	0.004 - 11.6	0.003 - 0.680	0.001 - 0.267	2.91 - 14.4	0.008 - 0.903
TKN	3.17 - 32.5	0.617 - 147	1.94 - 49.7	1.87 - 7.60	0.823 - 11.90	20.4 - 37.8	0.704 - 7.701
TSS	1.45 - 941	5.26 - 596	0.767 - 362	0.908 - 82.8	12.10 - 7010	27.1 - 307	13.4 - 869

[a] Numbers represent levels associated with batch type flow.

[b] Pollutant levels are expressed in pounds per 1,000 pounds of product produced.

wash waters and rinses for process solution makeup; and recirculation of noncontact cooling water, such as for vacuum driers, are some of the water conservation methods which may be used to reduce the volume of tannery wastes.

- *Repair and Replacement of Faulty Equipment.* Tannery waste problems may be complicated or intensified if faulty or obsolete process equipment is allowed to remain in service. Elimination of careless or accidental spills and excessive drainage of liquids from hides during transfer from one process to another is also desirable.
- *Automatic Monitoring Devices.* Abnormal levels of selected constituents can be detected by automatic monitoring equipment. For example, abnormal and accidental concurrent discharges of concentrated, highly alkaline lime-sulfide unhairing liquors and highly acid chromium tanning or pickle liquors are immediately detectable by pH meters and alarms installed on effluent lines from these processes.
- *Recovery and Reuse of Process Chemicals.* Reuse or reduction of process solutions or recovery of process chemicals are demonstrated methods of waste constituent reduction. A number of vegetable tanneries are using recycle streams to reduce the amounts of tan liquor discharged into waste streams. Reuse or recovery of chrome tan liquors also reused. Many tanneries recycle their pasting frame water, either wholly or partially.
- *Control of Specific Waste Constituents.* Lime reduction, sulfide removal, chrome reduction, and ammonia nitrogen reduction are examples of in-plant control of specific waste constituents in tanneries. The chrome content can be reduced by using one of a number of technologies. One technique is to increase the uptake of chrome by the leather in tanning. A second is to reuse chrome liquors in some part of the beamhouse, tanyard, or retan process without first recovering the chrome from the solution. A third way is to precipitate the chrome with an alkaline chemical, producing a chrome sludge either for disposal or chrome recovery.

End-of-Pipe Treatment Technology: Stream segregation is often the first step in implementing most end-of-pipe treatment technologies available to tanneries. It is the physical separation of at least two major wastewater streams, one originating in the beamhouse which is highly alkaline and the other originating in the tanyard which is acidic and has a measurable chrome content. These two streams are the specific process waste streams which respond more completely and cost effectively to separate treatment.

Preliminary treatment consists of one or a combination of the following operations and processes:

- Screening - Fine screening removes hair particles, wool, fleshings, hide trimmings, and other large particles which have a potential for damaging plant equipment and clogging pumps or sewers.
- Equalization - In order to produce optimum results in subsequent treatment operations, equalization of flow, strength, and pH of strong liquors may be necessary.
- Sulfide oxidation - Sulfides in beamhouse waste are a potential problem because they release hydrogen sulfide if mixed with wastes, thereby reducing the pH of the sulfide-bearing waste. Various methods for oxidizing sulfides include air oxidation, direct chemical oxidation, and catalytic air oxidation.
- Carbonation of beamhouse waste stream - Carbonation is effective in the treatment of alkaline wastes. Carbon dioxide reacts with lime to form calcium carbonate, whose crystalline structure provides an effective surface for adsorption of organic matter. Ammonia introduced into tannery wastewaters by the de-liming process can be removed by segregation of this stream and treatment by physical-chemical methods.

The effluents resulting from the preliminary treatment of segregated streams are combined for equalization, followed by coagulation-clarification. Alum, lime, iron salts, and polymers are used as coagulants. The effectiveness of a technology scheme which includes in-plant controls, preliminary treatment of segregated streams, and primary treatment of combined streams is shown in Table 31. Table 32 shows the effectiveness of various secondary treatment technologies in reducing pollutants from leather tannery wastewaters. The data shown in Table 32 originated from a field sampling and analysis of wastewaters from 22 tanneries.

Table 31: Leather Tanning and Finishing–Pollutant Removal Using In-Plant Controls and Primary Treatment

Wastewater parameter	Percent removal for waste stream segregation,		Percent removals for treatment levels 1 and 2,[a]		Percent removals for combined waste streams, for treatment level 3[a]
	Beamhouse	Tanyard	Beamhouse	Tanyard	
Ammonia	0.0	100	0.0	67	0.0
BOD_5	65	35	60	0.0	60
Chemical Oxygen Demand	56	44	60	0.0	65
Chromium (Total)	0.0	100	0.0	80	to 2 ppm
Oil and Grease	49	51	70	0.0	70
Phenol	0.0	100	0.0	0.0	0.0
Sulfide	100	0.0	100	0.0	-
Total Kjeldahl Nitrogen	46	54	65	-[b]	0.0
Total Suspended Solids	69	31	65	0.0	60

[a] Treatment level 1 is in-plant control. Treatment level 2 is preliminary treatment of segregated streams. Treatment level 3 is coagulation-clarification of combined waste effluent streams from levels 1 and 2.

[b] Reduction of 67 percent of the ammonia content.

Table 32: Leather Tanning and Finishing–Pollutant Removal Using Secondary Treatment Technologies

Wastewater parameter	Activated Sludge or Extended Aeration								Aerated Lagoon		Physical/Chemical	
	Hair save, chrome tan, retan-wet finish B		Hair save, chrome tan, retan-wet finish B		Hair save, chrome tan, retan-wet finish C		Shearling G		Hair, pulp, chrome tan, retan-wet finish A		Retan-wet finish D	
	Influent	Effluent	Influent	Effluent	Influent	Effluent	Influent	Effluent	Influent	Effluent	Influent	Effluent
Conventional - mg/l												
BOD	1,241	917	2,000	297	1,530	49	1,015	27	1,867	21	617	6.7
COD	2,557	1,777	4,033	893	5,953	553	2,367	488	5,530	217	1,900	28
NH_3	98	60	150	123	437	236	11	17	263	64	183	4.4
Oil and grease	171	91	553	17	247	35	413	25	720	17	180	15.4
pH (units)	11.0	10.5	8.4	7.6	8.6	7.6	5.2	7.7	8.4	6.3	4.3	4.4
Sulfide	50	30	16	6	19	17	0.16	0.13	204	0.4	0.5	0.3
TKN	252	186	287	163	750	277	49	27	500	105	183	4.4
TSS	1,098	557	2,252	129	6,377	227	768	108	2,907	155	522	7.7
Inorganics - ug/l												
Chromium	31,000	20,000	170,000	17,000	6,400	170	53,000	22,000	160,000	1,100	16,000	20
Copper	57	37	220	8	200	25	120	7	50	5	260	8
Cyanides	20	40	50	40	100	400	10	trace	60	150	trace	trace
Lead	100	30	3,100	60	100	50	80	30	1,100	80	300	
Nickel	5	34	75	30	60	30	27	19	60	30	6	4
Zinc	230	140	2,100	170	460	59	500	68	500	49	150	61
Organics - ug/l												
Antrhacene	56	trace	2.9	1.4	7.6	ND	36	6	ND	ND	133	7.3
Benzene	ND	ND	trace	trace	trace	trace	5	ND	trace	trace	trace	trace
Bis(2-ethylhexyl) phthalate	ND	ND	32	6	ND	26	93	34	51	2	ND	ND
Chloroform	41	trace	trace	ND	ND	ND	12	10	ND	ND	trace	trace
1,2-Dichlorobenzene	215	69	ND	trace	49	ND	ND[b]	ND	255	ND	ND	ND
1,4-Dichlorobenzene	99	21	ND	trace	19	ND	54	20	ND	ND	ND	ND
Ethylbenzene	ND	ND	100	trace	43	trace	ND	ND	88	ND	100	12
Naphthalene	49	15	ND	2.3	19	ND	35	ND	24	ND	ND	8.5
Pentachlorophenol	9,500	3,100	ND	12	2,900	200	400	130	ND	ND	ND	ND
Phenol	480	435	5,500	1,400	845	ND	91	ND	4,400	ND	3,200	60
1,1,2,2-Tetrachloroethane	ND	ND	ND	ND	trace	ND	18	ND	ND	ND	ND	ND
2,4,6-Trichlorophenol	10,500	4,300	ND	12	1,700	38	ND	ND	880	ND	573	trace
Toluene	trace	trace	100	trace	trace	trace	9	ND	100	trace	11	trace

[a] ND - compound not detected in samples.

[b] Compounds detected in trace amounts were generally less than 10 ug/l.

Type of Residue Generated

Three principal types of residue are generated from treatment of wastewaters from this industry: coarse leather debris, hair, trimmings, and screenings; sludge; and skimmings. Coarse material will have absorbed substantial amounts of chromium and other tanning chemicals. Sludge may be wasted secondary sludge from biological treatment, or it may be from chemical coagulation and clarification. The major portion of the sludge is organic material originating as blood, flesh, or manure particles from the hide. Inert materials in the sludge, such as grit, come from the hide; other inert materials, such as $CaCO_3$, originate as lime used in the dehairing process which then precipitates during recarbonation of the beamhouse waste stream. Except where nonchrome tanning methods are used, both chemical and biological sludges will have trivalent chromium. Skimmings consist mostly of large amounts of animal oils and greases released during the tanning process. Lesser amounts of oil and grease result from the fatliquoring step that follows tanning.

Residual Management Options

The bulk of the dischargers in this industry are relatively small, indirect dischargers. Because wastewaters normally contain chromium, phenols, benzenes, and other substituted aromatic compounds, the screenings, skimmings, and sludges produced by pretreatment will probably be classified as toxic and hazardous. Current final disposal practice would be land burial at a suitable site.

Although some pilot experimental work has been done in applying pretreatment sludge to crops, with careful control of the Cr(III) concentration therein, the results have been mixed. Although the nitrogen and organic content of the sludge is significant and would otherwise be expected to promote growth, the Cr(III) content is usually enough to counteract this. Other experimental work involves oxidizing the organic matrix in which the Cr(III) is held — the leather trimmings, dust, hair, and biological sludge — at a carefully controlled, elevated temperature. The chromium is then recovered from the ash, by acidification, and reused in the process.

Production process changes that will avoid the generation of some or all of these offensive residues are the major alternative in the management of residues from this industry.

MECHANICAL PRODUCTS

Sufficient data is not available on any aspect of this categorical industry.

NONFERROUS METALS MANUFACTURING

General Industry Description

Nonferrous metal manufacturers include smelters and refiners of ore concentrates and processors of scrap metals. This industry does not include mining of these materials or manufacturing of final products. Approximately 800 nonferrous metal plants currently process some 65 different metals in the United States. Approximately 30 percent of these plants discharge wastewaters directly to receiving waters, while 20 percent are indirect dischargers. The remaining 50 percent discharge no process wastewater.

Industrial Categorization

For the purpose of establishing effluent limitation guidelines, the nonferrous metals industry has been subcategorized as shown in Table 33. USEPA has excluded many of the small metal subcategories (one or two plants), as well as all subcategories with zero discharge of process wastewater. Metal subcategories which have a fairly large number of plants and which discharge wastewater, or those subcategories which produce some of the metals listed in the NRDC Consent Decree and discharge wastewater, have been included.

Table 33: Nonferrous Metals

Subcategories	Designation
Primary aluminum	A[a]
Secondary aluminum	B
Primary columbium and tantalum salts	C
Primary columbium and tantalum metal	D
Primary copper smelters	E[a]
Primary copper refineries	F[a]
Secondary copper	F
Primary lead smelters	H[a]
Primary lead refineries	I[a]
Secondary lead	J
Secondary silver from photographic wastes	K
Secondary silver from nonphotographic wastes	L
Primary tungsten salts	M
Primary tungsten metal	N
Primary zinc	O
Metallurgical acid plants	P

[a] These subcategories are all direct dischargers, and have been excluded from further consideration herein.

Process Description

The manufacturing process first separates the desired metal from all other metals and compounds with which it is found. Often the other metals and compounds are important byproducts, and the wastes from one plant may become the source of raw material for another metals manufacturer. A description of processes associated with subcategories B, C, D, G, J, K, L, M, N, O and P is given below. Subcategories A, E, F, H, and I are all direct dischargers and have been excluded from further consideration herein.

Secondary Aluminum (B): In the secondary aluminum subcategory, scrap aluminum, clippings, forgings, borings, and residues from primary reduction plants are used to produce aluminum metal or alloys. The varied nature of the raw materials requires two operations, presmelting and smelting.

Presmelting is the physical separation of aluminum scrap from other materials. The process varies with source of raw materials, and may include drying, magnetic separation, or furnace "sweating," in which aluminum is separated by allowing molten aluminum to flow down a sloped surface, leaving other metals and residuals behind. These processes generate dust and gases which require removal by scrubbing. In addition to wet scrubbing wastewaters, some processes use washings to remove impurities from the scrap, and thus add to the wastewater load.

Smelting of secondary aluminum consists of six steps: charging of scrap into the furnace, addition of fluxing agents, addition of alloying agents, mixing, demagging or degasssing, and skimming. Charging of scrap aluminum may be either batch or continuous. Fluxing agents employ a chloride or fluoride salt or sodium, potassium, calcium, or cryolite to form a semi-solid barrier on the melt which optimizes oxidation of the metal and aids the removal of impurities. Mixing is accomplished by injecting nitrogen gas or a mixture of nitrogen and chlorine. The nitrogen-chlorine mixture generates fumes which must be removed by scrubbing.

Demagging, or magnesium removal, employs chlorine or aluminum fluoride and generates heavy fumes which require scrubbing. Skimming removes flux and slag, along with impurities, for recycling, further processing, or disposal.

Sources of process wastewater within the secondary aluminum industry include demagging air pollution control, wet milling of residues, and contact cooling water.

Primary Columbium and Tantalum Salts (C): Columbium and tantalum are typically produced in two stages: processing of ore to produce a metal salt, and reduction of those metal salts to columbium and tantalum metal. This subcategory includes only those plants engaged in the first stage. Figure 22 shows a process flow diagram for a typical columbium and tantalum facility.

Columbium and tantalum are commonly found together in tin slags. Columbium is also found as an ore in combination with iron. Initial processing of ore concentrates and slags consists of leaching with hydrofluoric acid to solubilize the columbium and tantalum salts. Solvent extraction is the most commonly used purification technique, while ion exchange is less frequently applied. Methyl isobutyl ketone is often used to preferentially extract columbium and tantalum from solution.

Tantalum oxide is stripped from solution by deionized water. Hydrofluoric acid is added and the solvent is again stripped, yielding a columbium solution. Tantalum and columbium are then chemically precipated as salts.

Process wastewater sources include discharge from washings, overflow from process liquids, and wet scrubbing or gaseous emissions.

Primary Columbium and Tantalum Metals (D): This subcategory includes plants capable of reducing columbium and tantalum metal salts to metal by various methods. These include aluminothermic reduction, sodium reduction, carbon reduction, and electrolysis.

In aluminothermic reduction, purified salts of columbium and tantalum or ferrocolumbium ores are mixed with aluminum potassium chlorate and magnesium. Ignition of this mixture leads to the reduction of columbium and tantalum, and the oxidation of aluminum.

Sodium reduction entails charging a pressurized furnace with alternating layers of tantalum salts and sodium. The reduced metal is then purified by magnetic removal of iron, leaching with water, and leaching with nitric or hydrochloric acid.

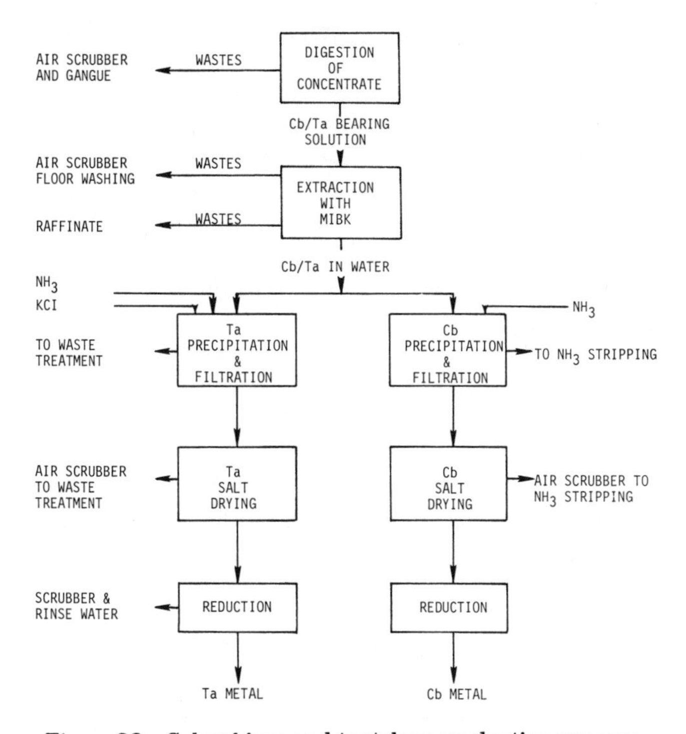

Figure 22: Columbium and tantalum production process.

Carbon reduction achieves a high purity columbium by a dry two-stage process. Electrolytic reduction of tantalum, employing a fused salt technique, is also used, though less frequently than the above methods.

Following the reduction processes, special techniques further purify these highly reactive metals. These processes include resistance heating in a vacuum furnace, cold-crucible ore melting, and electron beam melting.

Wastewater sources associated with the reduction of columbium and tantalum from salts to metals are: salt calcination wet air pollution scrubbing, reduction leaching, reduction wet air pollution scrubbing, and tantalum sizing.

Secondary Copper (G): The secondary copper subcategory consists of plants which process copper bearing scrap, refinery residues, and recycled copper to produce copper metals and alloys. The manufacturing process normally includes three operations: preparation of scrap, smelting, and refining. Figure 23 shows a process flow diagram for the three operations used by secondary copper facilities. Most secondary plants do not go beyond the smelting process. Raw materials are grouped into three classifications: low grade, which includes slags, skimmings and other highly contaminated wastes; intermediate grades consisting mostly of brass or bronze alloys; and high grade which is basically pure copper.

Process wastewater sources within the secondary copper subcategory include slag milling and washing; smelting wet air pollution scrubbers; contact cooling waters; spent electrolyte; and slag granulation operations.

Secondary Lead (J): The secondary lead industry processes lead scrap to produce various grades of lead metal. The main source of recovered lead is battery storage plates with minor amounts derived from solder, babbitt, cable coverings, type metal, soft lead and antimonial lead. Figure 24 shows a typical process flow diagram for a secondary lead facility.

Preparation of battery storage plates includes the sawing off of battery tops, removing of plates and paste for processing, and disposing of battery components.

Scrap lead is charged to a reverberatory or blast furnace to produce "soft" lead or "hard" lead, respectively. Hard lead contains significant amounts of antimony. Soft lead may be further refined by the Barton process to produce lead oxide for battery paste. Hard (antimonial) lead may be processed and alloyed on site to produce various products.

Process wastewater sources associated with the smelting and refining of secondary lead include furnace wet air pollution scrubbers, kettle wet air pollution scrubbers, and contact cooling waters.

Secondary Silver from Photographic Wastes (K): This subcategory includes plants that recover silver from photographic films, developing solutions, and manufacturing solutions. A process flow diagram is shown in Figure 25.

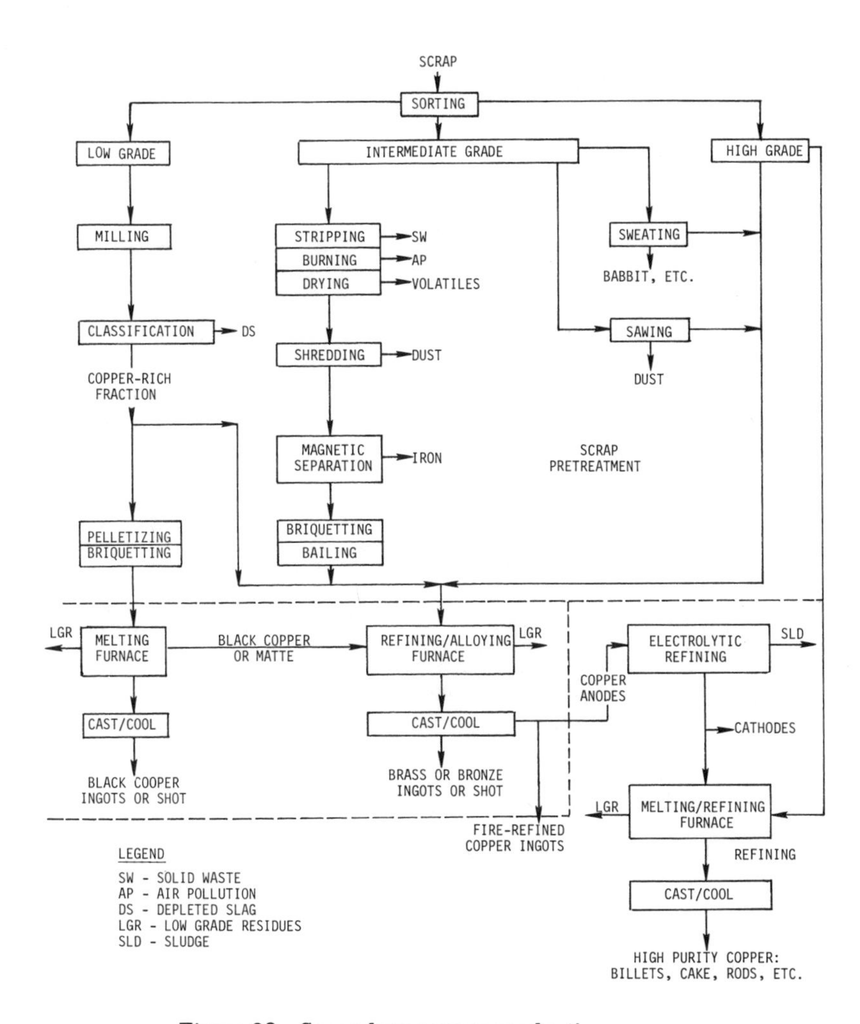

Figure 23: Secondary copper production process.

Photographic films account for 30 percent of secondary silver production. One process chops the film into small pieces and strips the silver by adding sulfuric acid, alum, and caustic soda. The liquid is clarified with the sediment being centrifuged and dried, and the resulting cake roasted and cast into ingots or dore plates. An alternate technique incinerates the photographic film followed by roasting of the silver bearing ash.

Electrolysis separates pure silver from dore plates along with a slime which may be processed for gold and platinum recovery. Silver bearing ash and ingots are refined by a furnace slagging process. Silver-rich solutions from film developing and manufacturing also undergo silver recovery processes such as chemical precipitation or electrolysis.

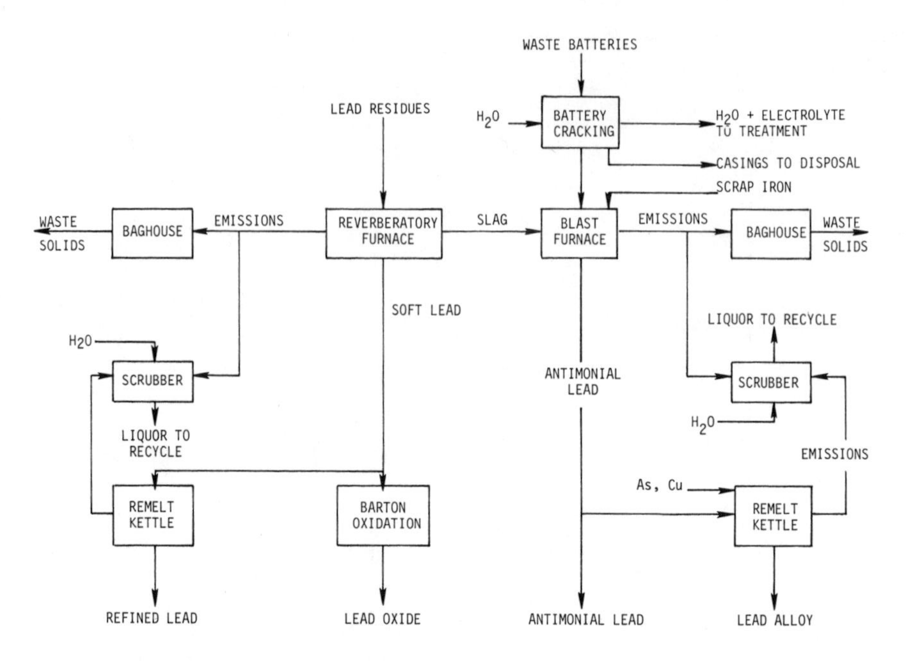

Figure 24: Secondary lead/antimony smelting process.

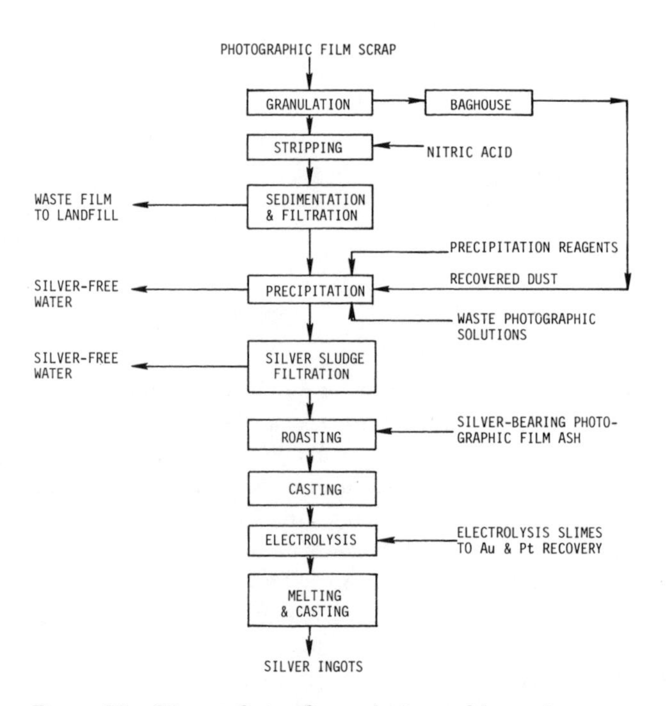

Figure 25: Silver refining from photographic wastes.

Sources of wastewater associated with the processing of photographic wastes include leaching and stripping operations, precipitation and filtration operations, furnace wet air pollution scrubbers, electrolysis operations, and contact cooling waters.

Secondary Silver from Nonphotographic Wastes (L): This subcategory includes plants that recover silver from jewelry, art objects, electrical components, industrial alloys, chemicals, and ceramics. Figure 26 shows a process flow diagram for a typical facility.

High purity waste silver, such as art objects and jewelry, are typically remelted with little preparation needed. Other silver wastes may require more extensive preparation to facilitate recovery. These may include waste silver from industrial alloys, chemicals, plating baths, ceramics, and electrical components.

Electrical components scrap not suitable for immediate electrolysis may undergo smelting in a reverberatory furnace to produce lead bullion, copper matte, and slag. Lead bullion is remelted in a cupola furnace to produce a lead bearing litharge and layers containing precious metals. The precious metals are cast into anodes for electrolytic separation. Copper matte is crushed, ground, roasted, and leached with sulfuric acid to release silver.

Sources of process wastewater associated with the production of silver from nonphotographic scrap include leaching or stripping wet air pollution scrubbers, precipitation and filtration operations, furnace wet air pollution scrubbers, and electrolysis operations.

Primary Tungsten Salts (M): This subcategory includes those plants capable of converting tungsten ore to a metal salt, ammonium paratungstate (APT). Tungsten metal production ordinarily occurs in two stages: ore concentrate conversion to metal salt, and metal salt conversion to metal. Approximately 30 percent of the tungsten processors in the United States perform both stages at one site. The remaining 70 percent perform only one of the two refining stages.

The first step in the production of APT involves grinding and digesting the ore with a sodium salt to produce soluble sodium tungstate. Waste solids are filtered, settled, and either landfilled or sold to other processors. The addition of calcium chloride precipitates calcium tungstate, leaving sodium chloride to be rinsed from the solids. Calcium tungstate is then leached with hydrochloric acid to form tungstic acid. The tungstic acid is dissolved in ammonium hydroxide producing ammonium tungstate. Solids are removed and the solution is heated to form APT.

Process wastewater sources from production of APT include precipitation, filtration and leaching operations, and APT drying wet air pollution scrubbers.

Primary Tungsten Metal (N): This subcategory includes plants which process ammonium paratungstate (APT) to tungsten metal.

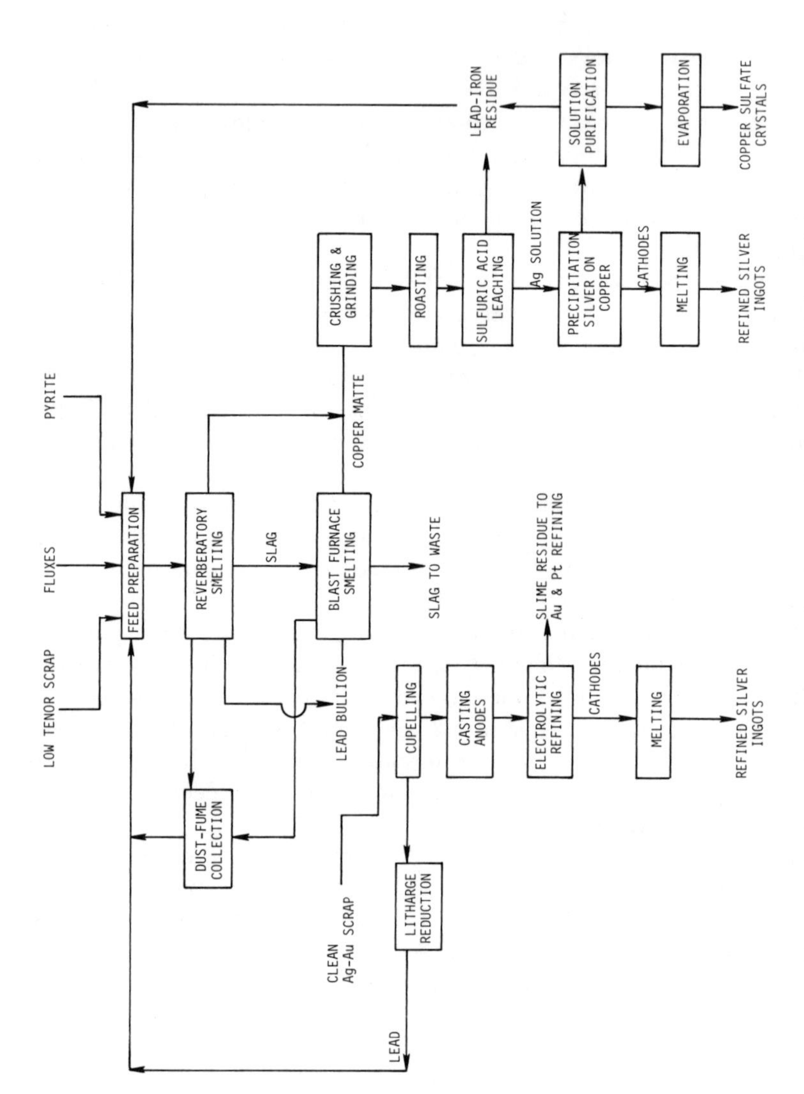

Figure 26: Secondary silver production process from nonphotographic scrap.

APT is used to produce refined tungsten metal. APT is dried, sifted, and converted to oxides in a nitrogen-hydrogen atmosphere furnace. A high temperature, hydrogen atmosphere reduces the oxide to tungsten powder for use as an alloying agent or conversion to metal. Monolithic tungsten metal is produced by sintering compacted tungsten powder in an inert atmosphere.

The main source of process wastewater from the production of metallic tungsten from APT are APT drying and reduction wet air pollution scrubbers.

Primary Zinc (O): This subcategory includes plants that produce zinc metal from zinc ores and from residues of the copper and lead refining industries. Figure 27 shows process flow diagrams typical of facilities in this subcategory.

There are two basic processes for production of zinc metal, pyrolytic and electrolytic. Pyrolytic zinc production begins with roasting of ores to remove sulfur as sulfur dioxide plus other volatile impurities such as mercury, lead, and cadmium. Roasted concentrate is blended with coke, water, and sometimes silica sand, and then pelletized. Pellets are sintered (bonded by heat) and crushed prior to feeding to a reduction furnace. Emissions are treated for dust collection, and cadmium and lead are recovered from the dust.

Either a vertical retort furnace or an electrothermic furnace drives off zinc as a vapor which is then condensed and cast into ingots. Wet scrubbers recapture uncondensed zinc and carbon monoxide emissions. Zinc is recovered from the scrubber water while exhausted carbon monoxide is used for fuel.

Electrolytic zinc production begins with the roasting of ore concentrates to remove sulfur and volatile impurities. If magnesium is present in significant amounts, preleaching with weak sulfuric acid is used to prevent buildup of magnesium in the electrolyte. After roasting, zinc and cadmium are leached from the roasted concentrate using spent electrolytics. Wastes are removed by clarification and filtration and sold to other processors for recovery of copper and lead. Further purification is achieved using zinc dust and scrap iron to precipitate copper, cadmium, and other impurities.

Electrolysis of the zinc solution deposits zinc on an aluminum cathode. Purified zinc is then stripped from the cathode and cast into desired shapes.

Process wastewater sources from the production of zinc include wet air pollution scrubbers, acid plant blowdown, preleaching operations, anode/cathode washing, and contact cooling waters.

Metallurgical Acid Plants (P): Primary copper, lead, and zinc smelters commonly operate by-product sulfuric acid plants. Most of the ores used in

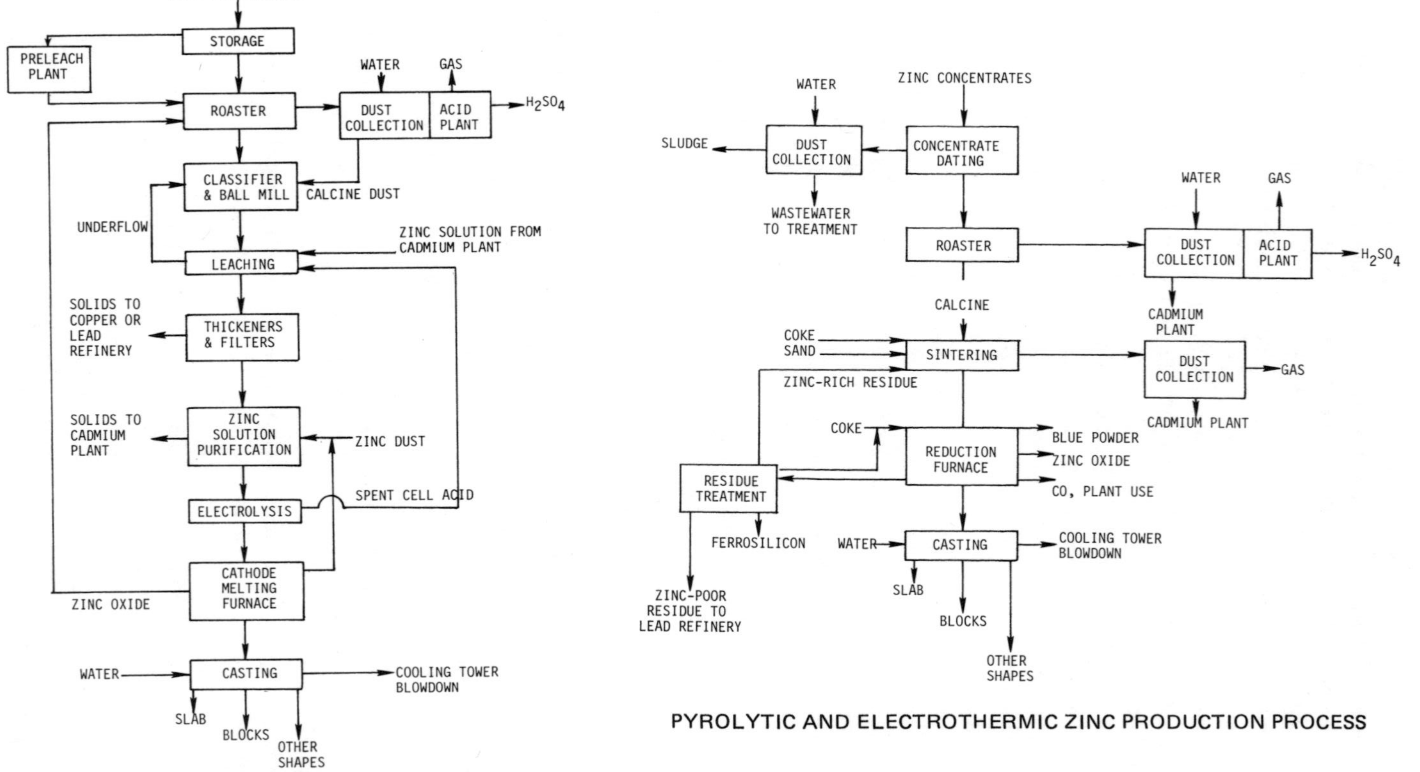

Figure 27: Zinc production processes.

these industries are sulfide ores, and the SO_2 in the off-gases from the roasting, sintering, or smelting operation usually needs to be removed before they are released to the atmosphere. Although not all plants using sulfide ores operate acid plants, most do. It is expected that all new smelters will have acid plants. Information on wastewater characteristics or control and treatment processes in this subcategory is not available.

Wastewater Characteristics

Table 34 shows raw wastewater characteristics.

Control and Treatment Technology

In-Plant Control Technology: Adoption or increased use of the following measures can result in both decreased wastewater volume and reduced pollutant loadings:

- Conservation and Reuse - Good housekeeping, tighter process control and wastewater recycle can all contribute to reductions in wastewater generation. Two types of recycle are used in nonferrous facilities: total stream and blowdown recycle. Total stream recycle, while more efficient, is often not practical due to resulting high solids concentrations.
- Dry Air Pollution Devices - Wet air pollution scrubbers are an important source of wastewater in the typical nonferrous plant, and substitution of dry devices can result in substantial wastewater volume reduction. Use of dry devices is limited by their inability to adequately remove gases as well as particulates.
- Dry Slag Processing - Slag is commonly processed by granulation, which involves the use of a water jet to produce the desired aggregates. Use of dry crushing and solidification procedures eliminates that contact-contaminated waste stream.
- Contact Cooling Water Reduction - Contact cooling of molten metal is widely used. Control of contact cooling water can be achieved by recycle, total evaporation, air cooling, or conversion to noncontact cooling.

End-of-Pipe Treatment Technology: Physical-chemical treatment and/or physical separation are the typical end-of-pipe technologies used in the nonferrous metals industry.

Physical-Chemical Treatment — Precipitation and coagulation/flocculation are common techniques used for removal of metals and fluorides. Adjustment of pH is also practiced to facilitate metals precipitation.

Ammonia which is introduced as a process reagent may be removed from the wastewaters by stripping with air or steam. Usually the wastewater is heated prior to delivery to the stripping tower and air is used at

Table 34: Nonferrous Metals Industry—Raw Wastewater Characteristics

Wastewater Parameter	Subcategory						
	Secondary aluminum B	Primary columbium and tantalum C and D	Secondary copper G	Secondary lead J	Secondary silver K and L	Primary tungsten M and N	Primary zinc O
Conventional - mg/l							
Ammonia nitrogen	< 0.10-240[a]	0.1-2,400		0-12	12-1,500	3.9-1,600	
Chemical oxygen demand	<5-930	22-11,000	5-900	65-380	230-15,000	22-880	20-59
Chloride	35-6,000	0-1,100		0-79	32,000	17,000-26,000	
Fluoride		1.6-3,500	0-0.29		1.2		
Oil and grease	2.5-98	4-22	1.7-30	6-59	8-132	2-17	10-14
Total organic carbon	3-220	2-1,800	5-99	4-330	24-13,000	4-270	7.3-9.33
Total suspended solids	4-20,000	1-28,000	4-11,000	0.05-10,000	110-3,700	1.9-6,700	13-15
Inorganics - ug/l							
Antimony	< 100-1,000	10-18,000	100-11,000	1,100-95,000	100-1.2E5[b]	100-800	2-2,100
Arsenic	< 10-4,000	10-27,000	10-4,200	3,800-16,000	50-2,000	20-7,200	3-3,000
Asbestos (crystalline fibers/l)	58E6	2.7E9	3.3E9	2.8E9	9.1E7-2.5E9	1.1E9	1.08E7-5.5E5
Asbestos (total fibers/l)	75E9	14E9	3.8E7	1.3E11	3.8E8-8.4E9	6.0E9	4.8E7-32E6
Beryllium	<1-310	1-500	1-160	2-80	1-20	2-30	2-20
Cadmium	<2-500	2-40,000	2-1,200	120-3,000	370-80,000	20-200	350-44,000
Chromium	<5-97	5-20,000	5-2,100	100-1,000	200-1E5	100-5,000	37-26,000
Copper	<6-700	60-5E5	6-2.1E6	160-10,000	720-70,000	100-5,000	37-26,000
Cyanide	1-11	2-10	1-26	2-10	1-6,000	1-140	2-3.0E6
Lead	<20-2,000	20-5E7	20-20,000	7,000-40E5	2,000-50,000	200-20,000	280-18,000
Mercury	< 0.1-6.4	0.1-63	0.1-0.53	0.6-13	0.1-17	0.3-3.0	2.9-52
Nickel	<5-50	5-10,000	5-3.1E6	180-2,000	500-8E5	50-1,000	50-4,300
Selenium	<10-200	2-45,000	10-270	2-10	10-900	10-1,000	24-1,200
Silver	<20-70	20-1,000	20-1,600	40-320	250-4,700	20-290	25-740
Thallium	<100	50-100	50-100	50-1,000	100-650	100-700	20-360
Zinc	<60-8,000	60-1E6	60-97,000	590-20,000	4,000-2.0E6	250-2,000	8.7E3-1.7E6
Organics - ug/l							
Phthalates							
Bis (2-ethylhexyl) phthalate	0-2,000	0-1,100	0-7,000	0-580	7-34	0-880	0-90
Butyl benzyl phthalate	0-98	0-47	0-56	0-85	0-53		0-30
Di-n-butyl phthalate	0-44	0-60	0-390	0-27	0-300	0-23	0-26
Di-n-octyl phthalate	0-25	0-95	0-67	0-27	0-58	0-1	
Diethyl phthalate		0-17	0-83		0-38		0-18
Dimethyl phthalate	0-56	0-39		0-13			0-22

(continued)

Table 34: (continued)

Wastewater Parameter	Subcategory: Secondary aluminum B	Primary columbium and tantalum C and D	Secondary copper G	Secondary lead J	Secondary silver K and L	Primary tungsten M and N	Primary zinc O
Nitrogen Compounds							
Benzidine				0-6			
3,3'-Dichlorobenzidine	0-2						
Aromatics							
Benzene	0-94	0-44	0-13	0-2	3-160	0-3.0	0-24
Chlorobenzene				0-5	0-9		
1,4-Dichlorobenzene	0-26						
2,4-Dinitrotoluene		0-16					
2,6-Dinitrotoluene		0-16					
Ethylbenzene			0-4	0-1.2	0-21	0-11	0-2
Hexachlorobenzene			0-5,000				0-100
Nitrobenzene		0-163		0-16			
Toluene			0-10		3-35	0-45	0-54
1,2,4-Trichlorobenzene		0-260					
Phenols							
Pentachlorophenol		0-17					0-80
Polycyclic aromatic hydrocarbons							
Acenaphthene		0-17	0-36		0-10	0-100	
Acenaphthylene	0-17	0-2	0-120	0-35		0-110	0-18
Anthracene	0-4	0-2	0-3,000	0-20	0-4	0-150	0-0.4
Benzo(a)anthracene		0-1	0-1				
Benzo(a)pyrene	0-12	0-1		0-10		0-1	
Benzo(ghi)perylene		0-2					
Benzo(k)fluoranthene				0-5.3			
3,4-Benzofluoranthene				0-5.3			
2-Chloronaphthalene		0-3					
Chrysene	0-190	0-45	0-10,000	0-540		0-240	0-11
Dibenzo(ah)anthracene		0-4					
Fluoranthene	0-12	0-7.2	0-3,000	0-27		0-1	0-15
Fluorene		0-20	0-94	0-2		0-55	0-14
Ideno(1,2,3-cd)pyrene		0-4					
Naphthalene	0-1	0-84	0-5,000	0-4	0-1	0-1,100	
Phenanthrene	0-10	0-2	0-3,000	0-20	0-4		
Pyrene	0-24	0-3	0-7,000	0-38	0-2,100		0-15

(continued)

Table 34: (continued)

Wastewater Parameter	Subcategory: Secondary aluminum B	Primary columbium and tantalum C and D	Secondary copper G	Secondary lead J	Secondary silver K and L	Primary tungsten M and N	Primary zinc O
Halogenated aliphatics							
Bromoform		0-21		0-49	0-65	0-48	
Carbon tetrachloride	0-10	0-74	0-120		0-2,300		
Chlorodibromomethane		0-81			0-64	0-38	
Chloroform	0-31	0-140	0-1,000	0-31	0-890	0-1,800	0-71
Dichlorobromomethane	0-19	0-13					
1,1-Dichloroethane							0-180
1,2-Dichloroethane	0-1	0-150	0-32	0-10	0-560	0-8	0-32
1,2-Trans-dichloroethylene	0-57	0-480	0-5			0-2	
1,1-Dichloroethylene		0-22	0-530	0-10	0-6,100	0-19	0-23
Methylene chloride	0-93	0-88,000	0-510		0-3,100		0-2,600
Hexachloroethane		0-23					
1,1,2,2-Tetrachloroethane		0-6		0-4	0-32	0-35	
Tetra chloroethylene	0-310	0-65	0-72	0-5	0-109	0-69	0-8
1,1,1-Trichloroethane		0-40			0-22	0-10	
1,1,2-Trichloroethane		0-29					
Trichloroethylene	0-530	0-230	0-72	0-6	0-900	0-19	0-160
Trichlorofluoromethane							0-101
Pesticides and metabolites							
Aldrin	0-6	0-4	0-0.2	0-0.1	0-1.1	0-7	
BHC (alpha)	0-0.1	0-0.04	0-0.2	0-0.2		0-0.6	
BHC (beta)	0-0.4	0-4.5	0-0.02	0-0.3	0-0.02	0-0.2	
BHC (gamma)	0-0.1	0-0.03	0-0.04	0-0.1		0-0.2	
BHC (delta)		0-4.0	0-0.2		0-1.1		
Chlordane	0-0.3	0-0.8	0-0.7	0-0.2	0-0.1	0-1.2	
4,4'-DDT	0-0.02	0-1	0-0.03	0-0.1	0-0.01	0-0.1	
4,4'-DDE	0-0.01	0-0.4	0-0.02	0-0.2	0-0.01		
4,4'-DDD			0-0.1			0-0.1	
Dieldrin	0-0.2	0-0.1	0-0.03	0-0.2	0-0.01	0-0.1	
-endosulfan		0-0.01	0-0.3	0-0.2		0-15	
-endosulfan			0-0.3			0-15	
Endosulfan sulfate		0-0.03					
Endrin	0-0.01	0-5.4	0-0.4	0-4	0-2	0-0.8	
Endrin aldehyde	0-0.4	0-0.2	0-0.03	0-0.16		0-0.9	
Heptachlor	0-0.4	0-0.5	0-0.02	0-0.3	0-0.02	0-0.2	
Heptachlor epoxide	0-0.2	0-0.1		0-0.2		0-0.2	
Isophorone	0-3	0-29					0-18
PCB-1254	0-0.9	0-52	0-3.0	0-2.6	0-0.7	0-5.4	
PCB-1248	0-0.3	0-32	0-2	0-3.1	0-0.5	0-1.0	
Toxaphene		0-0.1	0-0.4				

[a] "L" indicates levels below detection limits.
[b] Exponential notation: "X" "E" "Y" is the equivalent of X times 10^Y.

ambient temperature. Evaporation of water and the volatilization of ammonia generally produces a drop in both temperature and pH, which ultimately limit the removal of ammonia that may be achieved in a single air stripping tower.

Steam stripping offers better ammonia removal (99 percent or better) than air stripping for the high ammonia concentrations found in some wastewaters of the nonferrous metals industry. Tray towers are used, and the pH is adjusted to 12 or more with lime.

Physical Separation — Filtration by various means (sand, dual- and multi-media, pressure, and cloth filters) is generally used as a polishing step to further reduce suspended solids after sedimentation. Vacuum filtration is commonly used to dewater clarifier sludges.

Sedimentation is widely used throughout the nonferrous metals industry. Settling ponds and clarifiers are used, both as separate treatment units and in conjunction with chemical or pH induced precipitation.

Table 35 shows removal efficiencies of two of the commonly used treatment technologies in the nonferrous metals industry.

Table 35: Nonferrous Metals Industry–Percent Removal Efficiencies of Two Treatment Technologies

	Treatment Technology[a]	
Wastewater parameter	Lime precipitation	Multimedia filtration
Conventional		
Chemical oxygen demand	50-85	
Total suspended solids	30-99	
Fluoride	60-99	
Inorganics		
Antimony	95-99+	
Arsenic	85-99	
Cadmium	95-99	
Chromium	95-99+	
Copper	85-99+	
Lead	70-99+	
Mercury	90-95+	
Nickel	60-99+	
Silver	70-99+	
Zinc	70-99+	
Organics		
Acenaphthene	80	32
Benzene	80	53
Benzo(a)pyrene	93	32
Bis (2-Ethylhexyl) phthalate	32	54
Butyl benzyl phthalate	60	54
Chrysene	94	32
Dichlorobromoethane	76	59
1,1-Dichloroethylene	76	59
1,2-Dichloroethane	91	59
Diethyl phthalate	60	54
Di-n-butyl benzyl phthalate	89	54
Di-n-octyl phthalate	80	54
Fluoranthene	82	38
Fluorene	80	32
Methylene chloride	99	20
PCB-1248	39	80
PCB-1254	41	62
Pyrene	56	25
Tetrachloroethylene	50	62
1,2,4-Trichlorobenzene	90	62

[a] Removal efficiencies are ranges of reported values except single values, which are averages.

Type of Residue Generated

Information on the types and characteristics of residues generated by this industry is insufficient and cannot be discussed at this time.

Residual Management Options

Lack of sufficient information on types and characteristics of residues generated does not permit discussion of specific management options at this time.

ORE MINING AND DRESSING

General Industry Description

Ore mining industries include establishments engaged in mining ores for the production of metals, and includes all ore dressing and beneficiating operations, whether performed at mills operating in conjunction with the mines served or at mills operated separately. These include mills which crush, grind, wash, dry, sinter, or leach ore, or perform gravity separation or flotation operations.

Industrial Categorization

Based on similarities in types of processing, technology, wastewater, end products, and other factors, the following subcategories of the Ore Mining and Dressing industry were established: iron ore; aluminum; base and precious metals; uranium; ferroalloy; mercury; and metal ore not elsewhere classified.

Process Description

As mined, most ores contain the valuable metals whose recovery is sought, disseminated in a matrix of less valuable rock called gangue. The purpose of ore beneficiation is the separation of the metal-bearing minerals from the gangue to yield a product that is higher in metal content. To accomplish this, the ore must generally be crushed and/or ground small enough for each particle to contain either the mineral to be recovered or mostly gangue. Separation of the particles on the basis of some difference between the ore mineral and the gangue can then yield a concentrate high in metal value, as well as waste rock (tailings) containing very little metal. Separation is never perfect, and the degree of success attained is generally described by two parameters: (1) percent recovery, and (2) grade of the concentrate. Widely varying results are obtained in beneficiating different ores; recoveries may range from 60 percent or less to greater than 95 percent. Similarly, concentrates may contain less than 60 percent or more than 95 percent of the primary ore mineral. In general, for a given ore and process, concentrate grade and recovery are inversely related. (Higher recovery is achieved only by including more gangue, yielding a lower grade concentrate.)

Many properties are used as the basis for separating valuable minerals from gangue, including specific gravity, conductivity, magnetic permeability, affinity for certain chemicals, solubility, and the tendency to form chemical complexes. Separation processes in general use are gravity concentration, magnetic separation, electrostatic separation, flotation, and leaching. Amalgamation and cyanidation, which are variants of leaching, deserve special mention. Solvent extration and ion exchange are widely applied techniques for concentrating metals from leaching solutions, and for separating them from dissolved contaminants. All of these processes are discussed in general terms in the paragraphs that follow. This discussion is not meant to be all inclusive; rather, its purpose is to discuss the primary processes in current use in the ore mining and milling industry.

Gravity-concentration processes utilize the differences in density to separate valuable ore minerals (values) from gangue. Several techniques (e.g., jigging, tabling, spirals, and sing/float separation) are used to achieve the separation. Each is effective over a somewhat limited range of particle sizes, the upper bound of which is set by the size of the apparatus and the need to transport ore within it, and the lower bound by the point at which viscosity forces predominate over gravity and render the separation ineffective. Selection of the particular gravity-based process for a given ore will be strongly influenced by the size to which the ore must be crushed or ground to separate values from gangue, as well as by the density difference and other factors.

Ores can be leached by dissolving away either gangue or values in aqueous acids or bases, liquid metals, or other special solutions. The examples below illustrate various leaching possibilities:

- Water-soluble compounds of sodium, potassium, and boron can be mined, concentrated, and separated by leaching with water and recrystallizing the resulting brines.
- Vanadium and some other metals form anionic species that occur as insoluble ores. Roasting of such insoluble ores with sodium compounds converts the values to soluble sodium salts. After cooling, the water-soluble sodium salts are removed from the gangue by leaching in water.
- Uranium ores are only mildly soluble in water, but they dissolve quickly in acid or alkaline solutions.
- Native, finely divided gold is soluble in mercury and can be extracted by amalgamation (i.e., leaching with a liquid metal). One process for nickel concentration involves reduction of the nickel us-

ing ferrosilicon at a high temperature and extraction of the nickel metal into molten iron. This process, called skip-lading, is related to liquid-metal leaching.

- Certain solutions (e.g., potassium cynaide) dissolve specific metals (e.g., gold) or their compounds, and leaching with such solutions immediately concentrates the values.

In the amalgamation process, mercury is alloyed with some other metal to produce an amalgam. The process is applicable to free milling precious-metal ores, those in which the gold is free, relatively coarse, and has clean surfaces. Lode or placer gold/silver that is partly or completely filmed with iron oxides, greases, tellurium, or sulfide minerals cannot be effectively amalgamated. Hence, prior to amalgamation, auriferrous ore is typically washed and ground to remove any films on the precious-metal particles. Although the amalgamation process has been used in the past extensively for the extraction of gold and silver from pulverized ores, it has largely been superseded in recent years by the cyanidation process owing to environmental considerations.

In the cyanidation process, gold and/or silver are extracted from finely crushed ores, concentrates, tailings, and low-grade mine-run rock in dilute, weakly alakaline solutions of potassium or sodium cyanide. Gold is dissolved by the solution and subsequently sorbed onto activated carbon ("carbon-in-pulp" process) or precipitated with metallic zinc. The gold particles are recovered by filtering, and the filtrate is returned to the leaching operation.

Ion exchange and solvent extraction processes are used on pregnant leach solutions to concentrate values and to separate them from impurities. Ion exchange and solvent extraction are based on the same principle: polar organic molecules tend to exchange a mobile ion in their structure [typically, Cl^-, NO_3^-, HSO_4^-, or $CO_3^=$ (anions) or H^+ or Na^+ (cations)] for an ion with a greater charge or a smaller ionic radius.

Wastewater Characteristics

The wastewater situation evident in the mining segment of the ore mining and dressing industry is unlike that encountered in most other industries. Usually, industries (such as the milling segment of this industry) utilize water in the specific processes they employ. This water frequently becomes contaminated in the process and must be treated prior to discharge. In the mining segment, process water is not normally utilized in the actual mining of ores, except where it is used in placer mining operations (hydraulic mining and dredging) and in dust control.

Water is a natural feature that interferes with mining activities. It enters mines by groundwater infiltration and surface runoff and comes into contact with materials in the host rock, ore, and overburden. An additional source of water in deep underground mines is the water that results from the backfilling of slopes with the coarse fraction of the mill tailings. Transportation of these sands underground is typically accomplished by sluicing. Mill wastewater is usually the source of the sluice water. Mine water then requires treatment depending on its quality before it can be safely discharged into the surface drainage network. Generally, mining operations control surface runoff through the use of diversion ditching and grading to prevent, as much as possible, excess water from entering the working area. The quantity of water from an ore mine thus is unrelated, or only indirectly related, to production quantities.

Water is used in the ore mining and dressing industry for several principal uses under three major categories: noncontact cooling water; process water (i.e., wash water, transport water, scrubber water, and process and product consumed water); and miscellaneous water (i.e., dust control, domestic/sanitary uses, washing and cleaning, and drilling fluids).

Noncontact cooling water is defined as cooling water that does not come into direct contact with any raw material, intermediate product, by-product, or product used in or resulting from the process. Process water is defined as that water which, during the benefication process, comes into direct contact with any raw material, intermediate product, by-product, or product used in or resulting from the process.

Wastewater characteristics for the Ore Mining and Dressing industry, in general, reflect the diversity of the mining and milling operations associated with the various ores mined and processed. Each ore exhibits its own particular set of waste characteristics, as shown in Table 36. The peculiarities were, in part, criteria used to determine the various subcategories.

Control and Treatment Technology

In-Plant Control Technology: In-process recycling of wastewater after thickening or dewatering is utilized at some mines and has reduced the amount of wastewater discharged by five to seven percent. Some mines have also started to recover wastewater from their tailings ponds and in some cases have reduced wastewater discharge as much as 50 percent.

End-of-Pipe Treatment Technology: Removal of particulate matter is generally by clarification with or without chemicals (lime, alum, or polyelectrolyte) or with multi-media filtration. Suspended solids removals of 85 percent to 98+ percent have been achieved.

Beneficiation (dressing) of ores generally contributes significant amounts of cyanide or phenols. Currently these pollutants are being removed by alkaline chlorination, and research is being conducted into the use of ozone.

Table 36: Ore Mining and Dressing Industry–Raw Wastewater Characteristics

Wastewater parameters[b]	Subcategory[a]								
	Iron ore		Base and precious metals[b]		Uranium[c]		Ferralloy		Mercury[e]
	mine	mill	mine	mill	mine	mill	mine	mill	mine
Conventional - mg/l									
COD	1.0-48	<1-23	4.0-630	16-222	240-600	28-630		24-170	6.5-8.2
Oil and grease			0-29	<0.05-1			1.0-14	1-15	
pH (units)	7.2-8.4	7.2-9.5	3.3-9.6	7.9-11			4.5-7.3	3.5-8.6	6.5-8.6
TDS	120-1,300	200-2,400	260-29,000	400-4,300				210-2,600	
TOC			<1-31	12-29	16-25	<1-450			
TSS	<1-48	12-55	<2-1,000	2-550,000	300	110,000-530,000		2.3-500,000	
Inorganics - mg/l									
Aluminum			0.143-<0.2	<0.5[d]		18-1,600			
Antimony			<0.2-<0.5	<0.2-1.85					<0.5-3.8
Arsenic			<0.01-0.08	<0.07-3.7		0.13-2.3	<0.01-<0.07	0.01-0.1	0.02
Boron			0.01-0.20						
Cadmium			<0.002-1.3	<0.01-16			<0.005-0.07	<0.005-0.74	0.42
Calcium			45-46		93-120	220-3,200		206[d]	
Chromium			<0.02-0.42	9.8-40				0.02-0.03	
Cobalt			<0.05-1.9	1.68[d]					
Copper			<0.02-92	0.03-910		<0.5-1.7	<0.02-3.8	0.03-51	1.3
Cyanide			<0.01	<0.01-81				<0.01-0.45	
Gold				<0.05[d]					
Iron (total)	<0.02-4.5	0.4-1,200	<0.02-1,000		0.23-0.47	0.92-330		0.44-1,500	<0.5-2,900
Iron (dissolved)	<0.02-0.08	<0.02-0.16							
Lead			<0.05-4.9	<0.01-560		<0.005-2.1	0.06-0.19	0.02-9.8	0.58
Magnesium			28-32		36-45	190-550			
Manganese	<0.02-3.2	0.032-330	<0.02-100	300-570		<0.2-210	0.21-6.8	0.19-57	7-50

(continued)

Table 36: (continued)

Wastewater parameters[b]	Subcategory[a]								
	Iron ore		Base and precious metals[b]		Uranium[c]		Ferralloy		Mercury[e]
	mine	mill	mine	mill	mine	mill	mine	mill	mine
Mercury			< 0.0001-0.78	0.0006-0.15					28
Molybdenum			<0.2-< 0.5	28[d]	0.5-0.53	<0.3-16	0.1-0.5	0.5-18	
Nickel			< 0.05-0.24	0.05-0.14		0.52-1.4			2.4
Potassium			8-15						
Radium			44[d]		2,700-3,200	110-19,000			
Selenium			< 0.003-0.126	0.144-0.154					
Silver			<0.02	< 0.02					
Sodium			7-12						
Strontium			0.09-120	1.2[d]					
Tellurium			<0.3						< 0.08
Thallium			<0.1						
Titanium			<0.5			0.4-3			
Uranium			<0.5[d]		12	3.9-170			
Vanadium			<0.2		< 0.5-1.0	< 0.5-130	< 0.5-0.5	<0.5	
Zinc			< 0.1-170	0.02-3,000		<0.5	0.05-7	< 0.02-77	0.14-1.10

[a] Subcategory 2, "Aluminum" and Subcategory 7, "Other Metal Ores not Elsewhere Classified", have been excluded from this list. Data is not currently available on wastewater characteristics in the aluminum subcategory. The wide diversity of, and limited production figure on metals contained in Subcategory 7 precludes a meaningful comparison of wastewater characteristics; thus, it also has been excluded.

[b] Data in the "Base and Precious Metals" subcategory contains ranges from Copper, Lead/Zinc, Gold and Silver Mining and Milling Industries.

[c] Includes ranges from alkaline and acid processes.

[d] Indicates single unit values.

[e] Data on mill wastewaters nonexistent at this time.

Type of Residue Generated

Information on the types and characteristics of residues generated by this industry is insufficient and cannot be discussed at this time.

Residual Management Options

Lack of sufficient information on types and characteristics of residues generated does not permit discussion of specific management options at this time.

ORGANIC CHEMICALS

General Industry Description

Organic chemicals are the raw materials for a multitude of products the public uses daily, including plastics, synthetic fibers, synthetic rubber, dyes, solvents, food additives, pharmaceuticals, lubricants, detergents, and cosmetics. Synthetic organic chemicals are derived as a result of the physical and chemical conversion operations from naturally occurring raw materials such as petroleum, natural gas, and coal. Approximately 50 percent of the plants in this industry discharge to municipal treatment works.

Industrial Categorization

This industry has been divided into the subcategories shown in Table 37.

Process Description

General: The process area of an organic manufacturing plant is referred to as the "Battery Limit", while the remainder of the plant is called the "Off-Sites". The Off-Sites can be broken down into their components: the storage and handling facilities, the utilities, and the services. This is illustrated in Figure 28.

Storage facilities associated with any chemical plant depend upon the physical state (i.e., solid, liquid, or gas) of the feedstocks and products. Storage equipment includes cone-roof tanks for liquids, cylindrical, or spherical tanks for gases, and concrete pads of silos and solids. Wastewater emanating from this part of the plant normally results from storm run-off, tank washing, accidental spills, and aqueous bottoms periodically drawn from storage tanks. These wastes are generally small in volume, but since they do come into contact with process chemicals, these chemicals will appear in the waste stream.

Utility functions such as the supply of steam and cooling water gener-

Table 37: Organic Chemicals Industry

Subcategory	Products		Designation
Nonaqueous processes	Benzene-Toluene-Xylene (BTX) Cumene Cyclohexane	Vinyl chloride P-Xylene	A
Processes with process water contact as steam dilutent or absorbent	Acetaldehyde Acetone Acetylene Adiponitrile Benzoic acid and benzaldehyde Butadiene Chlorobenzene Chloromethanes Chlorotoluene Diphenylamine Ethyl benzene Ethylene and propylene Ethylene dichloride	Ethylene oxide Formaldehyde Hexamethylenediamine Maleic anhydride Methanol Methyl amines Methyl chloride Methyl ethyl ketone Perchloroethylene Phthalic anhydride Styrene Tricresyl phosphate Vinyl acetate Vinyl chloride	B
Aqueous liquid phase reaction systems	Acetic acid Acrylates Acrylic acid Acrylonitrile Aniline Bisphenol A Calcium stearate Caprolactam Coal tar Cyclohexanone oxime Dimethyl terephthalate Ethyl acetate Ethylene glycol Formic acid Hexamethylenetetramine Hydrazine solutions	Isobutylene Isopropanol Methyl methacrylate Oxalic acid Oxo chemicals p-Aminophenol p-Cresol Pentaerythritol Phenol and acetone Propyl acetate Propylene glycol Propylene oxide Saccharin Sec-butyl alcohol Synthetic cresol Terephthalic acid Tetraethyl lead	C
Batch and sem-continuous processes	Citric acid Citronellol and geraniol Dyes and dye intermediates Fatty acids Fatty acid derivatives Ionone and methylionone Methyl salicylate Miscellaneous batch chemicals Napthtenic acid	o-Nitroaniline p-Nitroaniline Pentachlorophenol Pigments (lakes) Pigments (toners) Plasticizers Sodium glutamate Tannic acid Vanillin	D

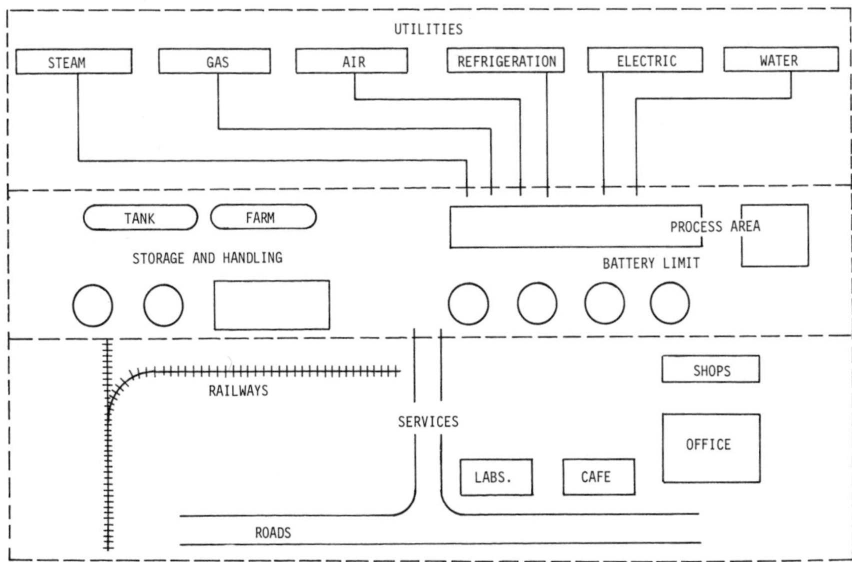

Figure 28: Plot plan for chemical plant illustrating four area layout.

ally are set up to service several processes. Noncontact steam, generated in the boiler house, is circulated through a closed loop, whereby varying quantities are made available to the different processes.

The uses for steam are as follows:

- For noncontact process heating.
- For power generation such as steam driven turbines, compressors, and pumps.
- For use as a diluent, stripping medium, or source of vacuum. This steam will become contaminated and will need treatment.

Wastes from non-contact use of steam come from purges of the system, boiler blowdowns, and water treatment systems for the steam generation system. Non-contact cooling water is also supplied to the processes. Once-through cooling systems constitute an uncontaminated waste stream, while cooling tower blowdowns from closed cooling systems contain water treatment chemicals.

The service area of the plant contains the buildings, shops, and laboratories, in which the personnel work. Waste streams are generated from laundry facilities, sanitary facilities, and wastes from laboratory and shop operations.

In regard to the "battery limits", most plants manufacture many different products. Each process is itself a series of unit operations which causes chemical and physical changes in the feedstock or products. In the commercial synthesis of a single product from a single feedstock, there generally are sections of the process associated with: preparation of the feedstock; chemical reaction; separation of reaction products; and final purification of the desired product. Each unit operation may have drastically different water usages associated with it. Type and quantity of contact wastewater are therefore directly related to the nature of the various processes. This in turn implies that the types and quantities of wastewater generated by each plant's total production mix are unique.

Production from a given process module is related to the design capacities of the individual unit operations within it. In many cases, unit operations are arranged as a single train in series. In other cases, some unit operations, such as the reaction, are carried in several small reactors operating in parallel. Flow of material between unit operations within a process may be either a continuous stream or through a series of batch transfers.

Facilities utilizing continuous processes manufacture products in much greater volumes than do batch operations. Batch processing is still extensively practiced, however, particularly when production is small or where safety demands that small quantities be handled at one time. Furthermore, batch operations are more easily controlled when varying reaction rates and rapid temperature changes are key considerations.

The feed preparation section may contain equipment such as furnaces where the liquid feed is vaporized or heated to reaction temperature, or large steam-driven compressors for compressing gaseous feed to the reaction pressure. It may contain distillation columns to separate undesired feed impurities which might damage the catalyst in the reactor or cause subsequent unwanted side reactions. Impurities may also be removed by preliminary chemical conversion (such as the hydrogenation of diolefins) or by physical means such as silica gel driers to remove trace amounts of moisture.

The reaction section of the process module is where the principal chemical conversions are accomplished. The reactor may be as simple as a hollow tube used for noncatalytic vapor-phase reactions. However, most industrial reactions are catalytic and generally require more complex reactor designs. Specific reactor design is usually governed by the required physical state of the reactants and catalyst.

Catalysts are of two types: heterogeneous and homogeneous. Heterogeneous catalysts are usually solids which may be composed of chemically inactive material such as finely ground aluminum, or contain metals such as cobalt, platinum, iron, or manganese which are impregnated on a solid support. In heterogeneous reaction systems, the reactants are usually in the vapor phase. Homogeneous catalysts exist in the same physical state as the reactants and products. This may require the use of an aqueous or nonaqueous solvent to provide a reaction media. Typical homogeneous catalysts include strong acids, bases, and metallic salts which may be in the form of a solution or a slurry. Recovery, reconcentration, or regeneration of these catylsts may require the use of processing equipment much more elaborate than the reactor itself.

Recovery of reaction products may involve a wide variety of processing operations. If the reactor effluent is a vapor, it may be necessary to condense and quench the products in a direct contact medium such as water. In many instances the desired products are absorbed in water and are subsequently stripped from the water by heating. Liquid reactor effluents are separated from solvents (and catalysts) by distillation. In almost all cases, the conversion of feed is not complete, so that continuous separation and recycle of unconverted feed to the reactor is necessary.

Final purification of the products is normally required both when they are to be sold and when they are used as intermediates. Most specifications restrict contaminant levels to the range of parts per million. Because of this, additional operations such as distillation, extraction, crystallization, etc., are necessary. Product is pumped from the battery limits to tanks in the storage area.

When chemical manufacturing is on a small scale, or when it is not adaptable to continuous procedures, a batch sequence is frequently used. Batch operations with small production and variable products, transfer equipment from the making of one chemical to that of another based on the same type of chemical conversion. Hundreds of specific products may be manufactured within the same building.

This type of processing requires the cleanout of reactors and other equipment after each batch. Purity specifications may also require extensive purging of the associated piping. Rapid changes in temperature during the batch sequence may also require the direct addition of ice or quench water, as opposed to slower, non-contact cooling through a jacket or coils.

Process waters from batch or continuous processes within the battery limits include not only water produced or required by the chemical reactions, but also any water which comes in contact with chemicals within each of the process modules. Although the flows associated with these sources are generally much smaller than those from non-contact sources, the organic pollution load carried by these streams is greater by many orders of magnitude.

Process water is defined as all water which comes in contact with chemicals within the process, and includes:

- Water required or produced (in stoichiometric quantities) in the chemical reaction.
- Water used as a solvent or as an aqueous medium for the reactions.
- Water which enters the process with any of the reactants or which is used as a diluent (including steam).
- Water associated with the catalyst system, either during the reaction or during catalyst regeneration.
- Water used as an absorbent or as a scrubbing medium for separating certain chemicals from the reaction mixture.
- Water introduced as steam to strip certain chemicals from the reaction mixture.
- Water used to wash, remove, or separate chemicals from the reaction mixture.
- Water associated with mechanical devices, such as steam-jet ejectors for drawing a vacuum on the process.
- Water used as a quench or direct contact coolant, such as in a barometric condenser.
- Water used to clean or purge equipment used in batch-type operations.
- Runoff or wash water associated with Battery Limits process areas.

The type and quantity of process water usage are related to the specific unit operations and chemical conversions within a process. The term "unit operations" is defined to mean specific physical separations such as distillation, solvent extraction, crystallization, adsorption, etc. The term "chemical conversion" is defined to mean specific reactions such as oxidation, halogenation, neutralization, etc.

Description of Subcategories: Four process subcategories have been established. The nonaqueous, process water contact as steam diluent or absorbent, and aqueous liquid-phase subcategories are related to continuous processes, while the batch and semi-continuous subcategory relates to batch processes. The subcategories are described as follows.

Nonaqueous Processes (A) — In this subcategory there is minimal contact between water and reactants or products within the process. Water is not required as a reactant or diluent and is not formed as a reaction product. The only water usage stems from periodic washes of working fluids or catalyst hydration. Raw waste loads should approach zero, with the only variations caused by spills or process upsets.

Processes with Process Water Contact as Steam Diluent or Absorbent (B) — Process water usage is in the form of dilution steam, a direct contact quench, or as an absorbent for reactor effluent gases. Reactions are all vapor-phase, and are carried out over solid catalysts. Most processes have an absorber coupled with steam stripping of chemicals for purification and recycle. Steam is also used for de-coking of catalyst.

Continuous Liquid-Phase Reaction Systems (C) — Liquid-phase reactions occur where the catalyst is in an aqueous medium, such as dissolved or emulsified mineral salt, or acid-caustic solution. Continuous regeneration of catalyst system requires extensive water usage. Substantial removal of spent inorganic salt by-products may also be required. Working aqueous catalyst solution is normally corrosive. Additional water may be required in final purification or neutralization of products.

Batch and Semi-continuous Processes (D) — Processes are carried out in reaction kettles equipped with agitators, scrapers, reflux condensers, etc., depending on the nature of the operation. Many reactions are liquid-phase with aqueous catalyst systems. Reactants and products are transferred from one piece of equipment to another by gravity flow, pumping, or pressurization with air or inert gas. Much of the material handling is manual, with limited use of automatic process control. Filter presses and centrifuges are commonly used to separate solid products from liquid. Where drying is required, air or vacuum ovens are used. Cleaning of noncontinuous production equipment constitutes a major source of wastewater. Waste loads from product separation and purification will be at least ten times those from continuous processes.

Wastewater Characteristics

Table 38 shows raw wastewater characteristics. The pollutants indicated in the table for the first three subcategories are based on contact process water only. Noncontact water is excluded for continuous processes since these plants have been able to achieve segregation of noncontact cooling water or steam. The "Batch and Semi-Continuous Process" subcategory includes all water usage associated with the process in that rapid cooling with direct contact is required.

Table 38: Organic Chemicals Industry—Raw Wastewater[a] Characteristics

Wastewater parameter	Non-aqueous processes	Process with process water as steam diluent or absorbent	Aqueous liquid phase reaction systems	Batch and semi-continuous processes
Ammonia nitrogen	1-150	1-50	0-1,000	5-1,000
BOD_5		100-500		
Chemical oxygen demand	100-10,000	200-5,000	10,000-50,000	1,000-10,000
Color				Present
Cyanide	0-1	0-0.3	0-1	0.02
Metals	Present	Present	Present	Present
Oil	10-250	2-200	0-1,200	10-4,000
Phenol	0-15	0-20	0-6,000	0-150
Total dissolved solids	20-50,000	400-100,000	23-100,000	40-300,000
Total organic carbon	100-3,000	100-2,000	3,000-5,000	
Total suspended solids	10-100	10-2,300	10-4,000	20-4,000

[a] All values in mg/l.

Control and Treatment Technology

In-Plant Control Technology: The following in-plant control measures are practiced in this industry:

- Substitution of surface heat exchangers for contact cooling water used in barometric condensers.
- Regeneration of contact process steam from contaminated condensate.
- Substitution of vacuum pumps for steam jet ejectors.
- Recycle of scrubber water.
- Recovery of insoluble hydrocarbons.
- Solvent extraction for recovery of phenols.

End-of-Pipe Treatment Technology: Biological treatment is the major treatment technology used in this industry. Both single stage and multiple stage plants are used, especially when phenol removal is required. Filtration is also used as a polishing step after biological treatment.

Activated carbon is becoming more accepted as an alternate treatment scheme to biological treatment. Contact times of 22 to 660 minutes are required, as opposed to 10 to 50 minutes for domestic waste. Pretreatment for suspended solids and oil removal is required to levels of at least 10 mg/l total suspended solids and 50 mg/l of oil.

Equalization is also good practice preceding all forms of treatment schemes. Table 39 gives removal efficiencies for the treatment systems described above.

Table 39: Organic Chemicals Industry–Removal Efficiencies of Various Wastewater Treatment Practices

	Treatment Technology Removal Efficiency - Percent		
Pollutant parameter	Biological treatment[a]	Filtration[b]	Pretreatment plus activated carbon
Biochemical oxygen demand	93	17	90
Chemical oxygen demand	69 (74)	20	69
Total organic carbon	60 (79)	20	87

[a] Number not in ( ) is for single stage treatment, number in ( ) is for multiple stage treatment.

[b] Percentage is between inlet and outlet of filter.

Type of Residue Generated

This industry is characterized by a large variety of compounds produced in a somewhat standard set of process vessels, ancilliary equipment, and utilities. Although production facilities may be of many different sizes, may be replicated in varying degrees in the same plant, and may be used either intermittently or continuously on the same product, much of the wastewater characteristics and treatment schemes are similar.

Biological treatment is the major treatment method used. The organic fraction of the waste biological sludge is biomass, with a fairly uniform composition. This sludge may also have small amounts of refractory organic compounds, and some adsorbed heavy metals. The latter may originate as process catalysts, or as excess process reactants. Frequently the wastewater is lacking in nutrients such as nitrogen and phosphorus.

Carbon adsorption is the main alternate method for treating wastewater flows from this industry. The residue which is generated by this treatment step is related to the form of activated carbon used — granular or powdered — and the method used for its regeneration, if it is regenerated.

Granular carbon is customarily used in beds, either downflow or upflow. The bed must be periodically regenerated. Common regeneration methods are: steam stripping the bed in place; thermal regeneration in a separate furnace, with movement of the carbon in and out of the carbon bed or column as a water slurry; and chemical regeneration in place. Steam stripping produces an enriched water/contaminant flow, which frequently can be returned to the production process. Thermal regeneration at a controlled temperature burns off the adsorbed contaminant from the carbon particle while leaving the particle itself essentially intact. Chemical regeneration is a less well-developed procedure that produces a brine that may be either reused in the production process or wasted, depending on specific circumstances.

Powdered carbon is usually added prior to a settling tank or an activated sludge aeration tank, and allowed to settle out with the sludge in the settling tank. Powdered carbon is usually disposed of along with the sludge, although some processes for recovering it from the sludge, regenerating, and reusing it are under development.

Filtration has also been used as a polishing step following biological treatment. Residue generated from this step will be relatively small in quantity. Its exact nature will be determined by the type of filter, method of backwashing, and any coagulant or precoat additives used.

Various inorganic brines with high total dissolved solids can also be generated, in relatively small amounts, in organic chemical plants. Reuse of these brines in a production process after removal of some organic contaminants has been discussed in the literature, but the general methods for brine disposal are unclear.

Numerous other wastewater treatment unit processes have been used or proposed for this industry, including evaporation, distillation, stripping, and reverse osmosis. However, these processes are economically justifiable only when some potentially valuable material can be recovered and recycled into the manufacturing process or sold, or as a last resort preparatory to disposal by incineration.

Residual Management Options

The amount of sludge generated by this industry is the second largest (after iron and steel) of the twelve largest sludge producing industries. The greatest amount is from biological treatment, with possible contamination with toxic organic compounds or heavy metals. Under ideal circumstances, the sludge might be spread on land for agricultural purposes in the same manner as municipal sludge, although there is little discussion of this in the literature. Under less ideal circumstances, burial in an acceptable landfill may be the only suitable method of disposal.

Incineration is another possibility. Organic chemical plants usually have an incineration unit to dispose of nonaqueous and aqueous, high strength, organic production process residues. Residues may be excess reactants, unusable by-products, or dumps of unsatisfactory-quality products, but are distinct from wastewater treatment residues. In certain cases — wastewaters with contaminants in a concentration range too high for wastewater treatment and too low or too impure for reuse in the production process, or wastewater with very toxic, nonbiodegradable contaminants — it may be most economical to concentrate the contaminants and then incinerate the concentrated fraction. Evaporation or distillation would be suitable concentration processes. Incineration would then be in the plant's general purpose incinerator.

In spite of the options available for wastewater treatment, there is considerable emphasis in this industry on reducing wastewater and sludge generation by production process modification, reuse of partially contaminated flows in other production processes, and consideration of waste generation and disposal problems and costs in choice of production technique.

PAINT AND INK FORMULATION

General Industry Description

The paint and ink manufacturing industry is essentially a product formulation industry, in that few, if any, of the raw materials are manufactured on site. The major products consist of interior and exterior paints, industrial finishes for such products as automobiles, appliances, furniture; varnish and lacquer; putty; caulking compounds; sealants; paint and varnish removers; and printing inks. The principal raw materials are oils, resins, pigments and solvents.

The vast majority of paint and ink plants discharge wastewater to POTWs.

Industrial Categorization

A useful subcategorization for the purpose of raw waste characterization and organization of pretreatment information is shown in Table 40.

Table 40: Paint and Ink Formulation

Major Category	Subcategory
Paint	Solvent Wash (solvent-base solvent wash) Water and/or Caustic Wash
Ink	Solvent Wash (solvent-base solvent wash) Water and/or Caustic Wash

Process Description

Both paint and ink can be either solvent-base or water-base, but there is little difference in the production processes used. The major production difference is in the carrying agent — solvent-base paints and inks are dispersed in an oil mixture, while water-base paints and inks are dispersed in water, with a biodegradable surfactant used as the dispersing agent. Another significant difference is in the cleanup procedures. Since the water-base products contain surfactants, it is much easier to clean up the tubs with water. The tubs used to make the oil-base products are generally cleaned with an organic solvent or with a strong caustic solution. Plants that use the caustic rinse systems usually rinse the caustic residue with water.

Paints and inks are generally made in batches. The major difference in the size of a plant is in the size of the batches. A small plant will make up batches from 100 to 500 gallons, while a large plant will manufacture batches of up to 10,000 gallons. There are generally too many color formulations to make a continuous process feasible.

Solvent-Base Paint and Ink: There are three major steps in the solvent-base paint and ink manufacturing process: (1) mixing and grinding of raw materials, (2) tinting and thinning, and (3) filling operations.

At most plants, mixing and grinding of raw materials for solvent-base paints and inks are accomplished in one production step. For high gloss paints, pigments and a portion of the binder and vehicle are mixed into a paste of a specific consistency. This paste is fed to a grinder, which disperses the pigments by breaking down particle aggregates rather than by reducing the particle size. Two types of grinders are ordinarily used for this purpose: pebble or steel ball mills, or roll-type mills. Other paints are mixed and dispersed in a mixer using a saw-toothed dispersing blade.

In the next stage of production, the material is transferred to tinting and thinning tanks, occasionally by means of portable transfer tanks, but more commonly by gravity feed or pumping. Here, the remaining binder and liquid, as well as various additives and tinting colors, are incorporated. Finished product is then transferred to a filling operation where it is filtered, packaged, and labeled.

The product remaining on the sides of the tubs or tanks may be allowed to drain naturally, and the "clingage", as it is called, is either wasted or the sides may be cleaned with a squeegee during the filling operation until only a small quantity of product remains. The final cleanup of the tubs generally consists of flushing with an oil-base solvent until clean. Dirty solvent is handled in one of three ways: (1) used in the next batch as a part of the formulation; (2) placed in drums that are sold to a company where it is redistilled and resold; or (3) collected in drums with the cleaner solvent being decanted for subsequent tank cleaning and returned to the drums until only sludge remains in the drum. Sludge is then sent to a landfill for disposal.

Figure 29 is a flow diagram for solvent-base paint manufacture.

Water-Base Paint and Ink: Water-base paints and inks are produced by slightly different methods as compared to solvent-base products. Pigments and extending agents are usually received in proper particle size, and the dispersion of the pigment, surfactant, and binder into a vehicle is accomplished with a saw-toothed disperser. In small plants, the product is thinned and tinted in the same tub, while in larger plants, the product is transferred to special tanks for final thinning and tinting. Once the formulation is correct, the product is transferred to a filling operation where it is filtered, packaged and labeled in the same manner as for solvent-base paints and inks.

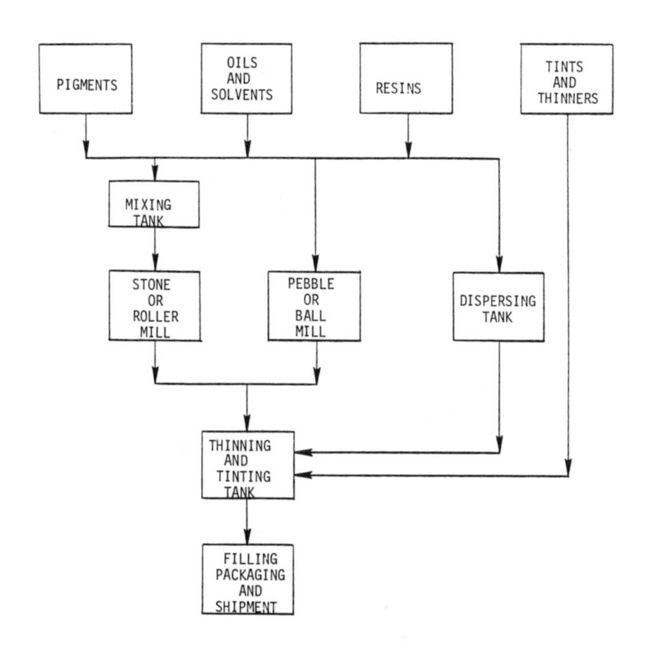

Figure 29: Flow diagram for solvent-base paint manufacturing.

As in the solvent-base paint and ink operation, as much product as possible may be removed from the sides of the tub or tank before final cleanup starts. Cleanup of the tubs is done simply by washing the sides with a garden hose or a more sophisticated washing device. Wash water may be: (1) collected in holding tanks and treated before discharge; (2) collected in drums and taken to a landfill; (3) discharged directly to a sewer or receiving stream; (4) reused in the next batch; or (5) reused in the washing operation.

Allied products manufactured by the paint portion of the industry include putty, caulking compounds, paint and varnish removers, shellacs, stains, wood fillers, and wood sealers. Manufacturing processes for these products do not generally utilize water, except for some water-base stains and paint removers. Wastes generated in allied product manufacture do not greatly differ from those generated in paint formulation. These categories are generally low in water use, and being very similar to paint manufacture, they have been placed in the same category.

Figure 30 is a flow diagram for water-base paint manufacture.

Wastewater Characteristics

Table 41 shows raw wastewater characteristics. Table 42 shows raw wastewater as a function of production based data. Since batches of solvent-base paints and inks that are rinsed with solvents ordinarily generate no wastewater, the most important factor affecting the volume of process wastewaters generated and discharged at paint and ink plants is the percentage of solvent-base and water-base products. The paint industry generates a total of approximately 1.5 MGD of process wastewater, about half of which is actually discharged. About 40 percent of the paint industry does not discharge any wastewater; about 30 percent of all paint plants discharge less than 100 gpd; and less than 10 percent of the plants discharge more than 1,000 gpd, but these plants account for at least 85 percent of total wastewater discharged from all paint plants. The ink industry generates about 40,000 gallons of wastewater daily, of which only 75 percent, or 30,000 gpd, are actually discharged. Because the flow from this industry is small but potent, and can therefore have a serious deleterious effect on the treatment plant and sludge of any municipal wastewater system that it enters, it has been recommended that all discharges from this industry into POTWs cease.

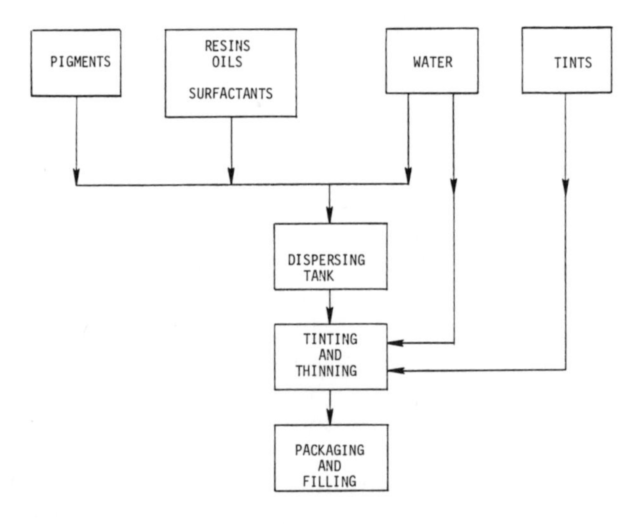

Figure 30: Flow diagram for water-base paint manufacturing.

Control and Treatment Technology

Virtually all paint and ink plants discharge wastewater to POTWs without any prior treatment. Ink process wastewater is often diluted with cooling water, sanitary wastewater, or other waste streams before discharge. Among paint plants, there is a wide variation in the amount, characteristics, handling, and treatment of process wastewater.

In-Plant Control Technology

Wastewater Reduction — There are methods that are used by some paint and ink plants to reduce overall water usage. The amount of water required to clean paint tanks or ink tubs can be reduced by cleaning the vessel surfaces with a squeegee prior to rinsing. The quantity of wastewater from vessel cleaning can also be reduced by the use of high pressure water. Several commercial systems consisting of booster pumps, flow regulators, and nozzles are available which can supply low volume, high pressure water sprays.

Another in-plant wastewater reduction measure is the sealing or elimination of floor drains and trenches. Plants that have no drains must collect all rinse water which may facilitate controlling the volume of water used for each purpose. Spills are picked up with shovels or squeegees, and floors are mopped, vacuumed, or cleaned by machine.

Table 41: Paint and Ink Formulation–Raw Wastewater Characteristics

Wastewater Parameter	Paint	Ink
Conventional - mg/l		
Biochemical oxygen demand	9,892	19,804
Chemical oxygen demand	54,956	39,819
Oil and grease	1,099	622
Total dissolved solids	10,619	11,244
Total organic carbon	10,601	9,874
Total solids	28,945	11,351
Total suspended solids	20,404	991
Total volatile solids	13,017	10,870
Total volatile suspended solids	7,789	341
Inorganics - ug/l		
Aluminum	196,758	40,926
Antimony	209	613
Arsenic	286	384
Barium	8,656	19,792
Beryllium	126	8
Boron	4,268	553
Cadmium	524	44
Calcium	2,277,000	962,000
Chromium	3,120	35,271
Cobalt	912	396
Copper	2,476	17,130
Cyanide	79	161
Iron	271,307	29,454
Lead	6,300	151,009
Magnesium	107,000	93,000
Manganese	2,901	505
Mercury	5,161	131
Molybdenum	674	55,990
Nickel	1,350	261
Selenium	165	384
Silver	15	8
Sodium	397	3,559
Thallium	151	371
Tin	1,111	349
Titanium	16,677	1,479
Vanadium	409	131
Yttrium	206	156
Zinc	74,746	4,080
Organics - ug/l		
Phthalates		
Bis (2-ethylhexyl) phthalate	418	12,520
Butyl benzyl phthalate	474	10
Di-n-butyl phthalate	5,745	188
Di-n-octyl phthalate		3,600
Diethyl phthalate	233	25
Dimethyl phthalate		10

(continued)

Table 41: (continued)

Wastewater parameter	Paint	Ink
Nitrogen Compounds		
N-nitrosodiphenylamine		10
Phenol	460	121
2-Chloronaphthalene	0.001	
Aromatics		
Benzene	1,933	368
Chlorobenzene	1,405	278
1,2-Dichlorobenzene		10
3,3'-Dichlorobenzidene		10
1,1,-Dichloroethylene	138	15
2,4-Dinitrotoluene		10
2,6-Dinitrotoluene		10
1,2-Diphenylhydrazine		3,805
Ethylbenzene	7,482	4,151
Hexachlorobenzene	92	
1,2,4-Trichlorobenzene		10
Polycyclic Aromatic Hydrocarbons		
Acenaphthene		10
Fluoranthene		10
Halogenated aliphatics		
Carbon tetrachloride	3,770	96
Chloroform	186	37
1,1-Dichloroethane	11	21
1,2-Dichloroethane	118	89
1,1,2,2-Tetrachloroethane	29	
1,1,1-Trichloroethane	141	560
1,1,2-Trichloroethane	562	10

Table 42: Paint and Ink Formulation–Raw Wastewater Characteristics–Production-Based Data

Wastewater Parameter	Paint[a]	Ink[b]
Conventional (kg/day)		
Biochemical oxygen demand	28,032	1,796
Chemical oxygen demand	155,737	3,611
Oil and grease	3,114	56
Total dissolved solids	30,093	1,020
Total organic carbon	30,042	895
Total suspended solids	57,878	90
Inorganics (kg/day)		
Aluminum	558	3.711
Antimony	0.59	0.056
Arsenic	0.81	0.035
Barium	24.5	1.795
Beryllium	0.36	0.001
Boron	12.1	0.050
Cadmium	1.5	0.004
Calcium	6,453	87
Chromium	8.8	3.198
Cobalt	2.6	0.036
Copper	7.0	1.554
Cyanide	0.22	0.015
Iron	768	2.671
Lead	17.9	13.694
Magnesium	303	9
Manganese	8.2	0.046
Mercury	14.6	0.012
Molybdenum	1.9	5.077
Nickel	3.8	0.024
Selenium	0.47	0.035
Silver	0.04	0.001
Sodium	1.1	0.232
Thallium	0.43	0.034
Tin	3.1	0.032
Titanium	47.3	0.134
Vanadium	1.2	0.012
Yttrium	0.58	0.014
Zinc	211.8	0.370

(continued)

Table 42: (continued)

Wastewater Parameter	Paint[a]	Ink[b]
Organic (kg/day)		
Acrolein	0.001	
Aldrin	0.001	
Alpha-BHC	0.002	
Anthracene	0.001	
Benzene	3.2	0.025
Benzo (A) pyrene	0.001	
Beta-BHC	0.001	
Beta-Endosulfan	0.002	
Butyl benzyl phthalate	0.16	
Carbon tetrachloride	2.7	0.001
Chlorobenzene	0.48	0.006
Chloroform	0.25	0.002
2-Chloronapthalene	0.001	
4-Chlorophenyl phenyl ether	0.02	
4,4'-DDD	0.001	
4,4'-DDE	0.001	
Delta-BHC	0.001	
3,3-Dichlorobenzidine	0.001	
Dichlorobromomethane	0.002	
1,1-Dichloroethane	0.002	
1,2-Dichloroethane	0.05	0.002
1,1-Dichloroethylene	0.06	0.001
Di (2-Chloroethyoxy) methane	0.0001	
Di (2-Chloroisopropyl) ether	0.27	
2,4-Dichlorophenol	0.002	
1,2-Dichloropropane	0.09	
1,3-Dichloropropylene	0.009	
Dieldrin	0.001	
Di (2-Ethylhexyl) phthalate	0.42	0.988
Di-N-Butyl phthalate	9.9	0.013
2,4-Dinitrophenol	0.04	
4,6-Dinitro-O-Cresol	0.001	
Di-N-Octyl phthalate		0.039
Endrin aldehyde	0.001	
Ethylbenzene	17.0	0.139
Fluoranthene	0.001	
Gamma-BHC	0.001	
Hexachlorobenzene	0.008	
Isophorone		0.479
Methylene chloride	52.4	0.053
Napthalene	2.4	0.001
Nitrobenzene	0.03	
Pentachlorophenol	3.2	0.015
Phenol	0.74	0.007
Phenols (total)	0.74	0.021
1,1,2,2-Tetrachloroethane	0.003	
Tetrachloroethylene	0.93	0.070
Toluene	44.3	0.128
1,2-Trans-Dichloroethylene	0.02	
1,1,1-Trichloroethane	0.20	
1,1,2-Trichloroethane	0.26	0.013
Trichloroethylene	0.12	0.083
2,4,6-Trichlorophenol	0.42	

[a] The paint industry generates a total of approximately 1.5 million gallons of process wastewater daily, of which half, or 750,000 gallons, is actually discharged. Mass loadings shown are based on an industry flow of 750,000 gallons daily, and the average raw wastewater characteristics as shown in Table 41.

[b] The ink industry generates a total of about 40,000 gallons of wastewater daily, of which 75 percent, or 30,000 gallons per day, is actually discharged. Mass loadings shown are based on industry flow of 24,000 gallons daily, and the average raw wastewater characteristics as shown in Table 41.

Wastewater Recycle — Although most paint plants produce a wide variety of paint colors and finishes, many plants' production consists of predominantly white and off-white batches. Good practice, already in use at some plants, is to attempt to segregate white paint production and to reuse the wastes from each batch in the subsequent batch.

Even where plants cannot dedicate tanks to a single product, the same recycle opportunities can be obtained by scheduling batches of the same or similar products back-to-back in the same tank. The rinse water from the first batch can be held in the tank and used in the next batch as part of the formulation, reducing raw material requirements and avoiding disposal costs.

Where paint rinse water cannot be reused immediately, there are several methods that are practiced by paint plants for eventually recycling this water. There are some plants that collect all paint wastewater in drums or tanks, label it by color and base, and reuse it in the next compatible batch of similar or darker color. The wastewater may have to be treated with a biocide, and is usually used as soon as possible.

Another method for reusing paint wastewater is to treat the wastewater by physical-chemical precipitation or some other method, and reuse the rinse water for subsequent paint batches or as rinse water. Wastewater recycle is also practiced by some ink plants.

End-of-Pipe Treatment Technology: The most common treatment methods used by paint plants are clarification (with or without chemical addition) and neutralization. Few plants employ biological treatment for paint wastes, and those that do usually have a combined treatment plant for wastes from other plant operations. Advanced wastewater treatment methods such as activated carbon or ultrafiltration are not used. Of the paint plants that discharge their wastewater to a municipal sewer, approximately 40 percent pretreat their wastes prior to disposal.

The most common methods used by ink plants for treating or pretreating wastewater prior to disposal are clarification and neutralization. Wastewater treatment is practiced by less than 15 percent of all ink plants. Few plants employ any physical-chemical treatment or secondary biological treatment. No ink plants use advanced wastewater treatment methods such as activated carbon or ultrafiltration. Of the ink plants that discharge their wastewater to a municipal sewer, less than one-third pretreat their waste prior to disposal.

Primary Treatment Systems — Many paint plants utilize physical treatment systems such as equalization or clarification. Clarification of paint wastewater will remove many of the suspended solids, but will still leave a supernatant layer high in solids and other pollutants. Primary treatment used by some ink plants may consist of clarification and/or neutralization. Treatment effectiveness obtained from sampling conducted at one ink plant is reported in Table 43. Treatment at this plant consisted of neutralization, oil skimming, and settling.

Physical-Chemical Treatment Systems — Physical-chemical (P-C) treatment systems in the paint industry are basically enhancements of gravity settling systems. Most plants utilizing P-C systems operate them on a batch basis. The plant's wastewater flow is collected in a holding tank

Table 43: Paint and Ink Formulation—Untreated and Treated Wastewater Concentrations and Percent Removals at One Ink Plant with Primary Treatment

Wastewater parameter[a]	Batch 1			Batch 2		
	Untreated[b]	Treated[c]	Percent removal	Untreated[b]	Treated[c]	Percent removal
Conventional - mg/l						
Biochemical oxygen demand	21,000	2,600	87			
Chemical oxygen demand	32,000	4,800	85	NR[d]	6,810	
Oil and grease	2,400	260	84	NR	5,384	
Total dissolved solids	21,000	5,500	73			
Total organic carbon	4,000	940	76			
Total solids	22,000	5,600	75	NR	15,311	
Total suspended solids	1,600	100	93	NR	1,830	
Total volatile solids	6,300	200	96			
Total volatile suspended solids	1,000	47	95			
Inorganics - ug/l						
Aluminum	20,000	600	97	31,800	6,710	78
Antimony	<25	<25	0	< 2,000	<2,000	0
Arsenic	<25	<25	0	< 2,000	<2,000	0
Barium	20,000	100	>99	120,000	4,020	96
Beryllium	<10	<10	0	<10	< 10	0
Boron	< 500	< 500	0			
Cadmium	90	<20	77	160	< 20	87
Calcium	71,000	<50,000	29	39,000	<5,000	87
Chromium	10,000	<50	>99	38,100	4,940	87
Cobalt	900	<50	94	3,110	670	78
Copper	10,000	< 60	>99	62,800	2,170	96
Cyanide	330	30	90	NR	1,300	
Iron	30,000	2,000	93	200,000	4,260	97
Lead	90,000	< 200	>99	150,000	32,500	78
Manganese	400	< 50	87	1,260	60	95
Magnesium	13,000	9,000	30	8,000	1,000	87
Mercury	<5	<5	0	NR	< 1	
Molybdenum	700	< 50	92	2,760	2,240	18
Nickel	< 50	< 50	0	2,410	< 50	97
Selenium	< 25	< 25	0	< 2,000	< 2,000	0
Silver	< 10	< 10	0	<10	< 10	0
Sodium	3,700	450	87	22	364	0
Thallium	< 10	< 10	0	2,000	2,000	0
Tin	< 50	< 50	0	460	50	89
Titanium	3,000	3,000	0	5,500	450	91
Vanadium	< 100	< 100	0	< 120	<120	0
Yttrium	< 200	< 200	0	< 160	<160	0
Zinc	1,000	1,000	0	2,530	720	71

(continued)

Table 43: (continued)

Wastewater parameter[a]	Batch 1			Batch 2		
	Untreated[b]	Treated[c]	Percent removal	Untreated[b]	Treated[c]	Percent removal
Organics - ug/l						
Acenaphthene				ND[e]	< 10	0
Anthracene				ND	< 10	0
1,2-Benzanthracene				ND	< 10	0
Benzene				220	96	56
Butyl benzyl phthalate				< 10	ND	>99
Chlorobenzene				530	ND	>99
Chlorodibromomethane				43	ND	99
Chloroform	10	NR				
Chrysene				< 10	ND	>99
3,3'-Dichlorobenzidene				ND	< 10	0
1,2-Dichloroethane	< 10	NR				
Di (2-chloroethyoxyl) methane				ND	< 10	ND
2,4-Dichlorophenol				ND	< 10	0
Di (2-ethylhexyl) phthalate	< 10	19	0	< 10	< 10	0
Di-n-butyl phthalate	ND	< 10	0			
Di-n-octyl phthalate				ND	< 10	0
1,2 Diphenylhydrazine	7,600	ND	> 99			
Ethylbenzene				6,700	2,400	64
Fluorene				< 10	ND	>99
Isophorone	44,000	ND	> 99	ND	46	0
Methylene chloride				45	29	35
Naphthalene	ND	110	0	17	<10	41
Pentachlorophenol				< 10	ND	>99
Phenanthrene	ND	12	0			
Phenol	< 10	18	0			
Tetrachloroethylene				22	ND	>99
Toluene	< 10	NR		3,600	1,100	69
2,4,6-Trichlorophenol				< 10	ND	>99

[a] Priority pollutants not measured in either stream are not indicated.

[b] Discharge from caustic washer.

[c] The plants' neutralization system malfunctioned during sampling.

[d] NR - not run.

[e] ND - not detected.

until a sufficient quantity is obtained to warrant treatment. If necessary, the pH is adjusted to an optimum level, a coagulant (often lime, alum, ferric chloride, or iron salts) and/or a coagulant aid (polymer) is added, mixed and the batch is allowed to settle from 1 to 48 hours. The supernatant is discharged and the sludge is generally disposed of as a solid waste. Often the sludge is left in the treatment tank for one or more subsequent batches, to reduce the overall sludge volume. Solvents, oils, and skins may float to the surface where they are manually removed.

Table 44 shows the pollutant removal efficiency obtained from sampling conducted at paint plants with batch P-C treatment systems. In total, 48 batches from 16 plants were analyzed for conventional pollutants and metals, and 23 batches were analyzed for priority pollutants. P-C treatment may be applicable to ink wastewater to reduce metals, solids, and some organics.

Other P-C wastewater treatment processes, such as ultrafiltration and activated carbon, have been mentioned as having potential application to the paint and ink industry. These systems are not currently used by any paint or ink plants.

Biological Treatment Systems. — Several paint plants that are part of multi-plant sites treat their wastewater by biological treatment system. These plants generally pretreat this wastewater and combine it with other plant wastewater, which is usually more dilute. Because of the exceptionally high solids and metals concentrations in paint wastewater, biological treatment must almost certainly be preceded by some kind of preliminary treatment (such as physical-chemical). The most common types of biological treatment systems are activated sludge, aerated lagoons, and trickling filters. In the paint industry, aerated lagoons predominate. It is likely that the shorter detention times of activated sludge and trickling filter plants may make these units more susceptible to failure from interferences and shock loading. Table 45 shows the pollutant removal efficiency obtained from sampling conducted at one plant with an aerated lagoon that treats primarily paint effluent after batch P-C treatment. There are essentially no data regarding the applicability of biological treatment to ink wastewater.

Type of Residue Generated

Sludge data obtained from sampling conducted at 22 paint plants are presented in Table 46. Of the 22 plants, 17 treated their wastewater by means of a batch physical-chemical treatment system, utilizing chemical addition, mixing and clarification. Three additional plants utilized a continuous system with the same unit operations as the batch physical-chemical plants. The remaining two plants utilized neutralization and clarification.

Table 44: Paint and Ink Formulation–Pollutant Removal Efficiency of Batch Physical-Chemical Treatment Systems at Paint Plants

Waterwater parameter	Influent[a]	Effluent[a]	Percent reduction[a]	Number of batches[b]	Median percent removal[c]
Conventional - mg/l					
Biochemical oxygen demand	10,778	5,585	48	41	20
Chemical oxygen demand	55,833	18,749	66	45	74
Oil and grease	1,158	118	89	38	97
Total dissolved solids	11,330	4,911	56	35	35
Total organic carbon	10,856	3,498	67	40	75
Total solids	31,967	5,905	81	40	80
Total suspended solids	23,831	1,310	94	38	99
Total volatile solids	13,224	2,499	81	35	88
Total volatile suspended solids	8,709	956	89	25	98
Inorganics - ug/l					
Aluminum	227,760	6,921	96	43	99
Antimony	281	226	19	10	60
Arsenic	334	194	43	22	80
Barium	6,770	842	87	42	94
Beryllium	162	8	95	13	50
Boron	2,206	1,874	15	31	16
Cadmium	674	31	95	25	80
Calcium	2,799,000	304,000	89	41	51
Chromium	3,399	1,403	58	38	54
Cobalt	1,018	533	47	35	75
Copper	3,083	1,894	38	44	69
Cyanide	75	57	24	4	54
Iron	334,466	110,795	66	42	90
Lead	6,033	1,062	82	36	90
Magnesium	129,000	21,000	83	42	60
Manganese	3,508	2,102	40	42	53
Mercury	4,022	322	91	32	88
Molybdenum	802	213	74	33	83
Nickel	1,729	3,386	0	19	77
Selenium	210	211	0	5	79
Silver	17	9	47	6	73

(continued)

Table 44: (continued)

Waterwater parameter	Influent[a]	Effluent[a]	Percent reduction[a]	Number of batches[b]	Median percent removal[c]
Sodium	397	803	0	39	0
Thallium	194	190	2	11	33
Tin	858	163	81	41	86
Titanium	19,300	738	96	42	97
Vanadium	495	79	84	19	72
Zinc	94,634	6,821	92	45	90
Organics - mg/l					
Benzene	1,190	563	52	17	65
Butyl benzyl phthalate	380	695	0	4	0
Carbon tetrachloride	19	16	15	5	100
Chloroform	144	283	0	18	68
1,2-Dichloroethane	81	20	75	5	69
1,1-Dichloroethylene	9	13	0	4	50
Di(2-ethylhexyl) phthalate	340	26	92	7	97
Di-n-butyl phthalate	6,474	90	98	12	99
Ethylbenzene	2,387	4,342	0	21	80
Methylene chloride	19,874	4,480	77	21	62
Naphthalene	3,278	335	89	5	70
Phenol	448	80	82	11	0
Tetrachloroethylene	545	90	83	14	98
Toluene	6,165	1,438	76	22	74
1,1,1-Trichloroethane	104	70	32	15	30
1,1,2-Trichloroethane	355	203	42	4	50
Trichloroethylene	59	50	15	11	9

[a] Average only of plants with batch physical-chemical treatment systems. Batches where both influent and effluent were not detected are not included in calculation of average concentrations.

[b] Individual percent removals were calculated only where both influent and effluent values were determined and where one or both values were above 10 ug/l (or other detection limit in the case of metals or nontoxic pollutants).

[c] Priority pollutants with less than four calculated removals are not listed.

Table 45: Paint and Ink Formulation–Pollutant Removal Efficiency of an Aerated Lagoon at One Paint Plant

Wastewater parameter	Untreated wastewater	Physical-chemical effluent	Lagoon effluent
Conventional - mg/l			
BOD	> 25,000	23,400	17
COD	70,000	260,000	675
pH (units)	7.4	7.0	8.3
TOC	7,500	25,000	200
TSS	46,000	400	42
Inorganics - ug/l			
Antimony	<1,000	170	30
Arsenic	440	< 100	< 20
Beryllium	7	2	< 1
Cadmium	130	58	< 2
Chromium	1,450	105	9
Copper	264	115	7
Lead	12,000	98	< 20
Mercury	1,010	142	0.1
Nickel	450	<5	< 5
Selenium	< 200	400	< 200
Thallium	< 200	100	< 20
Zinc	60,000	4,200	< 60
Organics - ug/l			
Benzene	280	200	< 10
Chlorodibromomethane	ND[a]	ND	ND
Chloroform	ND	23	ND
Dichlorobromomethane	ND	ND	ND
Di(2 ethyl-hexyl) phthalate	ND	ND	< 10
Di-N-butyl phthalate	< 10	<10	ND
Ethylbenzene	730	ND	ND
Methylene Chloride	6,300	31,000	1,000
Pentachlorophenol	< 10	<10	ND
Phenol	< 10	<10	ND
Tetrachloroethylene	110	25	ND
Toluene	290	200	ND
1,1,1-Trichloroethylene	120	560	22

[a] ND - not detected.

It will be noted that the sludges generated contain substantial amounts of heavy metals, particularly lead, mercury, nickel, and zinc. They also contain a wide variety of toxic organic compounds, a few in substantial concentrations, and moderate amounts of BOD, TSS, and other nontoxic materials.

Residual Management Options

Production procedures in the industry are usually batch processes with clarification, with or without chemical coagulation, as the main pretreatment process, incorporating long sludge-thickening retention times. The sludge produced should have a high solids content and be easy to dewater. The small amounts of skimmings removed from the sedimentation tanks during clarification may be landfilled along with the dewatered sludge.

Disposal of sludges generated by this industry is essentially limited to landfilling in an acceptable manner. However, the cost of pretreatment is expected to stress production procedure changes, and materials recovery and reuse in the production process.

Table 46: Paint and Ink Formulation–Sludge Data Summary Based on Sampling at 22 Paint Plants

Wastewater parameter	Number of samples analyzed	Number of times above detection limit	Concentration[a]			
			Average[a]	Median[a]	Minimum[a]	Maximum
Conventional - mg/l						
pH (units)	35	35		7	2	10
Biochemical oxygen demand	34	34	24,982	10,200	1	150,000
Chemical oxygen demand	38	38	171,641	130,000	7	950,000
Oil and grease	35	35	7,578	2,500	230	129,000
Total dissolved solids	30	27	13,122	9,800	< 1	100,200
Total organic carbon	36	36	35,126	29,500	12,000	108,000
Total solids	32	32	107,785	78,000	8	470,000
Total suspended solids	33	33	101,201	70,000	100	466,100
Total volatile solids	31	31	39,438	37,000	3,100	187,000
Total volatile suspended solids	20	20	23,002	14,133	880	89,000
Inorganics - ug/l						
Aluminum	37	37	867,154	600,000	50,000	3,000,000
Antimony	17	7	1,579	150	< 10	13,000
Arsenic	12	5	879	495	< 25	<2,000
Barium	36	36	9,831	4,345	72	50,000
Beryllium	39	25	192	20	2	3,760
Boron	34	29	2,965	2,000	175	14,200
Cadmium	39	28	840	200	< 8	14,700
Calcium	36	36	2,870,000	914,000	250,000	30,200,000
Chromium	39	37	7,050	700	< 50	90,000
Cobalt	36	32	2,131	600	< 50	15,600
Copper	39	39	7,121	1,000	216	80,000
Cyanide	34	3	1,151	< 20	< 1	36,500
Iron	36	36	886,452	200,000	22,600	8,000,000
Lead	39	37	10,770	3,000	100	80,000
Magnesium	36	35	156,000	79,000	<1,000	1,500,000
Manganese	36	36	7,249	5,000	300	50,000
Mercury	36	31	15,061	640	5	220,000
Molybdenum	35	34	1,680	1,000	< 50	15,000
Nickel	39	27	10,443	< 200	< 20	200,000
Selenium	36	2	547	< 250	8	2,000
Silver	38	8	22	< 10	< 2	<100
Sodium	36	32	583	260	1,100	3,500
Thallium	9	1	967	< 400	< 10	<2,000

(continued)

Table 46: (continued)

Wastewater parameter	Number of samples analyzed	Number of times above detection limit	Concentration[a] Average[a]	Median[a]	Minimum[a]	Maximum
Tin	36	36	2,640	2,000	200	14,500
Titanium	36	36	38,345	20,000	1,000	230,000
Vanadium	35	30	891	400	60	11,500
Yttrium	33	12	207	< 200	45	< 600
Zinc	39	36	230,946	90,000	<600	2,000,000
Organics - ug/l						
Aldrin	9	0	10	< 10	<10	< 10
Anthracene	9	1	210	< 210	<10	410
Benzene	9	4	414	30	<10	1,900
Beta-endosulfan	9	0	10	< 10	<10	< 10
Butyl benzyl phthalate	9	4	10,410	1,412	18	38,800
Carbon tetrachlorine	9	0	10	< 10	<10	< 10
Chlorobenzene	9	2	176	176	12	340
Chloroform	9	2	920	920	840	1,000
Delta-BHC	9	0	10	< 10	10	< 10
1,2-Dichloroethane	9	1	17	17	17	17
Diethyl phthalate	9	3	370	100	50	960
Di(2-ethylhexyl) phthalate	9	6	455	215	<10	1,940
2,4-Dimethylphenol	9	0	10	< 10	<10	
Dimethyl phthalate	9	0	10	< 10	<10	< 10
Di-n-butyl phthalate	9	4	3,622	70	<10	17,750
2,4-Dinitrophenol	9	1	18	< 18	<10	27
Ethylbenzene	9	8	14,277	237	26	99,000
Methylene chloride	9	8	120,201	1,735	300	900,000
Naphthalene	9	3	366	202	<10	1,050
Pentrachlorophenol	9	4	346	125	35	1,100
Phenol	9	3	325	150	<10	1,120
Pyrene	9	0	10	< 10	<10	<10
1,1,2,2-Tetrachloroethane	9	1	13	13	<10	17
Tetrachloroethene	9	4	2,142	170	<10	8,200
Toluene	9	8	44,740	905	130	350,000
2,3,6-Trichloroephenol	9	0	10	< 10	<10	<10
1,1,1-Trichloroethane	9	4	866	14	<10	3,200
Trichloroethylene	9	2	39	< 10	<10	130

[a] Average, median, and minimum values based only on number of times detected for organic priority pollutants.

PESTICIDES MANUFACTURING

General Industry Description

The pesticides manufacturing category includes facilities which produce phosphorus-based, nitrogen-based, and metallo-organic pesticides from raw materials. Also included are plants which formulate and package purchased pesticide compounds.

Industrial Categorization

This industry can be subdivided into the following subcategories: halogenated organic pesticides, organo-phosphorus pesticides, organo-nitrogen pesticides, metallo-organic pesticides, and formulating and packing.

Process Description

Pesticide plants which manufacture active ingredient products use many diverse manufacturing processes. Rarely does a plant employ all of the processes found in the industry, but most plants use several in series. The principal processes utilized include chemical synthesis, separation, recovery, purification and product finishing, such as drying.

Chemical synthesis can include chlorination, alkylation, nitration, and other substitutive reactions. Separation processes include filtration, decantation, and centrifuging. Recovery and purification are utilized to reclaim solvents or excess reactants as well as to purify intermediates and final products. Evaporation, distillation, and extraction are common processes in the pesticides industry. Product finishing can include blending, dilution, pelletizing, packaging, and canning.

In the manufacture of halogenated organic pesticides, principal sources of high organic wastes are decanting, distillation, and stripping operations. Sources of wastewater from the manufacture of organo-phosphorus pesticides include decanter units, distillation towers, overhead collectors, solvent strippers, caustic scrubbers, contact cooling, hydrolyzing, and product and equipment washing. Sources of wastewaters from the manufacture of organo-nitrogen and metallo-organic pesticides are similar to the products mentioned above.

Wastewater Characteristics

Generally, the wastewaters of this industry are high in BOD concentration. This high level of biodegradable organic matter contributes to numerous problems for POTWs and industrial treatment facilities. Table 47 provides raw wastewater characteristics for this industry.

Table 47: Pesticides Manufacturing–Raw Wastewater Characteristics

	Subcategory				
Wastewater parameters	Halogenated organic pesticides, mg/l	Organic phosphorus pesticides, mg/l	Organic nitrogen pesticides, mg/l	Mettalo-organic pesticides, mg/l	Formulating and packing, mg/l
Biochemical oxygen demand	125-8,500	140-750	1,200-2,500	20-800	150-1,600
Chemical oxygen demand	100-250	10-100	10-2,000	1,500-3,000	100-650
Total dissolved solids	850-16,000	350-1,800	800-15,000	1,500-2,200	500-6,000
Total organic carbon	650-8,400	100-4,000	450-5,300	80	
Total suspended solids			2,000-44,000		

Control and Treatment Technology

In-Plant Control Technology: Waste segregation is an important in-plant control measure, since high organic loading streams will require different treatment schemes than low organic loading streams. The use of dry cleanup systems can also reduce wastewater flows. Steam jet ejectors and barometric condensers can often be replaced in most cases with vacuum pumps and surface condenser systems, significantly reducing wastewater generation.

End-of-Pipe Treatment Technology: The pesticide manufacturing industry uses a multiplicity of chemical synthesis, separation, purification, and other processes. As a special class of organic chemicals, the starting materials, by-products, and dumps of unsatisfactory product are all wastes that are themselves concentrated organic compounds. Many of these are readily biodegradable, and biological treatment is the predominant wastewater treatment method in this industry. However, there are significant exceptions and qualifications. Chemical synthesis steps are frequently carried out at extremes of pH or temperature, such that the wastes must be neutralized before they can be biologically treated. Also, there are substituent groups — such as halogens and nitrates — and heavy metals — such as copper, zinc, lead, and mercury — that impair the biodegradability of wastes containing them. Segregation and separate handling of such flows can frequently lessen the overall treatment problem. Sometimes such wastes are biodegradable after acclimation. There are cases where the waste is more biodegradable if it is less concentrated. For wastewaters which cannot be made biodegradable, the following treatment techniques have been used: incineration, evaporation, filtration, chemical precipitation, adsorption, chemical oxidation, deep well injection, and ocean disposal. No data are available for showing removal efficiencies of this industry.

Type of Residue Generated

Residues generated by this industry can be classified as biological sludges, chemical sludges, and brines. The biological sludges are similar to domestic wastewater treatment sludges, but are likely to be "contaminated" with toxic organics or heavy metals. Chemical sludges are likely to have a high chemical oxygen demand or total organic carbon, and be toxic. Waste brines may have a substantial chemical oxygen demand, but have as their predominant feature a high total suspended solids, in part from chemicals added to neutralize excess acids or bases.

Although the basic manufacturing processes are well known, minor changes in reactants or reaction conditions greatly affect the product quality and yield, and the waste generation. Pesticide manufacture is something of an "art," and many of the details are looked upon as proprietary information. Because the manufacturing technique can be inferred from the waste generated, many details of wastewater flows and treatment techniques are also regarded as confidential. Public sensitivity to pesticide waste problems adds to this tendency toward secretiveness on pesticide wastewater treatment and residue generation.

Residual Management Options

The principal methods used for disposal of sludges from this industry are landfilling and incineration. Evaporation, usually in open lagoons, is frequently used for concentration of waste brines, or for thin sludges preparatory to incineration. Solidified brines may be removed from lagoons and landfilled, or the whole lagoon may be abandoned when it is full of solids. Deep well disposal of wastes from the industry is strictly regulated in most areas, is quite expensive, and will in general not be an available alternative for pretreatment residues of indirect dischargers. Ocean discharge of these residues is now prohibited.

All residues from treatment of wastewaters from this industry have the potential of being toxic. When landfilled, the landfill must be secure, be protected from surface runoff, seepage, erosion, and leaching. Attention must be given to the protection of the groundwater. When placed in a landfill that also receives other wastes, the portion receiving the pesticide manufacturing residues should be segregated and preferably fenced off. Careful records should also be kept of the origin, nature, and placement of such residues. Encapsulation may be required for some pesticide manufacturing residues. When incinerated, there is the potential for equipment corrosion and for incomplete combustion and the production of volatile organic fragments and/or heavy metal vapors that are also toxic. Generally pesticide wastes must be burned in special incinerators designed to insure a minimum retention time at a minimum temperature. Pesticide residues containing heavy metals with low boiling points — such as mercury, zinc, and lead — should not, however, be incinerated.

PETROLEUM REFINING

General Industry Description

The petroleum refining industry produces products such as propane, gasoline, jet fuels, heating oils, lubricating oils, asphalt, and coke. These materials are derived from crude oil by means of distillation, catalytic conversion, solvent extraction, and chemical conversion operations.

In 1979 the petroleum refining industry in the United States processed about 15 million barrels per day. Growth in refining capacity has averaged about five percent per year and has resulted largely from additions to existing refineries rather than construction of new ones. The combined crude throughput of refineries with indirect discharges amounts to about ten percent of the total capacity of all petroleum refineries operating in the United States.

Industrial Categorization

Refineries are complex manufacturing facilities. Over 100 distinct processes are used in the petroleum refining industry. Moreover, sizes (throughputs) of the processes significantly affect the effluent flow. It is not practical to group refineries by both process and size of process because each refinery would constitute a distinct subcategory. In fact, factors which impact wastewater generation in a refinery are so numerous that subcategorization of the petroleum refining industry, based on these factors, is not practical. Accordingly, for purposes of raw waste characterization and organization of pretreatment information, petroleum refining is treated as a single industrial category. A reasonable approximation of what occurs in a refinery can be formulated as a mathematical flow model which correlates achievable effluent flow with a relatively small number of process variables. The general form of the model is:

Flow = industry average flow + asphalt effect + crude effect + cracking effect + lube effect

For the purpose of effluent limitation guidelines, the specific effluent flow model for predicting current industry flow on a refinery-by-refinery basis is given by the equation:

$$\text{Flow} = 0.568 + 0.048A + 0.004(C - 59.2) + 0.046(K - 7.2) + 0.048L$$

Where the flow is in units of million gallons per day, constants are median values in units of million gallons per thousand barrels per day, and

A, C, K, L are in units of thousands of barrels per day throughput with the following definitions:

A = sum of asphalt processes
asphalt production
asphalt oxidizer
asphalt emulsifying

K = sum of cracking processes
hydrocracking
visbreaking
thermal cracking
fluid catalytic cracking
moving bed catalytic cracking

C = sum of crude processes
atmospheric crude distillation
crude desalting
vacuum crude distillation

L = sum of lube processes
hydrofining, hydrofinishing, lube hydrofining
white oil manufacture
dewaxing, deasphalting, fractioning, and deresining of propane
treating, extraction, dewaxing, deasphalting of solvent
manufacturing of oil, lube, grease and allied products
dewaxing of MEK, ketone, and MEK-toluene
separation, pressing, sweating and slabbing of wax
extraction of sulfur dioxide, furfural and phenol
miscellaneous blending and packing

Process Description

Figure 31 is a process flow diagram for the petroleum refining industry and shows the interrelationships among the different processes. A description of the various processes is given below.

Crude Oil and Product Storage: The storage area of the refinery serves to provide a working supply, equalizes process flow and also acts as a place for separation of water and suspended solids from the crude oil. Wastewaters associated with storage of crude oil are high in oil, suspended solids, and COD.

Crude Oil Desalting: The crude oil desalting process is a pretreatment step to remove impurities. Wastewaters containing inorganic salts and suspended solids are discharged.

Crude Oil Fractionation: Fractionation is the basic refining process for the separation of crude petroleum into intermediate fractions of specified boiling point ranges. Wastewaters contain sulfides, ammonia, chlorides, mercaptans, and phenols.

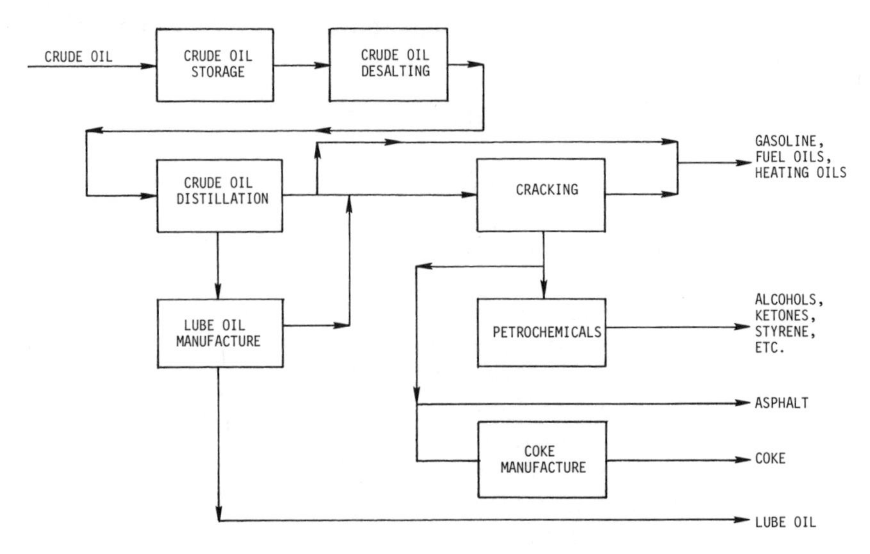

Figure 31: Petroleum refining–product manufacturing flow diagram.

Cracking: In this process, heavy oil fractions are converted into lower molecular weight fractions including domestic heating oils, high octane gasoline stocks, and furnace oils. Three types of cracking are used: thermal, catalytic, and hydrocracking. Thermal cracking is accomplished by heating (480°C to 603°C) without the use of a catalyst. Wastewaters usually contain oils and distillates, and are high in biochemical and chemical oxygen demand, ammonia, phenol, sulfides, and alkalinity. Catalytic cracking is operated at lower temperatures and pressures than thermal cracking because of the use of a catalyst. Catalytic cracking units are one of the largest sources of sour and phenolic wastewaters in a refinery. Major pollutants are oil, sulfides, phenols, cyanides, and ammonia. Regeneration of the catalyst may constitute an air pollution problem. Hydrocracking is a catalytic cracking process in the presence of hydrogen and has greater flexibility in adjusting operations to meet changing product demands. Wastewaters are high in sulfides and possibly in phenols and ammonia.

Lube Oil Manufacturing: Lube oil manufacturing processes include hydrofining, hydrofinishing, lube hydrofining, propane dewaxing, propane deasphalting, propane fractioning, propane deresining, white oil manufacture, solvent treating, solvent extraction, duotreating, solvent dewaxing, solvent deasphalt, oil fractionation, batch still (naphtha strip), bright stock treating, centrifuging, chilling, MEK dewaxing, ketone dewaxing, MEK-toluene dewaxing, deoiling (wax), naphthenic lubes production, sulfur dioxide extraction, furfural extracting, wax pressing, wax sweating, wax slabbing, clay contacting and percolation, acid treatment, phenol extraction, lube and fuel additives operations, sulfonate plant operations, MIBK, rust preventives operations, petroleum oxidation, manufacturing of grease and allied products, blending, product finishing, etc.

Petrochemical Operations: These include production of second generation petrochemicals (i.e., alcohols, ketones, cumene, styrene, etc.) or first generation petrochemicals and isomerization products (i.e., BTX, olefins, cyclohexane, etc.).

Hydrocarbon Rebuilding: Higher octane products for use in gasoline may be manufactured by two hydrocarbon rebuilding techniques: polymerization or alkylation. Wastewaters are high in sulfides, mercaptans, ammonia, suspended solids, and oils. Waste sulfuric acid is usually recovered.

Hydrocarbon Rearrangements: Isomerization and reforming are process techniques for obtaining higher octane gasoline blending stock. Isomerization, a molecular rearrangement process rather than a decomposition process, generates no major pollutant discharge. Reforming, a mild decomposition process, generates low volume discharges with small quantities of sulfides, ammonia, mercaptans, and oil present.

Solvent Refining: Various solvents are used to improve the quality of a particular feedstock component. Major pollutants are the solvents themselves, many of which can produce a high biochemical oxygen demand. Under ideal conditions, solvents are continually recirculated. Actually, some solvent is always lost. Oil and solvents are major wastewater constituents.

Hydrotreating: Hydrotreating processes are used to purify and pretreat various feedstocks by reacting with hydrogen. Contaminants, including sulfur and nitrogen compounds, odor, color, and gum-forming materials, are removed. Strength and quantity of wastewaters generated by hydrotreating depends upon the subprocess and feedstock used. Ammonia and sulfides are present. Phenols may also be present.

Grease Manufacturing: Grease is primarily a soap and lube oil mixture. A small amount of oil is lost to the wastewater system through leaks in the pumps. The largest waste loading occurs when the units are washed.

Asphalt Production: Asphalt feedstock is contacted with hot air at 203°C to 280°C to obtain a desirable asphalt product. Wastewaters contain high concentrations of oils which have high biochemical oxygen demand. Small quantities of phenols may also be present.

Product Finishing: Drying and sweetening processes are used to remove sulfur compounds, water and other impurities from gasoline, kerosene, jet fuels, domestic heating oils, and other middle distillate products. Spent caustic, large quantities of high sulfides, and phenols are generated. Phenolic caustic streams are usually sold for recovery of phenolic materials. Clay and acid treatment to remove color forming and other undesirable materials further refine lube oil stocks. Acid wastes high in dissolved and suspended solids, sulfates, sulfonates, and stable oil emulsions are generated. Managing acid sludge can create additional problems. Some refineries neutralize the sludge and discharge it to the sewer resulting in organic and inorganic pollution.

Blending various gasoline stocks and additives, and packaging the products are relatively clean processes. The primary source of waste material is from tank car washing. These wash waters are high in emulsified oil. Tetraethyllead, a gasoline additive, is highly toxic and may be washed into the sewer.

Wastewater Characteristics

The wastewaters generated by the petroleum refining industry are diverse and complex, representing a full range of organic and inorganic substances.

Wastewater characteristics of effluents from the oil-water separator in the petroleum refining industry is shown in Table 48. Table 49 shows raw wastewater characteristics of effluents from the oil-water separator as a function of refinery production. These characteristics were compiled based on a sampling program conducted in 1977 to 1979, involving 23 refineries.

Table 48: Petroleum Refining Industry–Wastewater Characteristics of Oil/Water Separator Effluent[a]

Wastewater parameter	Concentration
Conventional - mg/l	
Ammonia nitrogen	1-168
BOD_5	10-556
Chemical oxygen demand	75-3,330
Oil and grease	3-775
pH (units)	6-11
Sulfide	0.1-32
Total organic carbon	25-240
Total suspended solids	6-490
Inorganics - ug/l	
Antimony	1-360
Arsenic	5-480
Beryllium	1-20
Cadmium	1-200
Chromium	1-2,000
Copper	4-380
Cyanide	0-1,500
Lead	10-600
Mercury	0.4-78
Nickel	1-500
Selenium	4-682
Silver	1-250
Thallium	1-15
Organics - ug/l	
Acenaphthene	37-3,000
Acenaphthylene	4-655
Aldrin	0.3-12
Alpha-BHC	0.1-2.2
Anthracene	30-660

(continued)

Table 48: (continued)

Wastewater parameter	Concentration
Benzene	100-5,800
Benzo (a) pyrene	3
Beta-BHC	0.1-5
Beta-Endosulfan	13
Bis (2-etheylhexyl) phthalate	290-700
Butyl benzyl phthalate	16
Chlorobenzene	31
Chloroform	5-100
2-Chlorophenol	315-830
Chrysene	0.1-20
Chrysene/benzo (a) anthracene	30-40
4,4'-DDE	7
4,4'-DDT	0.1-5
Delta-BHC	5-12
Dichlorobromomethane	24
1,2-Dichloroethane	16-54
Diethyl phthalate	1-14
2,4-Dimethylphenol	16-9,300
Di-n-butyl phthalate	1-40
4,6-Dinitro-o-cresol	60
2,4-Dinitrophenol	110-11,000
2,4-Dinitrotoluene	20
1,2-Diphenylhydrazine	23
Ethylbenzene	28-9,000
Fluoranthene	3-9
Fluoroanthene/pyrene	20-40
Fluorene	80-270
Heptachlor	< 5
Heptachlor epoxide	0.3-5
Isophorone	12-3,500
Methylene chloride	3-1,600
Naphthalene	18-3,200
2-Nitrophenol	1,350
4-Nitrophenol	20-5,800
N-nitrosodiphenyl amine	41
PCB-1016	0.2-10
PCB-1221	0.1-10
PCB-1232	0.5-10
PCB-1239	0-9
PCB-1242	0.5-10
PCB-1248	<10
PCB-1254	<10
PCB-1260	<10
Pentachlorophenol	850
Phenanthrene	30-660
Phenanthrene/anthracene	5-1,100
Pyrene	7-21
Phenol	0-200,000
Tetrachloroethylene	50
Toluene	10-15,000
1,2-Trans-dichloroethylene	10-20

[a] Wastewater data based on a sampling and analysis program conducted in 1977-1979, involving 23 refineries.

Table 49: Petroleum Refining Industry—Wastewater Characteristics of Oil/Water Separator Effluent Based on Production[a]

Wastewater parameter	Range of values
BOD_5	0.56 - 34.66
TSS	1.86 - 11.45
COD	5.91 - 86.50
Oil and grease	1.32 - 19.08
Phenols	0.005 - 1.32
Ammonia nitrogen	0.19 - 5.45
Total hexavalent chromium	0.001 - 0.08
Sulfides	0.002 - 0.32

[a] kg pollutant/1,000 barrels of crude oil.

Control and Treatment Technology

In-Plant Control Technology: Many newer refineries are being designed or modified, with reduction of water use and pollutant loading as a major part of the design criteria. These advances include:

- Substitution of improved catalysts that require less regeneration.
- Replacement of barometric condensers with surface condensers or air fan coolers, reducing a major oil-water emulsion source.
- Newer hydrocracking and hydrotreating processes which produce less wastewater loadings than the older processes.
- Increased use of improved drying, sweetening, and finishing procedures to minimize spent caustics and acids, water washes, and filter solids requiring disposal.
- Reuse practices involving the use of water from one process in another process. Examples of this are: using stripper bottoms for makeup water to crude desalters; using blowdown from high pressure boilers as feed to low pressure boilers; and using treated effluent as makeup water wherever possible.
- Recycle systems that use water more than once for the same purpose. An example of recycle system is the use of steam condensate as boiler feedwater.
- Good housekeeping practices, including dry cleaning methods, to clean up oil spills, minimizing leaks, and treating segregated waste streams such as spent cleaning solutions.
- Cooling towers enable recycling of cooling water many times, eliminating large volumes of once-through cooling water.

Many waste streams are routinely treated at the source, including stripping of sour waters, neutralization and oxidation of spent caustics, ballast water separation, and slop oil recovery. Sour water stripping removes 85 to 99 percent of the sulfides before it enters the sewer. Spent caustics are treated, and occasionally products are extracted and sold. Slop oil or separator skimmings are treated and reused.

Treatment Technology: Wastewater treatment technologies for the petroleum refining industry are divided into three broad classes: preliminary treatment of sour water, oil-water separation, and end-of-pipe treatment.

Preliminary Treatment of Sour Water — Removal of sulfides and ammonia is accomplished by preliminary treatment of sour water before the waste enters the oil-water separator. Sour waters are produced in a refinery when steam is used as a stripping medium in various cracking processes, and are generally treated by stripping and oxidation as described below.

Sour Water Stripping - This technology involves a gas/liquid separation process using steam or flue gas to extract the bases (sulfides and ammonia) from the wastewater. The stripper is a distillation-type column containing either trays or packing material.

Sour Water Aeration - Oxidation of sour water by aeration involves injection of compressed air into the waste followed by sufficient steam to raise the reaction temperature to at least 190 degrees F. Reaction pressure of 50 to 100 psig is required.

Sour Water Oxidation - Oxidation of sour water by peroxide or chlorine occurs in open tanks, without the use of steam.

The use of sour waters as makeup to the desalter has been a proven technology for phenol removal for some time in the industry.

Oil/Water Separation — The following treatment technologies can be used for this purpose:

API Separator - Oil and water are separated by gravity and corrugated plates in the separator.

Dissolved Air Flotation - Oil and water are separated by air bubbles generated in a dissolved air flotation unit.

Other flotation systems and physical-chemical methods can also be used for oil/water separation.

End-of-Pipe Treatment — This includes all wastewater treatment systems that follow an API separator or a similar oil/water separation unit. The treatment technology relies heavily upon the use of biological treatment methods, after appropriate pretreatment, to insure the proper conditions. Various end-of-pipe treatment methods are described below:

Biological Treatment - Organic waste substances are converted to carbon dioxide and biomass. The process includes activated sludge, trickling filter, rotating biological contactor, aerated lagoon, and stabilization pond.

Filtration - Solids and other associated waste substances are removed by the filter media. The most common process is rapid sand filtration.

Granular Activated Carbon - The granular activated carbon system consists of one or more trains of carbon columns, each train having three columns operated in series. Granular carbons are the absorbents used for removing pollutants.

Powdered Activated Carbon - The technology consists of the addition of powdered activated carbon to biological treatment systems. The need for a filter press system or acid cleaning system as well as a carbon regeneration furnace should be determined on a case-by-case basis.

Other end-of-pipe treatment technologies include chemical precipitation/clarification, ion exchange, and reverse osmosis. Table 50 shows the removal efficiencies of various wastewater treatment methods practiced by the petroleum refining industry.

Zero Discharge: Although the use of flow reduction, recycle methods, and other control measures reduce the quantity of water discharged or subject to end-of-pipe treatment, none of these techniques will eliminate the discharge of all wastewater. Zero discharge of refinery wastewater is

Table 50: Petroleum Refining Industry–Summary of Effectiveness of Treatment Technologies

Wastewater parameter	Preliminary treatment of sour water, percent removal			Primary treatment, percent removal		End of pipe treatment,[d] percent removal				
	Stripping	Oxidation[a]	Desalter	Separator[b]	Flotation[c]	Clarifier	Filter[e]	GAC[f]	PAC[g]	BIO[h]
Ammonia nitrogen	6.8-99.8	15-88								
Antimony									0-5	
Arsenic					0-20					
BOD_5				5-40	0-15	30-60	0-60	0-80	25-96	40-99
Beryllium									0	
Cadmium							0-80	0	0-66	
Chemical oxygen demand				5-39	0	20-50	8-72	27-87	59-99	30-9
Chromium (+3)					0-39		0-86	0-76	70-97	
Chromium (+6)							0	50	0-82	
Copper					0-92		0-81	0-67	0-98	
Cyanides	0-90.9				0		0-10	0-92	0-75	
Lead					0-63		0-50	0-42	0-80	
Mercury					0-44					
Nickel					0-41				0-76	
Oil and grease				60-99		60-95	0-88	0-50	0-96	50-99
Phenolics	0-91	0-11	73-85	0-50	0-4	0-50	0-8	0-95	95-100	60-99
Selenium					0-44		0-10	0-19	0-13	
Silver									0-50	
Sulfide	17-100	100			0-18		0-40	0-75	0-91	70-100
Total organic carbon					0-17		0-57	0-76	52-94	
Total suspended solids				10-50	0-40	50-80	6-97	0-88	0-97	20-80
Zinc					0-62		0-60	0-65	0-99	

[a] Oxidation - Chemical oxidation using peroxide and chlorine.

[b] Sour waters are pretreated before being discharged to an oil/water separator, such as API separator.

[c] Dissolved air flotation (DAF) can be used either as a primary treatment unit, or as an end-of-pipe treatment unit following an API separator.

[d] End-of-pipe treatment follows primary treatment mainly by an API separator.

[e] Rapid sand or dual media filtration systems.

[f] GAC - Filter and granular activated carbon system.

[g] PAC - Powdered activated carbons are added to biological treatment systems for treatment.

[h] BIO - Biological treatment processes include oxidation ponds, aerated lagoons, and activated sludge systems.

technically achievable, however. Fifty-five existing refineries have reported zero discharge. Of the 55 plants, 32 use evaporation or percolation ponds, ten use disposal wells, five use contract disposal, two use leaching beds, one uses surface spray, and six reported no wastewater generated at all. Some of these zero discharge methods are discussed below:

- Percolation and evaporation ponds are attractive disposal methods when evaporation losses exceed rainfall. These ponds can be sized according to the annual flow, so that the inflow plus the incidentally added water (i.e., rainfall) are equal to percolation and evaporation losses.
- Deep well injection is also being practiced by the petroleum refinery industry. This method can be used only with the stipulation that extensive studies be conducted to insure ground water protection.
- Irrigation (or a similar land disposal technique) is a viable end-of-pipe treatment alternative which can eliminate discharge of all or a portion of the process wastewaters.
- Forced evaporation with heat is a zero discharge technology. The steam is condensed and reused as makeup water to the refinery, while the brine (slurry) stream is transformed into a solid state in a flash dryer.

Type of Residue Generated

Residues generated by this industry consist of a wide variety of skimmings, sludge, and brines from various unit processes in the refinery. Hydrocarbons of all kinds are the main contaminants, including many of those on USEPA's toxic pollutants list. Heavy metals are also frequently present.

Oil/water separation and sour water treatment are almost universally used in treating oil refinery wastewaters. Much of the oils that are skimmed off in a separator are returned to some part of the refining process, and are not ordinarily thought of as waste residues.

Data on removals for this industry by various wastewater treatment processes emphasize how much is in the influent versus how much is in the effluent, not the precise mechanism of removal. Thus, in biological processes involving aeration, removal of particular organic compounds — including toxic ones — may occur by biodegradation, stripping, chemical oxidation (with O_2 from the air), photodecomposition, precipitation, sedimentation, flotation, or adsorption on solids or liquid droplets which float or settle out. Only the last four will cause the organic contaminant to end up in the residues. The first four either transform the compounds to some other chemical form, or transfer it to another physical medium, the air. Similarly, sour water treatment will either strip out ammonia and hydrogen sulfide, or oxidize them to innocuous dissolved aqueous products, but will not transfer them to residues such as skimmings or sludges. For heavy metals, however, they will, if removed, end up in the residue.

Biological treatment is the predominant wastewater treatment technique for this industry. Therefore, the bulk of sludge to be disposed of is biomass. "Contaminants," organic and inorganic, toxic and nontoxic, will usually be present.

Other types of sludges or components of sludges that may be generated in refinery wastewater treatment include:

- Unremoved and unreacted pollutant;
- Chemical precipitants, such as lime, ferric chloride or sulfate, or organic polymers;
- Grit, fine sand, clay, asbestos fibers, and other mineral matter; sometimes added as adsorbents or filtering media;
- Powdered activated carbon, added as an adsorbent; and
- Precipitates resulting from acid-base neutralization or oxidation-reduction reactions.

When such materials are added as precipitants, adsorbents, or filtering media, the contaminants bound to them after removal from the wastewater may be of greater significance than the added materials themselves. Bound materials are likely to be toxic or objectionable wastewater pollutants which necessitated the use of the adsorbents or filters in the first place, but in a much more concentrated form.

Toxic materials detected by an EPA survey in the partially treated (Oil/Water Separation and Sour Waste Treatment) wastewaters from two refineries discharging to POTWs are listed in Table 51.

Table 51: Toxic Pollutants Detected in Discharges to POTWs from Partially Treated Refinery Wastewater

Metals	
Arsenic (total)	Mercury (total)
Chromium (total)	Nickel (total)
Copper (total)	Selenium (total)
Lead (total)	Zinc (total)
Organics	
Acenaphthene	2,4-Dimethylphenol
Aldrin	2,4-Dinitrotoluene
Alpha-BHC	1,2-Diphenylhydrazine
Anthracene	Ethylbenzene
Benzene	Fluorene
Benzo (a) anthracene	Heptachlor epoxide
Beta-BHC	Isophorone
Butyl benzyl phthalate	Methylene chloride
Chlorobenzene	Naphthalene
Chloroform	N-nitrosodiphenylamine
Chrysene	Pentachlorophenol
4,4'-DDE	Phenanthrene
4,4'-DDT	Phenol
1,2-Dichloroethane	Pyrene
Diethylphthalate	Tetrachloroethylene
Di-n-butyl phthalate	1,1,1-Trichloroethane
Di-n-octyl phthalate	Toluene

Brines can be produced from application of ion exchange, reverse osmosis, or evaporation in refinery wastewater treatment processes. Waste brines can also be produced from some of the refining processes themselves.

Residual Management Options

Landfilling is the predominant method of residue disposal for this industry. A 1976 survey sponsored by USEPA reported that 63 percent of all refinery wastes were landfilled, 28 percent lagooned; 7 percent land farmed; 2 percent incinerated, and small fractions were disposed of by ocean dumping and deep-well disposal. Sludge quantities for 1977 were estimated at 100,000 metric tons of hazardous solids in 1.8 million metric tons of wet sludge. The key question in landfilling sludges from this industry is whether the particular sludge is hazardous or not. If it is, then numerous extra precautions are necessary, including coverage of the placed residue with a layer of impervious material to reduce rainwater seepage; insulation from contact with groundwater by a floor and dike lined with a layer of impermeable material or a membrane liner; grading to prevent surface water erosion; collection and treatment of leachate and seepage from the landfill; and extensive recording-keeping of the origin and placement of every batch of residue received.

As to the other methods, ocean dumping is no longer permitted. Deep-well disposal requires ideal geological conditions, is somewhat expensive to build and operate, and is closely regulated by State and Federal laws. Incineration is not widely used — because of its high initial and operating costs — and is becoming less so as improved oil recovery in the plant lessens the fuel value of the wastewater residues. Further, it has been found that a single type of incineration does not necessarily function well on all the various refinery residues, and that more than one type of incinerator may be needed in a single refinery.

Lagooning is also declining as an option for this industry, because of the scarcity and expense of land adjacent to refineries, increasing waste flows and space needs of the refineries themselves, and increasingly stringent regulatory requirements. Usability of such land for other purposes may also be permanently impaired, unless the residue is later picked up and removed.

Land farming is perhaps the only promising technique for residues from this industry. In this technique, the residue is spread on cleared land and disked into the ground, both with regular farming equipment. Residue is degraded over a period of time, principally by bacterial action. Active cultivation of crops on the land is not part of the technique, although the soil structure is said to be improved by the organic degradation products. The attitude of regulatory authorities toward the technique is not yet clear, though the potential presence of toxic "contaminants" in the residue might be a source of concern. Surface runoff and possible groundwater contamination would have to be monitored and controlled in any application of this technique.

PHARMACEUTICAL MANUFACTURING

General Industry Description

The pharmaceutical manufacturing category includes facilities engaged in the production of biological products, medicinal chemicals and botanical products, pharmaceutical products, pharmaceutical active ingredients, cosmetic preparations which function as a skin treatment, pharmaceutical research facilities, and products which are similar to pharmaceuticals in the type of manufacturing processes used and wastewater generated.

Industrial Categorization

In order to facilitate raw waste characterization and the organization of pretreatment information, the industry is subcategorized as follows: fermentation, biological and natural extraction, chemical synthesis, formulation, and pharmaceutical research.

Process Description

Each of the pharmaceutical subcategories is a particular process or type of process. Many of the facilities in the pharmaceutical category employ several of these processes and thus do not fall into any one subcategory. A description of each subcategory and the processes employed is given below.

Fermentation (A): Although only six percent of pharmaceutical manufacturers are in this subcategory, fermentation is considered a basic production method for the pharmaceutical industry. Most steroids and antibiotics are produced by fermentation plants. Fermentation production involves three steps: inoculum and seed preparation, fermentation, and product recovery.

Like all pharmaceutical production, fermentation is generally a batch process. In the first step, activated spores are used to culture a sufficient quantity of seed. This seed is introduced, along with selected raw materials, into a batch vessel where fermentation is allowed to occur. A typical fermentation vessel is shown in Figure 32. When fermentation is complete, the broth is drained from the batch vessel for product recovery and the vessel is cleaned and sterilized.

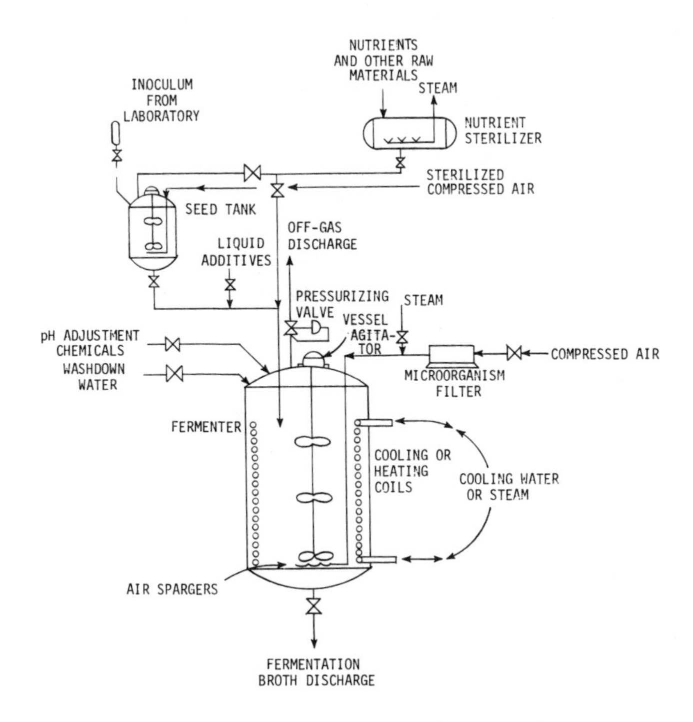

Figure 32: General fermentation tank.

There are three commonly used product recovery schemes: solvent extraction, direct precipitation, and ion exchange. In each case, the most significant waste load comes from the broth left over after recovery, which is known as "spent beers". Solvent extraction will also contribute relatively small amounts of organic solvents to the waste load, while direct precipitation results in increased metallic ion concentrations (particularly copper and zinc).

In addition to spent beers, wash water and fermentation gas scrubbers also contribute significant, although much more dilute, wastewater loads. An infrequent, but potentially troublesome, waste load results from the infection of a batch or facility by phage virus. When this occurs, the infected batches must be immediately dumped and the vessels thoroughly cleaned and sterilized. This results in large hydraulic and organic loads on subsequent treatment facilities.

Biological and Natural Extraction (B): Many pharmaceutical products are derived from natural sources such as plants, animal glands and fungi. Products range from allergy medications to powerful drugs like morphine and insulin. Extraction techniques basically consist of methods to concentrate a particular compound which is found in either plant or animal tissue. To accomplish this, sequential batch operations are used. Because of the many concentration steps required, the amount of product is usually insignificant in comparison to the amount of raw material. Due to this, waste material generated is practically equal to the amount of raw material processed. Most of this waste is disposed of through dry methods (i.e., incineration or landfilling), but some does appear in process wastewater.

Wash water is a major portion of the wastes generated by extraction operations. Ammonia appears in wastewater, as it is a common extraction solvent. Spills from batch operations are a principal source of wastewater. In general, extraction wastewaters are characterized by small flows and low pollutant concentrations.

Chemical Synthesis (C): As in other subcategories, batch processing is the standard method of operation for chemical synthesis facilities. Using multipurpose reaction vessels, raw materials and reagents are combined in a series of chemical reactions to produce the final products. It is common for a plant to produce a large stock of one product in a short "campaign" and then switch to an entirely different product.

This flexibility results in a wastewater which varies in characteristics, as well as in flow. A large number of toxics are used in chemical synthesis plants, and a correspondingly high incidence of toxic pollutants in plant wastewater has been observed. Organic solvents are commonly used as carriers in reactions, and although solvent recovery is generally practiced, some solvents do appear in process wastewaters.

In general, chemical synthesis plants produce a complex, variable wastewater which is often difficult to treat by conventional methods.

Formulation (D): Bulk pharmaceuticals produced from raw materials must be reduced to dosage form for consumer use. Pharmaceuticals can be formulated into tablets, capsules, liquids or ointments.

Most water usage is as cooling water, which generates no contact wastewater. Wastewater generally originates from cleanup, spills and breakage of packaged products. Some wastewater may also come from the use of dust scrubbers, which are sometimes needed to control dust from tablet and capsule production.

Formulation wastewaters typically have low flows and are very dilute in nature.

Pharmaceutical Research (E): Pharmaceutical research laboratories are engaged in attempts to improve existing process technologies or search for alternate methods to produce pharmaceutical products. In doing so, they may use all four of the previously described methods of pharmaceutical production.

Wastewater Characteristics

Table 52 shows raw wastewater characteristics. Facilities engaged in fermentation or chemical synthesis operations tend to generate large wastewater flows with high levels of BOD, COD, and TSS. Variation of pH over a broad range, particularly in the chemical synthesis subcategory, is also common. Wastewaters from chemical synthesis also contain significant levels of a large number of toxic pollutants.

Table 52: Pharmaceutical Industry—Raw Wastewater Characteristics[a]

Wastewater parameter	Fermentation	Biological and natural extraction	Chemical synthesis	Formulation	Industry-Wide
	A	B	C	D	
Conventional - mg/l					
BOD_5	1,330-4,350	11-520	561-10,000	90-950	
COD	3,260-9,420	77-834	2,250-17,000	304-1,670	
TSS	360-2,180	15-64	146-773	14-103	
Toxic - ug/l[b]					
Benzene					100-10,000
Chloroform					100-100,000
Chromium					50-1,000
Copper					5,000-6,500
Cyanide					150-5,500
1,2-Dichloroethane					100-15,000
Ethylbenzene					100-50,000
Lead					50-500
Methylene chloride					100-100,000
Phenol					100-15,000
Toluene					100-100,000
Zinc					150-2,500

[a] Data is not available for Subcategory E, "Pharmaceutical Research".

[b] Only pollutants found in significant concentrations are listed.

Plants in the extraction, formulation, and pharmaceutical research subcategories tend to produce low flows, with low levels of biochemical oxygen demand, chemical oxygen demand, and total suspended solids.

Control and Treatment Technology

In-Plant Control Technology: As a result of the competitive climate and strict FDA regulations, general operating practices within the industry are quite efficient. Nevertheless, adoption or increased usage of the following measures can reduce wastewater flows and concentrations.

Improved Housekeeping — Although housekeeping practices are generally good, improvements are possible. Practices such as spill cleanup and prevention, and centralized cleaning facilities can minimize accidental discharge of both solid and liquid chemicals.

Water Conservation and Reuse — By modifying processing procedures or auxiliary equipment, water usage, and therefore wastewater generation, may be significantly reduced. Examples are: use of surface rather than barometric condensers, reuse of noncontact water, concentration of reaction mixtures to limit waste volume, and combining several processes.

Concentration and Segregation of Wastes — In order to minimize the impact of intermittent hydraulic surges, especially in fermentation operations, wastewater may be concentrated. Various evaporation or dewatering methods may be utilized. Segregation of waste streams for individual treatment often allows more efficient removal of particular pollutants. Both new plants and individual process units are now commonly designed with allowance for waste stream segregation.

Waste Solvent Recovery — Recovering waste solvents is generally used as an economical in-plant control. It is also possible for facilities to recover previously unrecoverable solvents and solvent "bottoms" from recovery units. These methods can significantly reduce solvent discharges from solvent extraction facilities.

Segregation of Storm Water Runoff — Storm water from manufacturing areas can contain high levels of pollutants. Separation and treatment can eliminate the discharge of contaminated runoff.

Specific Pollutant Removal — Cyanide destruction, metals removal, and steam stripping to remove ammonia and organic solvents are demonstrated treatment technologies in the pharmaceutical industry. Although all three may be utilized as end-of-pipe treatment, they can be used more effectively as in-plant controls on segregated streams. Cyanide destruction, in particular, is much more efficient when used on a concentrated waste stream.

End-of-Pipe Treatment Technology: Approximately 30 percent of the facilities in the pharmaceutical industry have some form of wastewater treatment. Of that 30 percent, about half are indirect dischargers.

A basic end-of-pipe treatment scheme would involve screening, equalization, neutralization, and clarification, followed by secondary treatment — either biological or physical/chemical. Of the two, biological treatment is by far the more widely used. Activated sludge, trickling filters, lagoons and ponds, rotating biological contractors, and sand filtration (intermittent) have all been used in pharmaceutical facilities. Activated sludge is being used in over 60 percent of the plants employing biological treatment. Lagoons are also widely used. The remaining treatment methods find only limited use. Physical/chemical processes have been used in only 20 percent as many plants as biological systems. Thermal oxidation is the most widely used of the physical/chemical processes.

Table 53 gives typical results of various end-of-pipe treatment technologies for removal of biochemical oxygen demand, chemical oxygen demand, and total suspended solids. Table 54 lists the effectiveness of various treatment technologies in reducing toxic pollutants from pharamaceutical wastewaters.

Type of Residue Generated

Numerous process wastes are generated by this industry. A distinguish-

Table 53: Pharmaceutical Manufacturing–Conventional Pollutant Treatment Effectiveness[a]

Major end-of-pipe treatment	Subcategory[b]	Biochemical Oxygen Demand			Chemical Oxygen Demand			Total Suspended Solids		
		Influent, mg/l	Effluent, mg/l	Removal, percent	Influent, mg/l	Effluent, mg/l	Removal, percent	Influent, mg/l	Effluent, mg/l	Removal, percent
Activated sludge Aerated lagoon	B, C, D	1,195	331	72.3				116	251	
Physical/chemical	D		218		256	456		53	88	
Activated sludge	D		4						12	
Activated sludge Aerated lagoon	C, D		129			683			328	
Trickling filter	C		653		1,950				124	
Activated sludge with pure oxygen	A, B, C, D	1,220	146	88.0	2,628	407	84.5	2,000	320	84
Physical/chemical	C, D		3,636			8,481			286	
Activated sludge	A, B, C, D	3,000	120	96.0				950	500	47.4
Aerated lagoon	B, C, D	7,100	869	87.8	15,700			369	1,793	
Activated sludge	B, D	7,520	4,636	38.4	12,032	7,418	38.3	4,923	4,048	17.8
Trickling filters	B, D		288							
Aerated lagoon	B, C, D	1,500	150	90.0				500	150	70

[a] Values reported are based on individual plant samplings of indirect dischargers.

[b] A - fermentation, B - biological natural extraction, C - chemical synthesis, D - formulation.

Table 54: Pharmaceutical Manufacturing–Toxic Pollutant Removal Effectiveness[a]

Wastewater parameters	Biological[b]			Primary[c]			Chemical Precipitation[d]		
	Influent, g/l	Effluent, g/l	Removal, percent	Influent, g/l	Effluent, g/l	Removal, percent	Influent, g/l	Effluent, g/l	Removal, percent
Metals									
Antimony	0-78	0-50	0-100					0-90	
Arsenic	0-115		0-100		0-20			0-7,200	
Cadmium	0-32		100		0-40				
Chromium	0-680	0-190	0-100					0-9	
Copper	0-3,110	0-63	0-98		0-100			0-35	
Cyanide	0-1,980	0-7,700	0-100		0-60				
Lead	0-80	0-24	0-100		0-400				
Mercury	0-.9	0-6.4	0-44						
Nickel	0-630	0-200	0-70		0-300				
Selenium	0-860	0-300	0-100		0-26			0-310	
Silver					0-40				
Thallium	0-234	0-58	0-100						
Zinc	0-540	0-507	0-94		0-23			0-70	
Volatile Organics									
Acrolein									
Benzene	0-4,000	0-44	0-100		0-24				
Bromoform									
Carbon tetrachloride	0-50	0-61	0-100						
Chlorobenzene	0-19		0-100						
Chlorodibromomethane									
Chloroform	18-3,170	14-1,000	0-100		0-60				
Dichlorobromomethane									
1,1-Dichloroethane									
1,2-Dichloroethene	12-9,000	44-400	0-98					0-20	
1,1-Dichloroethylene	0-230	0-90	0-100				0-30	0-370	
1,2-Dichloropropylene	0-100		0-100						
1,2-Trans-dichloroethylene									
Ethylbenzene	0-1,600	0-160	0-100						
Methyl bromide	0-30		0-100						
Methyl chloride	0-3,000		0-100						
Methylene chloride	0-40,000	0-349	0-200		0-360				
1,1,2,2-Tetrachloroethane	0-20		0-100						
Tetra chloroethylene	0-36		0-100						
1,1,1-Trichloroethane	17-261	10-33	0-95					0-20	
1,1,2-Trichloroethane	0-19		0-100				0-1,250	0-890	0-29
Trichloroethylene	0-2,100	0-14	0-100						
Trichlorofluoromethane	0-970	0-280	0-100					0-80	
Toluene	0-33,000	0-1,350	0-100		200				

(continued)

Table 54: (continued)

Wastewater parameters	Biological[b] Influent, g/l	Biological[b] Effluent, g/l	Biological[b] Removal, percent	Primary[c] Influent, g/l	Primary[c] Effluent, g/l	Primary[c] Removal, percent	Chemical Precipitation[d] Influent, g/l	Chemical Precipitation[d] Effluent, g/l	Chemical Precipitation[d] Removal, percent
Extractables									
Acenaphthene	0-135		100						
Anthracene	0-14		100						
Bis (2-chlorisopropyl) ether	0-448	0-181	0-100						
Bis (2-chloroethyl) ether	0-10	0-20	0						
Bis (2-ethylhexyl) phthalate	0-200	0-400	0-100						
Butylbenzyl phthalate	0-720		100						
1,1-Dichlorobenzene	0-290		100						
1,4-Dichlorobenzene	0-10	0-2	80-100						
Diethyl phthalate	0-61		100						
2,4-Dimethylphenol	0-65		100						
Di-n-butyl phthalate	0-20		100						
1,2-Diphenylhydrazine	0-20		100						
Fluoranthene									
Fluorene	0-27		100						
Isophorone	0-1,014		100						
Napthalene	0-28	0-14	50						
Nitrobenzene									
N-nitrosodiphenylamine	0-12		100						
Phenanthrene	0-14		100						
Acid Extractables									
2-Chlorophenol	0-240	0-107	0-100						
2,4-Dichlorophenol	0-9	0-8	11						
2,4-Dimethylphenol	0-62	0-15	0-100				0-13,000	0-4,100	0-68
4,6-Dinitro-o-cresol		0-15							
2-Nitro phenol	0-119	0-7	0-100				0-3,600	0-1,100	0-69
4-Nitro phenol	0-1,600	0-19	0-100						
Pentachlorophenol	0-62		100						
Phenol	0-3,100	0-48	0-100		100		0-16,500	0-16,500	
2,4,6-Trichlorophenol	0-14	0-12	14-100						

a Values shown in table are based on a sampling program involving 26 facilities. A blank space indicates that the pollutant was not detected, or that insufficient data exists to determine removal efficiency.

b Activated sludge, aerated lagoons. Approximately 50 percent had some form of primary treatment.

c Equalization, neutralization, solids removal.

d Thermal oxidation, evaporation.

ing feature of this industry is that a large proportion of the material input to the manufacturing process ends up as process waste. This low-product-yield feature is especially conspicuous for the fermentation and the biological and natural extraction manufacturing techniques. These process wastes may be in either solid or liquid form, and generally would not be considered residues from wastewater treatment. However, these latter residues are similar in character to the corresponding process wastes. Also, the wastewater residue can frequently be disposed of in the same manner as the corresponding process waste.

Biological treatment is the prevalent wastewater treatment process in this industry. Wastewater residues will typically be waste biological sludges "contaminated" in varying degrees by potentially toxic materials, primarily organics rather than heavy metals. Organic contaminants of the sludge are either: traces of solvents used as part of the fermentation, chemical synthesis, or biological extraction manufacturing techniques; or are reactants or by-products of the chemical synthesis technique. Fats and oils may also occur in the biological extraction manufacturing technique and be skimmed-off in settling or flotation tanks.

Production methods also affect wastewater sludge generation through the almost universal use of batch production methods, which makes the wastewater vary greatly due to changes in product runs, batch dumps, and/or cleanup.

Residual Management Options

Manufacturing process wastes are generally disposed of either by incineration or by landfilling. Disposal sometimes occurs on-site, but can also be done off-site by private contractors. Wastewater treatment residues are similar in character and are usually combined with, and treated in the same manner.

Some large manufacturers have demonstrated that manufacturing and/or wastewater residues can be spread on land used for agricultural purposes in the same manner as municipal wastewater sludge. Industrial process wastes have also been successfully concentrated and sold as an animal feed supplement. However, these two residue disposal methods require a substantial capital investment and have not been widely adopted. The possibility of contamination with toxic materials, even intermittently, would have to be considered when determining the suitability of any of these techniques to specific cases. Ocean dumping, formerly widely used by this industry for residue disposal, has now been phased out.

Other management methods for pharmaceutical sludges are generally the typical ones used on biological sludges.

Fats and oils may also be incinerated or landfilled along with the sludges, or they may sometimes be sold as by-products to other industries, such as soap manufacturers.

PHOTOGRAPHIC EQUIPMENT AND SUPPLIES

Sufficient data is not available on any aspect of this categorical industry.

PLASTICS PROCESSING

Sufficient data is not available on any aspect of this categorical industry.

PLASTICS AND SYNTHETIC MATERIALS

General Industry Description

The plastics and synthetic material industry is composed of three segments: the manufacture of the raw material or "monomer"; the conversion of this monomer into a resin or plastic material; and the formation of the plastic resin, or polymer, into a plastic item such as a toy, synthetic fiber, packaging film, adhesive, paint, etc.

This summary addresses the manufacture of the plastic or synthetic resin and the manufacture of synthetic fibers, such as nylon, rayon, cellulose film, and others described in each industry subcategorication.

Industrial Categorization

The industry has been categorized according to waste characteristics and subcategorized along project lines, as shown in Table 55.

Process Description

Plastics and synthetics are produced by the conversion or polymerization of plastic resins. Polymerization is the formation of long chain molecules from a single-type molecule, or "monomer". For example:

Monomer A + monomer A ⟶ A-A-A-A-A (polymer)

Copolymers are formed by combining two different monomers. For example:

Monomer A + monomer B ⟶ A-B-A-B-A (copolymer)

Polymerization takes place in reactors which can be either a batch or continuous process. Many reactions require a catalyst in order for the reaction to occur. A brief description of the processes used by each subcategory follows.

Polyvinyl Chloride (A), Polyvinyl Acetate (B), Polystyrene (C), and ABS/SAN (H): Materials in subcategories A, B, C, and H can be manufactured by an Emulsion and Suspension Polymerization process in which the monomer is dispersed in an aqueous, continuous phase during the course of the reaction. The batch cycle consists of the continuous introduction of a water-monomer emulsion to a stirred, temperature controlled, reactor ranging in size from 5,000 to 30,000 gallons. On completion of a batch, a short "soaking" time is allowed for completion of reaction. Water is added to dilute to the desired end composition, and the batch is screened and stored.

In some cases, the water-polymer emulsions are marketed in this latex form, thus, no wastewater is generated. When the polymer is isolated and sold, a wastewater contaminated with polymer is discharged. Monomers that are protected by an inhibitor are subject to washing prior to polymerization. This contributes to the wastewater load. Figure 33 shows a flow diagram describing this process.

Atmospheric or Low-Pressure Mass Polymerization is another process used to manufacture subcategories A, C, and H. Catalysts and modifiers are used and remain in the product. Inhibitors, which are usually added to the monomer for protection during storage, are removed by washing, thus generating a wastewater. Vacuum-stripping is used to separate the unreacted monomer and contaminants from the product, resulting in a waste stream containing these chemicals. Figure 34 shows a flow diagram for this process.

Table 55: Plastic and Synthetic Materials

Main Category[a]	Subcategory	Designation
I	Polyvinyl Chloride	A
I	Polyvinyl Acetate	B
I	Polystyrene	C
I	Polypropylene	D
I	Polyethylene	E
II	Cellophane	F
II	Rayon	G
II	ABS/SAN (acrylonitrile, butadiene, styrene/ styrene, acrylonitrile)	H
III	Polyester Resin and Fibers	I
III	Nylon 66 Resin and Fibers	J
III	Nylon 6 Resin and Fibers	K
III	Cellulose Acetate Resin and Fibers	L
IV	Acrylics	M
I	Etylene-vinyl Acetate Copolymers	N
I	Polytetrafluoroethylene	O
I	Polypropylene Fibers	P
III	Alkyds and Unsaturated Polyester Resins	Q
III	Cellulose Nitrate	R
III	Polyamides	S
III	Polyester Resins (thermoplastic)	T
III	Silicones	U
III	Epoxy Resins	V
IV	Phenolic Resins	W
IV	Urea and melamine	X

[a]

I	Generates a low raw waste load (less than 10 units per 1,000 units of product produced); low BOD concentrations attainable (less than 20 mg/l).
II	Generates a high waste load (greater than 10 units per 1,000 units of product); low BOD concentrations attainable (less than 20 mg/l).
III	Generates a high waste load (greater than 10 units per 1,000 units of product); medium BOD concentrations attainable (30 to 75 mg/l).
IV	Generates a high waste load (greater than 20 units per 1,000 units of product); high BOD concentrations attainable (over 75 mg/l).

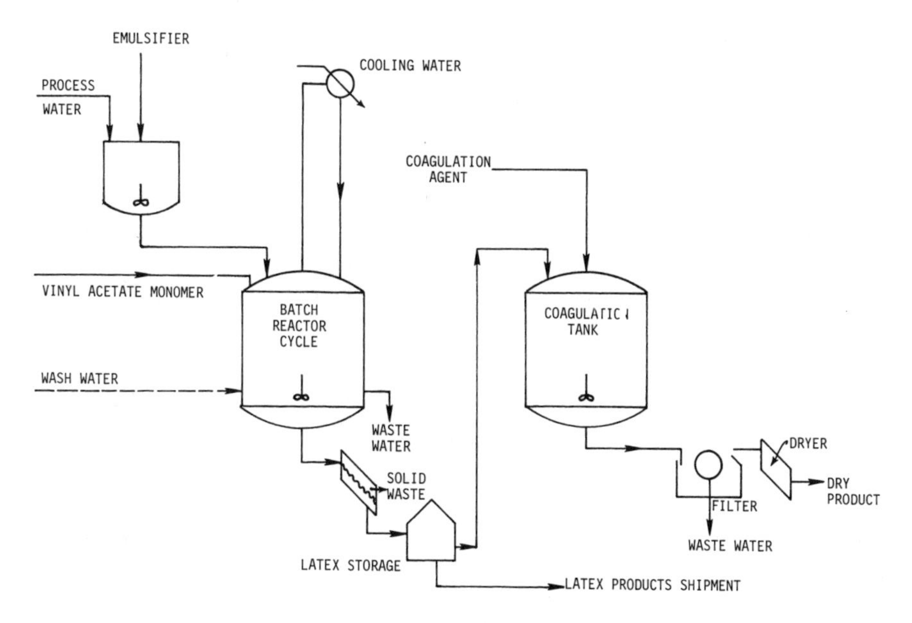

Figure 33: Process flow diagram for emulsion and suspension polymerization process.

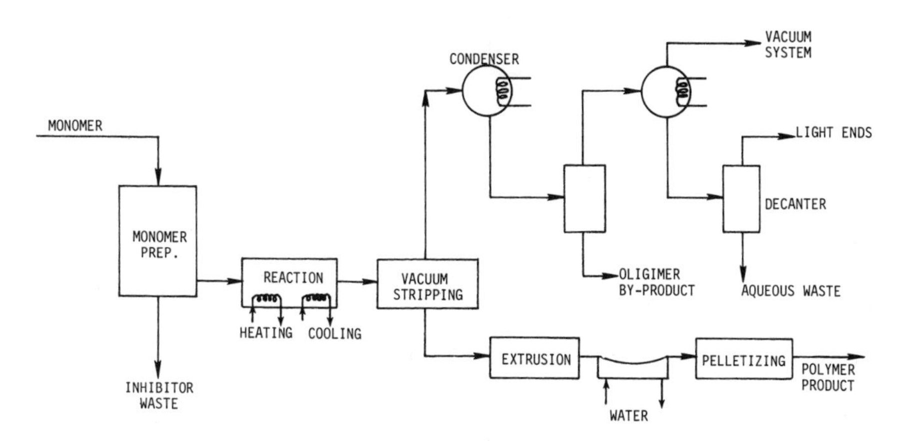

Figure 34: Process flow diagram for mass polymerization process.

Polyethylene (E) and Polypropylene (D): Ethylene gas is mixed with a very small quantity of catalyst and raised to a high pressure in a process called High Pressure Mass Polymerization. At the appropriate pressure and temperature, polymerization is carried out in jacketed (cooled) tubular reactors. On completion of the reaction, the polymer is flash-cooled in drums containing water. The low density polyethylene is then formed into pellets and separated from the water. The water is recycled, but is periodically purged, producing a waste stream.

Polyethylene (E) and copolymers can also be manufactured by the Solution Polymerization Process, in which the polymer is dissolved in the reaction solvent as it is formed and the catalyst is present as a separate, solid phase.

The catalyst system uses activated chromium oxide deposited on a carrier, such as alumina. After reaction, the catalyst and solvent are separated from the polymer and then separated from each other. Figure 35 shows a flow diagram for the Solution Polymerization Process.

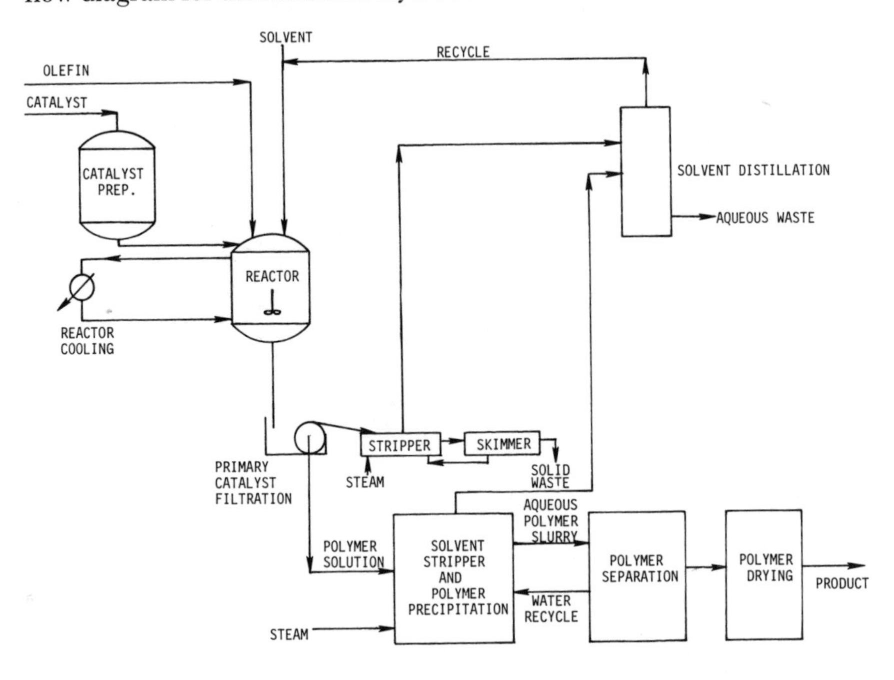

Figure 35: Process flow diagram for solution polymerization process.

Polyethylene (E), Polypropylene (D), and some copolymers can also be manufactured by the Ziegler Process. This processs is similar to the Solution Polymerization Process, except that the polymer precipitates as it is formed, rather than remaining in solution. The Particle Form Process is an improvement on the two processes previously described, using a continuous system whereby the product is drawn off continuously, using a "loop" reactor. Wastes from this process contain polymer fines and solvents (aqueous alcohols).

The water used in the separation process constitutes a waste stream which will contain quantities of catalyst, solvent, and polymer.

Cellophane (F) and Rayon (G): Cellophane and Rayon are both regenerated cellulose products that are produced by treating wood pulp and cotton linters (raw cellulose) using the Viscose process. Here, raw cellulosic polymer is treated to form a solution of viscose, then is processed and transformed back into cellulosic plastics of desired shapes.

Cellophane manufacture is performed in three steps: (1) viscose preparation, (2) film casting, and (3) film coating. Viscose is prepared in several batch operations in which the raw cellulose is depolymerized in a caustic solution, and is then reacted with carbon disulfide to make a solution of sodium cellulose xanthate called viscose. This solution is reacted with sulfuric acid in the next step to regenerate the cellulose as cellophane. Caustic solution from the first step is recycled, but is periodically purged, producing a waste stream.

Film is cast by pumping "viscose" through slit-dies into a spinning batch of sulfuric acid and sodium sulfate. The cellophane film is subsequently passed through finishing baths, dried, and wound into rolls. In the third step, coatings are applied to the film from organic solvent solutions. The solvent from these solutions is recovered and reused.

Waste liquors from the spinning bath are evaporated, crystallized, and recycled.

Rayon (G) is manufactured by the same process as cellophane except the regenerated cellulose is recovered in the form of fibers instead of as a film.

ABS/SAN (H): Included under Process Description for (A)(B)(C).

Polyester Resin and Fibers (I): In this process the monomer is generated first, then followed by polymerization. Although many plants still use the batch polymerization process, continuous polymerization with direct spinning of the fiber is more common for new facilities. The ester monomer is made by reacting an alcohol, usually ethylene glycol, with an ester, forming an "activated" ester.

When polymerization of the ester takes place continuously, the molten polymer is fed to spinning heads, forming the polyester fiber. Wastes associated with this process are primarily unused monomer and methanol. A process flow diagram is shown in Figure 36.

Nylon 66 Resin and Fibers (J): This process is similar to polyester, in that the monomer is generated first, followed by polymerization. Effluents from activated carbon filtration, evaporators, and scrubbers all produce waste streams containing small amounts of raw and intermediate chemicals.

Nylon 6 Resin and Fibers (K): Caprolactam is mixed with catalyst, acetic acid, and titanium dioxide, and polymerized. Formation of the resin into strands and chips, as well as monomer recovery processes, involve extensive water use.

Cellulose Acetate Resin and Fibers (L): Cellulose Acetate resin is produced by a batch operation in which wood pulp is dissolved with strongly acidic materials, and cellulose acetate "flakes" are recovered by an acid reaction to form a precipitate. Polymer "flakes" are washed to recover the acids used in the process. Wastewaters are high in dissolved solids.

Fibers are manufactured by dissolving in acetone the "flakes" produced by the previous process and by pumping the "dope" solution through spinnerettes. The acetone solvent recovery system is the major source of waste.

Acrylics (M): Acrylonitrile monomer is polymerized in a continuous reactor in the presence of a catalyst. Polymer is recovered, dried, and forced through spinnerettes. Solvent losses are a major contributor to the waste load.

Ethylene-Vinyl Acetate Copolymers (N): Ethylene-Vinyl Acetate (EVA) is manufactured in the same facilities as low and high density polyethylene (E). Monomers used in the reaction are vinyl acetate and ethylene. Polymerization is carried out in an autoclave. The reaction mass is then sent to a separator, to remove unreacted monomers. The EVA polymer is then fed to an extruder, which forms strands which are cut into pellets.

Cooling water used in the pelletizer is recirculated, but is purged periodically, producing a waste stream containing monomers and polymer fines.

Polytetrafluoroethylene (PTFE) (O): The PTFE monomer is produced in the gaseous phase. The product stream is scrubbed with water and then with dilute caustic solution to remove byproduct acid and other soluble components. The gas is then dried with concentrated sulfuric acid or ethylene glycol.

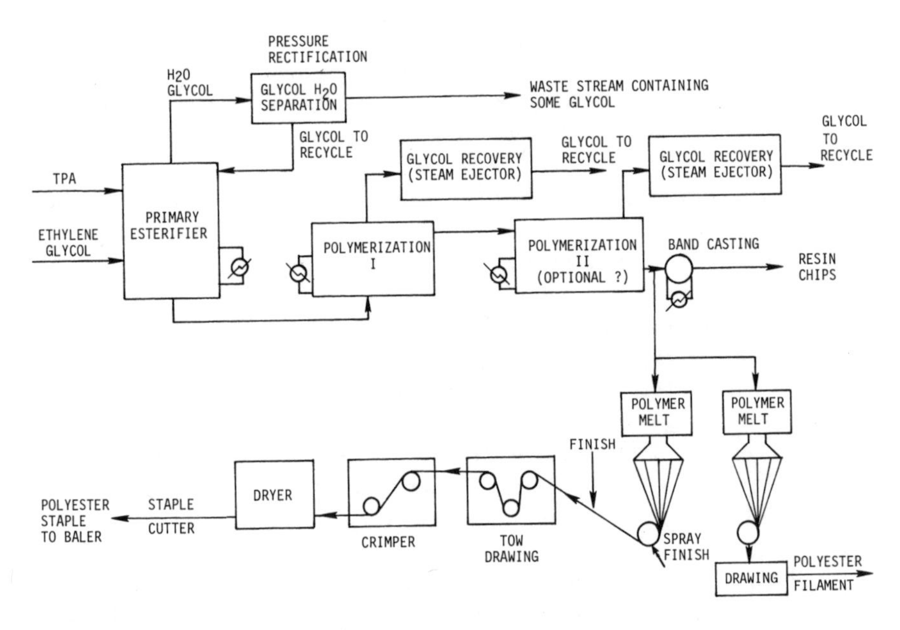

Figure 36: Process flow diagram for polyester resin and fiber production.

The waste stream produced from the scrubbing processes is acidic and contains fluoride. If glycol is used as the drying agent, it contributes to the waste load. PTFE polymer is produced in a batch operation and sold in a granular or pellet form, fine powder, or in an aqueous dispersion.

Polypropylene Fibers (P): The polymerization process was described under process descriptions for Plastics and Synthetic Materials. Polypropylene fibers are made by melt spinning. The process consists of coloring polypropylene flakes by dry blending the flakes with pigments, followed by a melting and extrusion process that forms colored polypropylene pellets. Pellets are then extruded through a spinnerette into a column of air which solidifies the molten filaments. Filaments are then stretched or spun into fibers.

Rinsewaters are generated from the blending process. Discharges from spinning wastes are very high in BOD.

Alkyds and Unsaturated Polyester Resins (Q): Unsaturated polyester resins are made in a batch process by an esterification reaction, involving several materials derived from petroleum fractions. Reinforced plastic is made by reinforcing the resin with glass or metallic fibers. Non-reinforced, unsaturated polyester resin is used for castings, coatings, and putty-like compounds.

Alkyds are often manufactured and used interchangeably with unsaturated polyesters, since they are chemically very similar. They are used for paint formulations and in molding compounds.

As a result of the polymerization reaction, and scrubber and equipment washouts, wastewaters contain a variety of contaminants and are high in BOD, COD, and oil and grease.

Cellulose Nitrate (R): Cellulose Nitrate is produced by reacting cellulose with a mixture of nitric and sulfuric acids, followed by washing to remove acid, stabilization by boiling with water, digestion (heating in water), and dehydration.

Wastewaters contain acids, unrecovered alcohols, and suspended solids.

Polyamides-Nylon 6/12 (S): Nylon 6/12 is produced in equipment used regularly for the production of Nylon 66. Wastewaters generated in this process are similar to those from the Nylon 66 process.

Polyester Resins (Thermoplastic) (T): These resins are produced by the same polymerization process used for fiber production (I). The two raw materials are ethylene glycol, and either dimethyl terephthalate (DMT) or terephtalic acid (TPA). The two reactants are polymerized in a reactor. Some integrated plants also produce polyester fibers.

Liquid wastes result from the condensation of steam ejector vapors. Process materials are present in the waste streams.

Silicones (U): Plants purchase silicon metal and react it with a wide range of chemicals to produce silicone. In general, silicones are produced by reacting chloride-containing compounds, such as methyl chloride or phenyl chloride, with silicon metal in the presence of copper catalyst to form chlorosilane, which is a silicone.

A significant amount of acid wastes (often containing copper) are generated due to the formation of hydrochloric acid (HCl) in the process. In addition, trace amounts of solvent may be present in the waste stream.

Epoxy Resins (V): The epoxy resin family should be regarded as intermediates rather than an end resin itself, since they require further reaction with a second component, or curing agent, in order to yield the final thermoset material. Almost all commercially produced epoxy resins are made by the reaction between epichlorohydrin and bisphenol A. The reaction takes place under alkaline conditions.

Epoxy resins fall into two broad categories: low molecular weight liquids and high molecular weight solids. Low molecular weight liquid resins can be manufactured by either batch or continuous processes, while solid resins are produced by batch processes. Wastewaters contain caustic and salt.

Phenolic Resins (W): These resins are based upon the reaction between phenol and formaldehyde. There are two broad types of resins produced by the industry: resols and novolaks. Resols are formed from a mixture containing an excess of formaldehyde; novolaks are formed from a mixture containing a deficiency of formaldehyde. These resins are generally produced by batch process.

Wastewaters are generated from the distillation process and the rinsing procedures.

Urea and Melamine (Amino Resins) (X): "Amino Resins" are a broad group of polymers formed by batch processes from formaldehyde and various nitrogen-containing organic chemicals such as urea and melamine. The product can be sold either as a thick syrup or as a solid.

Equipment used for the production of the "first-step" amino resins is often also used for the production of other resins such as phenolics. Between these different uses, and between production batches of melamine and urea resins, it is customary to clean the equipment with hot dilute caustic solution. This material is drained as process waste.

Wastewater Characteristics

The wastes of plastic and synthetic material plants are generally high in BOD and suspended solids. Table 56 shows raw wastewater characteristics. Table 57 shows raw wastewater characteristics as a function of production.

Control and Treatment Technology

In-Plant Control Technology: A major source of waterborne pollutants is

Table 56: Plastic and Synthetic Materials—Raw Wastewater Characteristics

Wastewater parameters[b] mg/l	Subcategory[a]											
	A	B	C	D	E[c]	F	G	H	I	J	K	L
BOD	350	1,500			400		200	1,200	4,400	1,300	400	
COD	1,600				1,500		500	2,100	5,800	2,100		

Wastewater parameters[b] mg/l	Subcategory[a]											
	M	N	O	P	Q	R	S	T	U	V	W	X
BOD	900	160			3,000				300	850	1,500	1,300
COD	1,700				4,000				700	2,300	5,000	6,500

[a] See Table 55 for names of industrial subcategories.

[b] Other pollutants which may be present in any of the subcategories are: color, oil and grease, mercury, chromium, copper, zinc, cyanide, cobalt, iron, titanium, cadmium nickel, and vanadium.

[c] Low density polyethylene.

Table 57: Plastic and Synthetic Materials—Raw Wastewater Characteristics—Production-Based Data

Wastewater Parameter[b]	Subcategory[a]											
	A	B	C	D	E	F	G	H	I	J	K	L
BOD	0.1-48	0-2	0-3	0-10	0-5	20-133	20-45	2-21	3-20	0-135	0-135	6-70
COD	0.2-100	0-3	0-6	0-20	0-54	40-334	33-100	5-34	6-45	0-300	0-300	11-100
TSS	1-30	0-2	0-8.4		0-4	6-70		0-30	0-12	0-8	0-8	2-20
Zinc							12-50					

Wastewater Parameter[b]	Subcategory[a]											
	M	N	O	P	Q	R	S	T	U	V	W	X
BOD	10-40	0.4-4.4	0-7	0.4-1	9-25	55-110		0-10	5-110	60-85	15-50	13
COD	10-70	0.2-54	4.4-44	1.8-3	15-80	75-275		1-30	15-200	30-127	90-65	60
TSS	0-2	0-4	2.2-6.6	0.2-2.2	1-2	35			50	5-25	0-7	
Zinc												

[a] See Table 55 for names of industrial subcategories.

[b] Units are in kilograms of pollutant per 1,000 kilogram of product produced.

attributable to spills, leaks, and accidents. The following list of spill prevention and control techniques applies to the Synthetics and Plastics industry:

- Dike areas around storage tanks.
- Install tank level indicators and alarms.
- Curb process areas.
- Install holding lagoons for general plant area.

End-of-Pipe Treatment Technology: Wastewater treatment technology in this industry relies heavily upon the use of biological treatment methods, preceded by pH adjustment, equalization, and primary solids removal.

COD/BOD ratios in the plastics and synthetic industry range from 4 to 12, which indicates the presence of substances which are not biodegradable. Removal efficiencies of these substances vary markedly from one subcategory to another. In general, longer residence times are required to treat wastes from this industry than for municipal wastes. Detention times of 500 to 900 hours have been reported for some plants. Of all the subcategories, acrylic wastes represent the most difficult treatment problems. Equalization prior to discharge to municipal treatment plant can help to maximize POTW efficiency.

Water recycle has not been practiced due to two factors: (1) the industry, except for cellulosics, is a relatively low user of water per unit of product, and (2) high purity process water is often required in order to maintain product quality.

Table 58 gives removal efficiencies for a common treatment scheme consisting of neutralization, equalization, clarification, and biological treatment.

Table 58: Plastics and Synthetics Industry—Percent BOD_5 and COD Removals for Various Subcategories Using a Common Treatment Scheme of Neutralization, Equalization, Clarification and Biological Treatment

Pollutant	Percent removal	Subcategories
BOD_5	80-87	G, K, M, R, U
	95-99	A, B, E, H, I, J, Q, V, X
COD	0-30	B, G
	60-70	D, M, U
	85	V, W
	90-95	A, E, H, I, J, N, Q, X

Type of Residue Generated

Information on types and characteristics of residues generated by this industry is insufficient and cannot be discussed at this time.

Residual Management Options

Lack of sufficient information on types and characteristics of residues generated does not permit a discussion of specific management options at this time.

PORCELAIN ENAMELING

General Industry Description

Porcelain enameling is the application of glass-like coatings to steel, cast iron, aluminum, or copper. The purpose of the coating is to improve resistance to chemicals, abrasion, and water, and to improve thermal stability, electrical resistance, and appearance. Porcelain enameled products include plumbing supplies, cookware, household appliances, and jewelry.

Industrial Categorization

For the purpose of establishing pretreatment guidelines, the porcelain enameling industry has been categorized into the following subcategories: porcelain enameling on steel; porcelain enameling on cast iron; porcelain enameling on aluminum; and porcelain enameling on copper.

Process Description

Regardless of the base metal being coated, the porcelain enameling process involves the preparation of the enamel slip, surface preparation of the base material, application of enamel, drying, and firing to fuse the coating. Figure 37 illustrates the processes used in the porcelain enameling operation. The processes are described below, followed by a description of the subcategories based on these operations.

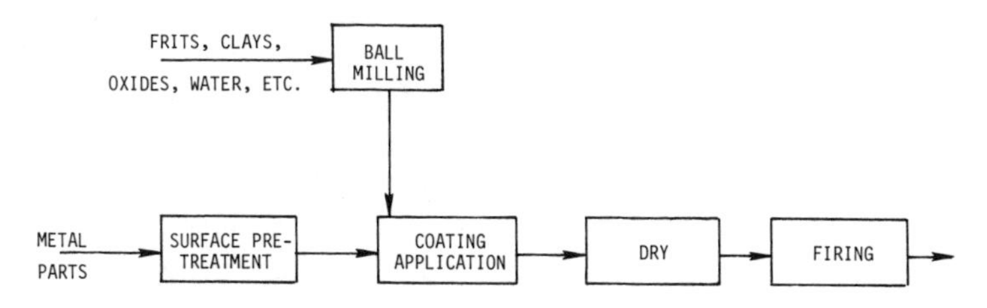

Figure 37: Typical porcelain enameling process sequence.

Frit Manufacture: Frit, the glassy raw material used in porcelain enameling, forms the basis of the enamel slip that is applied to the workpiece. Production of frit involves measuring and mixing of raw materials and smelting these into a uniform glass. The well mixed batch is loaded into a smelter and raised to a temperature in the range of 1040° to 1370°C, where the decomposition and fusion of the various components takes place. The final product is either packaged and sold or used within hours for porcelain enameling.

Ball Milling: Ball milling is the process of mixing and grinding frit and other raw materials to form an enamel slip of the appropriate consistency for a particular application. A typical enamel slip is composed of a combination of the following: frit, clays, gums, suspending agent (bentonites, colloidal silica), opacifiers (tin or zirconium oxide), coloring agents, electrolytes (borox, sodium carbonate), and water. Wastewaters generated by this process consist of cooling water and spent wastewater.

Basis Material Preparation: The base metal must be suitably prepared to enable the formation of a good bond with the porcelain. Depending on the type of metal being finished, one or more of the following surface preparation processes are used: solvent cleaning, alkaline cleaning, acid treatment, nickel deposition, neutralization, chromate cleaning, or grit blasting.

Solvent cleaning is used to remove oily dirt, grease, smears, and fingerprints from metal workpieces. Solvent cleaning is classified as either hot cleaning such as vapor degreasing, or cold cleaning which covers all solvent cleaning performed at or near room temperature. Vapor degreasing utilizes trichloroethylene or 1,1,2-trichloroethane at its boiling point to clean metal parts. It is very effective in removing non-saponifiable oils, and sulfurized or chlorinated components. It is also used to flush away soluble soil. In cold cleaning, the solvent or mixture of solvents is selected based on the type of soil to be removed. When soil removal requires the action of water and organic compounds, diphase cleaning provides the best method of cleaning. This approach uses a two layer system of water-soluble and water-insoluble solvents. Diphase cleaning is particularly useful when both solvent-soluble and water-soluble lubricants are used.

Alkaline cleaning by soak, spray, and electrolytic methods is used to remove oils, soils, or solid soil from workpieces. When aluminum is the metal being enameled, a stronger alkaline solution is often used to bring about a mild-etch or micro-etch of the metal. The purpose of the etch is to remove a thin layer of aluminum, thereby removing excess surface oxides.

Acid treatment is utilized to remove rust, scale, and oxides that form on a part prior to porcelain enameling. Acid treatment can be carried out in the form of acid cleaning, acid pickling, or acid etching, depending upon the strength of the acid solution.

Nickel deposition on the surfaces of many steels is sometimes employed in order to improve the bond between the porcelain enamel and the metal. A thin nickel deposit is obtained by immersing the steel in a nickel solution after the part has been acid treated and rinsed.

The neutralization step follows the acid pickling and nickel deposition (if used) steps, prior to the porcelain enameling of steel. Its function is to remove the last traces of acid left on the metal surface. There are two types of neutralizer: alkali and cyanide. The alkali neutralizer is made up of soda ash, borax or trisodium phosphate, and water. The cyanide neutralizer consists of sodium cyanide and water. Sodium cyanide forms ferrocyanides with iron salts left on the surface of the workpiece and holds them in solution. The result is a salt-free base metal surface ready for coating. Neutralization may or may not be followed by a rinse.

When particular aluminum alloys are being porcelain enameled, it is preferable to use a chromate cleaning or pickling solution to enhance adherence of the subsequent enamel application. Typical solutions contain a source of chromate (potassium chromate or sodium bichromate), sodium hydroxide, and water.

Grit blasting is a mechanical surface preparation in which an abrasive impacts the metal to be processed in order to produce a roughened, velvety surface. Cast iron, in particular, and certain grades of steel, are blasted to remove any surface irregularities. Parts which are grit blasted require no additional surface preparation since they are essentially clean, and their roughened surfaces provide a good base on which the porcelain enamel can adhere.

There are two types of abrasive blasting, dry, and wet. In a wet blasting system, abrasives are suspended in water or petroleum distillates along with rust inhibitors, wetting agents, and antisettling agents. In the process of abrasive blasting, the only wastewater created is when a light rinse is used to remove particles of the abrasive from the workpiece.

Coating Application Methods: Once the workpiece has undergone the proper base metal preparation, and the enamel slip has been prepared, the next step is the actual application of the porcelain enamel. Included among the application methods used are air spraying, electrostatic spraying, dip coating, flow coating, powder coating, and silk screening. After each coating is applied, the part is fired in a furnace to achieve a fusion between the enamel coating and the base metal or substrate. Each of these application methods are discussed below.

- *Air spraying* is the most widely used method of enamel application. In this process, enamel is atomized and propelled by air into a conical pattern, which can be directed by an operator or machine over the article to be coated.

  The major problem involved with air spraying is overspraying. Along with the expense of wasted enamel, this application method necessitates the installation of collection equipment, such as water wash spray booths. The booth generates process wastewater in the form of a curtain that captures particles carried into the booth by a stream of air. Oversprays can also be collected on a dry curtain, but this requires the use of a filter and fan system to prevent air contamination in the vicinity of the spray.
- *Electrostatic spray coating* incorporates the principles of air atomized spray coating with the atomized particles charged at

70,000 to 100,000 volts and directed toward the grounded part. Despite its increased efficiency over air spraying, spray booths are still necessary. Most of these use process water to capture enamel particles.

- *Dip coating* consists of submerging a part in a tank of enamel, withdrawing the part, and permitting it to drain to remove excess slip. Aside from water used for pretreatment rinses, dip coating requires no process water.
- *Flow coating* consists of pumping enamel slip from a storage tank to nozzles that are positioned according to the shape and size of the parts, so as to direct the flow of enamel onto the surface of the parts as the parts are conveyed past the nozzles. Excess enamel drains back to the storage tank for recirculation. No process water is used in flow coating.
- *Powder coating* is the application method used in cover coating cast iron. It is a dry process, thereby requiring no water.
- *Silk screening* is used by some companies to impart a decorative pattern onto a porcelain enameled piece. This is accomplished through the use of an oil-based porcelain enamel which is applied to the part through a silk stencil. Enamel is spread on the piece in a thin layer, using with a squeegee. After application, the workpiece is baked to achieve fusion of the enamel. Only one color can be applied and baked at a time.

The four subcategories of the porcelain enameling industry and the various combinations of processes which comprise these subcategories are described below.

Porcelain Enameling on Steel — Steel is by far the most widely used base metal for porcelain enameling. Parts to be coated are first cleaned with an alkaline solution, and rinsed to remove soils. This is followed by an acid treatment step and rinse, in which sulfuric acid, ferric sulfate in conjunction with sulfuric acid, or muriatic acid are used for oxide removal. Further preparation involves a nickel deposition step and rinse, followed by a neutralization operation to remove any remaining traces of acid.

Porcelain Enameling on Cast Iron — In this process, water is not generally used for metal preparation, but is sometimes used for coating application. The casting to be coated is blasted with sand or a combination of grit and sand to produce a smooth, velvety surface. The ground coat is then applied by spraying, dipping, or flow coating.

Porcelain Enameling on Aluminum — Porcelain enameling on aluminum is used in the cookware and housewares industry. Although all aluminum parts can be coated in a similar fashion, the surface preparation method can vary from company to company. Nearly all aluminum parts are first treated in an alkaline solution. In some cases, this is only a cleaner for removing grease and soils; sometimes it is a mild etchant to remove a layer of metal and its oxides. Frequently, this is all the surface preparation that is necessary. Any further preparation steps involve removing residual oxides (e.g., chemical deoxidizing with nitric acid), or imparting a thin protective layer on the metal (alkaline chromate treatment).

Porcelain Enameling on Copper — Porcelain enameling on copper is only used in a very small part of the porcelain enameling industry — mostly for ornamental purposes, such as jewelry, decorative ware, and metal sculpture. The part is first alkaline-cleaned, degreased, or annealed. After cleaning, the part is then typically pickled for oxide removal, and enamel coated.

Wastewater Characteristics

Table 59 shows raw wastewater characteristics. Principal wastewater sources are from metal preparation, enamel preparation and application, clean-up and dust control operations, wet scrubbers for air pollution control, and equipment cooling.

Control and Treatment Technology

In-Plant Control Technology

Water Reuse — Several porcelain enameling plants practice water reuse. Water which is employed for noncontact cooling or air conditioning can be reused for rinses in the base metal preparation line and as washdown water. One plant reported an overall water savings of 22 percent per year by recirculating acid pickling rinse waters to the alkaline cleaner rinses, and by using cooling water from air compressors as makeup water for the acid pickle rinses.

One common method for reusing rinse water in the porcelain-on-aluminum category is by using a closed loop, deionized rinse water system. Some plants rinse their work pieces in a deionized water final rinse and recirculate rinse water through an ion exchange unit, to remove the impurities picked up in the rinse process. Purified water is then returned to the rinse tank for further process work.

Process Materials Conversion — During the nickel deposition process, nickel ions come out of the solution and iron ions go into solution. It is good practice from a process standpoint to filter the nickel bath to prevent iron concentrations from building up to a level of contamination. Several types of filters are available for this purpose, including filter leaf, filter bag, flat bed filter, and string wound "cartridge" type filters. Filter installation extends the life of the process solution and decreases the frequency of discharges. In some cases, bath life can be increased as much as six months to one year. This represents a direct decrease in the pollutant load on the waste treatment system from the nickel deposition process. A similar filtration scheme can be used on neutralizer baths.

Table 59: Porcelain Enameling—Raw Wastewater Characteristics

Pollutants	Subcategory[a]							
	Steel		Cast iron		Aluminum		Copper	
	mg/l	mg/m^2 [b]	mg/l	mg/m^2	mg/l	mg/m^2	mg/l	mg/m^2
Aluminum	0.5-208	8-4,860	0.3-1,220	0-364	0.08-10	6-550	8-196	243-822
Antimony	0-22	2-1,036	6	1.3	0.15-0.26	7-9.3	0.12-2.35	4-9.8
Arsenic	0-2.5	0-59	1.8-2.8	0.4-0.8			0.42	1.8
Beryllium	0.001-0.008	0.041-0.10	0.002-0.120	0.001-0.151			0.005-0.035	0.147-0.153
Cadmium	0-0.6	0.011-6.2	0.01-9.57	0.005-12	0.007-5.2	0.1-422	0.008-0.220	0.234-0.923
Chromium (total)	0.02-0.8	0.1-39.2	0-1	0-1.4	0.001-0.013	0.015-0.694	0.023-0.63	0.704-2.640
Cobalt	0.2-9.2	0.7-251	0.4-95	0-119	0.006	0.443	4-64	123-269
Copper	0.03-2.2	0-51	0.001-8.750	0-11	0-0.14	0.005-6	7-11.5	30-371
Cyanide (total)			0.009	0.002	0.005-0.141	0.194-10.90	0.004	0.143
Fluoride	1.3-30	5-584	2-115	0.6-144	0.7-1.0	16-85	4-56	122-234
Iron	52-666	481-6,235	18-150	5-188	0.02-0.71	0.357-32	1.425-48.3	45.927-1,450
Lead	0-0.7	0-8.2	0.5-877	0-192	0-12.3	4-650	0.2-4.8	5.6-20.2
Manganese	0.5-61	3-1,410	0.003-65	0-82	0.002-0.131	0.09-4.5	5-118	166-495
Mercury			0.001	0				
Molybdenum			0.037	0.013				
Nickel	1.4-32	12-372	0-67,000	0-84				
Oil and grease	3.8-37.9	23-534	1-9.5	0.3-2.3	1.7-10.8	60-315	2-188	8-5,650
pH (units)	2-12.5		7.9-11.4		6.3-10.4		6.2-10.1	
Phenols	0.006-0.294	0.05-2	0.008-0.038	0.002-0.031	0-0.015	0.008-0.064	0.006	0.178
Phosphorus	3.7-13.3	73-282	0.5-2	0.2-1.9	0.9-24	37-1,050	0.08	2.6
Selenium	0.001-13	0.001-276	0.4-161	0.5-38	0.1-0.63	8-55	0.04-0.81	1.3-3.4
Temperature (°C)	26-104		20.6-24		18-33.5		19-25	
Tin		2.0	0.033	0.011				
Titanium	0-1,190	6-27,800	0-102	0-128	0.1-6	2.2-469	10-555	311-2,330
Total suspended solids	65-15,800	115-36,700	6,600-81,000	4,080-35,000	12-192	1,070-11,600	1,130-9,390	34,500-394,000
Zinc	0.8-44	0-1,040	0.7-645	0.2-810	0.12-0.53	2.5-45	5-196	155-822

[a] Values represent the range of results of analyses from sample estimates, and have been rounded.

[b] mg/m^2 = milligrams of pollutant generated per square meter of surface prepared and coated.

Replacement of wet spray booths with dry spray booths eliminates the possibility of oversprayed enamel entering the wastewater stream. Enamel overspray is allowed to dry on the floor, then is collected and reused. This type of spray booth is efficient only when color changes are infrequent, since dry enamel that is not cleaned up is easily blown around and could contaminate slip of other colors.

Reclamation of Waste Enamel — In those plants where one color is consistently used, enamel slip can be recovered and reused.

Process Modifications — Process modifications can reduce the amount of water required for rinsing, and in some cases, even eliminate waste loads from this category. For example, a number of plants have omitted the nickel deposition step for enameled surfaces hidden from view, such as the inside of water heater tanks. Another plant reported finding a new basis material preparation process called NPNN (No-Pickle, No-Nickel). This process consists of seven steps: (1) solvent clean, (2) detergent clean, (3) gold rinse, (4) acid clean (50 percent phosphoric acid), (5) acid clean (30 percent cleaner, 70 percent phosphoric acid), (6) cold rinse, (7) neutralizer (soda ash and borax). After this treatment, enamel is applied using normal application techniques.

Another process line modification may involve replacing a wet process with a dry one. For example, dry surface blasting can sometimes replace chemical cleaning, a wet process. This can only be employed with certain types of steel, since highly abrasive blasting may damage light gauge steel. Another water-saving process modification involves the method of enamel application. Electrostatic spray coating achieves the same results as normal spray coating, but at a much higher coverage efficiency. Consequently, electrostatic spray coating has much less overspray to be caught in a water curtain, and thus generates a fraction of the waste load of normal spray coating. Work is also being done in Europe using electrostatic dry powder application, a system which generates no wastewater.

Changes in production schedule can also directly or indirectly lighten the load on a wastewater treatment system. Decreasing the storage periods of the raw basis material reduces the degree of corrosion; consequently, preparation requirements are reduced. Another consideration is the timing of batch dumps. If an alkaline bath can be dumped safely with an acid bath, it reduces the consumption of treatment chemicals needed when baths are dumped separately. Holding tanks can be installed to facilitate this dumping method.

Material Substitutions — Substitution of nontoxic, or easily treatable, materials for toxic materials is another method of easing the load on, and increasing the effectiveness of an end-of-pipe treatment system. Acid in the pickling process might involve replacement of sulfuric acid with hydrochloric acid. Although sulfuric acid is cheaper to purchase, hydrochloric acid is easier to regenerate. This substitution has the added bonus of reducing the affect of iron concentration on pickling baths. However, acid regeneration done on a small scale has been shown to be economically not feasible. A final material substitution may involve the use of phosphate-free and biodegradable cleaners, which reduces the organic waste load on the wastewater treatment systems.

Rinse Techniques — Reductions in the amount of water used in porcelain enameling can be realized through installation and use of efficient rinse techniques. Five basic modes of rinsing that are potentially applicable to porcelain enameling are:

1. *Single Running Rinse* - This arrangement requires a large volume of water to effect a large degree of contaminant removal. Although in widespread use, single running rinse tanks should be modified or replaced by a more effective rinsing arrangement to reduce water use. A countercurrent rinse or a "dead rinse" followed by a single running rinse is more efficient than a single running rinse used alone. The dead rinse can potentially be returned to the process tank for reuse.
2. *Countercurrent Rinse* - For a countercurrent rinse system, there is only one freshwater feed for a series of tanks. Feed water is introduced in the last tank of the arrangement; overflow from each tank becomes the feed for the tank preceding it. Consequently, the concentration of contaminants decreases rapidly from the first to the last tank.
3. *Series Rinse* - In a series rinse system, each tank reaches its own equilibrium condition, the first rinse having the highest pollutant concentration, the last rinse having the lowest concentration. The major advantage of the series rinse over the countercurrent system is that water and heat levels in the tanks can be individually controlled, since each has a separate feed.
4. *Spray Rinse* - Spray rinsing is well suited for flat sheets. The impact of the spray also provides an effective mechanism for removing dragout from recesses having a large width-to-depth ratio.
5. *Dead, Still, or Reclaim Rinses* - This form of rinsing is particuarly applicable for initial rinsing after nickel deposition, since the high nickel concentration in the dead rinse allows easier metal recovery using reverse osmosis.

Porcelain enameling plants should review their rinse techniques to consider the efficiencies of substituting other rinse methods. Once a rinse system is selected, the rinse water feed rate must be controlled to minimize water use. Control devices include fixed orifice, conductivity controllers, or manually operated valves.

Good Housekeeping Practices — Good housekeeping and proper maintenance of coating equipment are required to reduce wastewater loads to the treatment systems. Ball milling and enamel application areas need constant attention in order to maintain cleanliness and to avoid the waste of cleanup water. Hoses should be shut off when not in use. Pressure nozzles can be installed on hoses to increase cleaning effectiveness and reduce water use.

Periodic inspection of the basis material preparation tanks and tank liners will reduce the chance of a catastrophic failure which could cause slug load in the waste discharge. Periodic inspections should also be performed on all auxiliary porcelain enameling equipment. This includes inspections for leaks in pumps, filters, process piping, and immersion steam heating coils.

Neutralizer and nickel filter cleaning should be done in curbed areas, or in a manner such that solution retained by the filter is dumped to the appropriate waste stream.

Good housekeeping practices also are applicable in chemical storage areas. Storage areas should be isolated from high hazard fire areas, and arranged so that if a fire or explosion occurs in such areas, loss of the stored chemicals due to deluge quantities of water would not overwhelm the treatment facilities or cause excessive groundwater pollution. Good housekeeping practices also include the use of drain boards between processing tanks. Bridging the gap between adjacent tanks by using drain boards allows for recovery of dragout, which drips off the parts while they are being transferred from one tank to another. The board should be mounted so that dragout drains back into the tank from which it originated.

End-of-Pipe Treatment Technology: The following treatment processes can be used for treating porcelain enameling wastewaters: ion exchange, ultrafiltration, dissolved air flotation, electrochemical chromium reduction, chemical precipitation, peat adsorption, membrane filtration, and clarification. Table 60 shows the efficiencies of these technologies in removing pollutants generated in the porcelain enameling industry.

Type of Residue Generated

Information on the types and characteristics of residues generated by this industry is insufficient and cannot be discussed at this time.

Residual Management Options

Lack of sufficient information on types and characteristics of residues generated does not permit a discussion of specific management options at this time.

PRINTING AND PUBLISHING

Sufficient data is not available on any aspect of this categorical industry.

Table 60: Porcelain Enameling—Percent Removal Efficiencies of Various Treatment Technologies

Wastewater parameter	Ion exchange[a]	Ultrafiltration[b]	Dissolved air flotation[c]	Electro-chemical chromium reduction[d]	Chemical precipitation[e]	Peat adsorption[f]	Membrane filtration[g]	Clarification[h]
Aluminum	96							80-100
Antimony								89-100
Arsenic					98			100
Cadmium	100				94			66-100
Chromium (+3)	99.6				99.9			
Chromium (+6)	99.9			99.5		99.9	97.8-99.9	
Chromium (total)							99.6-99.9	32-100
Cobalt								88-100
COD		98.3						
Copper	98-99.8				99.8	99.9	97-99.8	79-100
Cyanide	97.3-99.6							
Gold	95.6					98		
Iron	99.9				99		98.8-99.8	66-100
Lead	99.4				96	98.8	96.5-98.5	92-100
Manganese	100							87-100
Mercury					99.9	98		
Nickel	99.3-100				99.6	97.2	99.8	25-100
Oil and grease		99.7	72-100					
Selenium								97.8-100
Silver	99.9-100				99.6	95		
Sulfate	99							
Tin	91-100							
Titanium								90-100
TSS		99					38-99.9	92
Zinc	97			96.7	99.6	83	97.8-98.9	83-99.7

[a] Range of test results from electroplating and printed circuit plant effluent.
[b] Bench scale tests.
[c] Typical data.
[d] Plant data.
[e] Pilot plant data.
[f] Effluent analysis from two porcelain enameling plants.
[g] Effluent analysis from five porcelain enameling plants.

PULP, PAPER AND PAPERBOARD

General Industry Description

Paper is made from raw materials, including: wood pulp, cotton, linen rags, and straw. These materials all contain adequate amounts of the basic component, cellulose fiber. Today, wood accounts for over 98 percent of the virgin fiber used in paper making.

The pulping process separates cellulose fibers from a pulp. Commonly, pulps are derived from wood chips, recycled paper, rags, etc. The fibers are then deposited from a dilute suspension of the pulp onto a screen. Water passes through the screen, leaving layers of fibers to be pressed and dried.

Approximately 65 million tons of paper pulp were produced in the United States in 1977. Wastewater generated by the process varied from 1,800 gallons per ton to 70,000 gallons per ton of paper product, depending upon raw material, end product, plant process, and water conservation and reuse practices. Wastewater from pulp-paper plants contains conventional as well as toxic pollutants. Approximately 230 of the 730 pulp and paper mills in the United States discharge to POTWs.

Industrial Categorization

For the purpose of establishing pretreatment guidelines, the pulp, paper and paperboard industry has been subcategorized as shown in Table 61.

Process Description

A general process flow diagram for the pulp, paper and paperboard industry is shown in Figure 38. A brief description of the various processes follows.

Wood Preparation: This process removes bark from the logs by washing and by machinery (drum, pocket, or hydraulic debarkers). The process generates large quantities of wastewater with high BOD and suspended solids concentrations. Once debarked and cleaned, the logs are chipped and sent to the pulping process.

Pulping: Pulping involves chemical or mechanical processes, or a combination of the two, to release cellulose fibers into a dilute suspension. The fiber suspension then enters the papermaking stage in a concentration of approximately 0.25 to 0.50 percent. Many of the characteristics of the end product are determined by the pulping process. For example, mechanical pulping does not remove the natural wood binder (lignin) and resin acids. Lignin and resin can cause paper to deteriorate rapidly by yellowing and weakening of the oxidized material.

Most pulp mills recover and reuse chemicals and waste liquors from each stage of the pulping process. There is potential for sudden discharge of these chemicals and waste liquors when an upset in the process, such as shutdown or startup, causes flows to exceed recovery and available storage capacity. Mills accommodate process upsets and imbalances by placing sewage outlets at all potential overflow points.

Bleaching: After pulping, unbleached pulp is brown or deeply colored, due to lignins, resins, or inefficient washing of the cooking liquor from the pulp. Plants achieve desired brightness by various combinations of washing, chlorination, alkaline extraction, chlorine dioxide addition, and hypochlorite application which are used before sending the pulp to the papermaking process. Bleaching is often a major contributor of plant effluent quantity.

Stock Preparation: In the stock preparation area, pulps are blended with materials such as alum and rosin for sizing the paper sheets. Fillers such as clay can be added to give improved brightness, smoothness and opacity. Dyes are added for color and shade control. Waste discharges from the stock preparation area are usually minimal, and may occur from washups, order changes, shutdowns, and other upsets to the normal production process.

Paper Making: Suspended cellulose fibers are deposited onto a cylinder or wire screen, which allows most of the water to pass through. The fiber layer is then pressed, dried, and treated for desired surface characteristics. It is possible in this process to recover stock and finishing chemicals, thus leaving very little waste discharge. Again, process imbalance and upsets may cause unexpected wastewater discharges.

Table 61: Pulp, Paper and Paperboard Industrial Subcategorization

Subcategory	Designation
Integrated Mills	A
Alkaline-Dissolving	011
Alkaline-Market	012
Alkaline-BCT (for paperboard, coarse and tissue)	013
Alkaline-Fine	014
Alkaline-Unbleached	015
Semi-Chemical	016
Alkaline-Unbleached and Semi-Chemical	017
Alkaline-Newsprint	019
Sulfite-Dissolving	021
Sulfite-Papergrade	022
Thermo-Mechanical Pulp (TMP)	032
Groundwood-CMN	033
Groundwood-Fine	034
Secondary Fiber Mills	B
Deink-Fine and Tissue	101
Deink-Newsprint	102
Wastepaper-Tissue	111
Wastepaper-Board	112
Wastepaper-Molded Products	113
Wastepaper-Construction Products	114
Nonintegrated Mills	C
Nonintegrated-Fine	201
Nonintegrated-Tissue	202
Nonintegrated-Lightweight	204
Nonintegrated-Filter and Nonwoven	205
Nonintegrated-Paperboard	211
Miscellaneous Mill Groupings	D
Integrated-Miscellaneous, including	
Alkaline-Miscellaneous	
Groundwood Chemi-Mechanical	
Nonwood Pulping	
Secondary Fiber-Miscellaneous	
Nonintegrated-Miscellaneous	

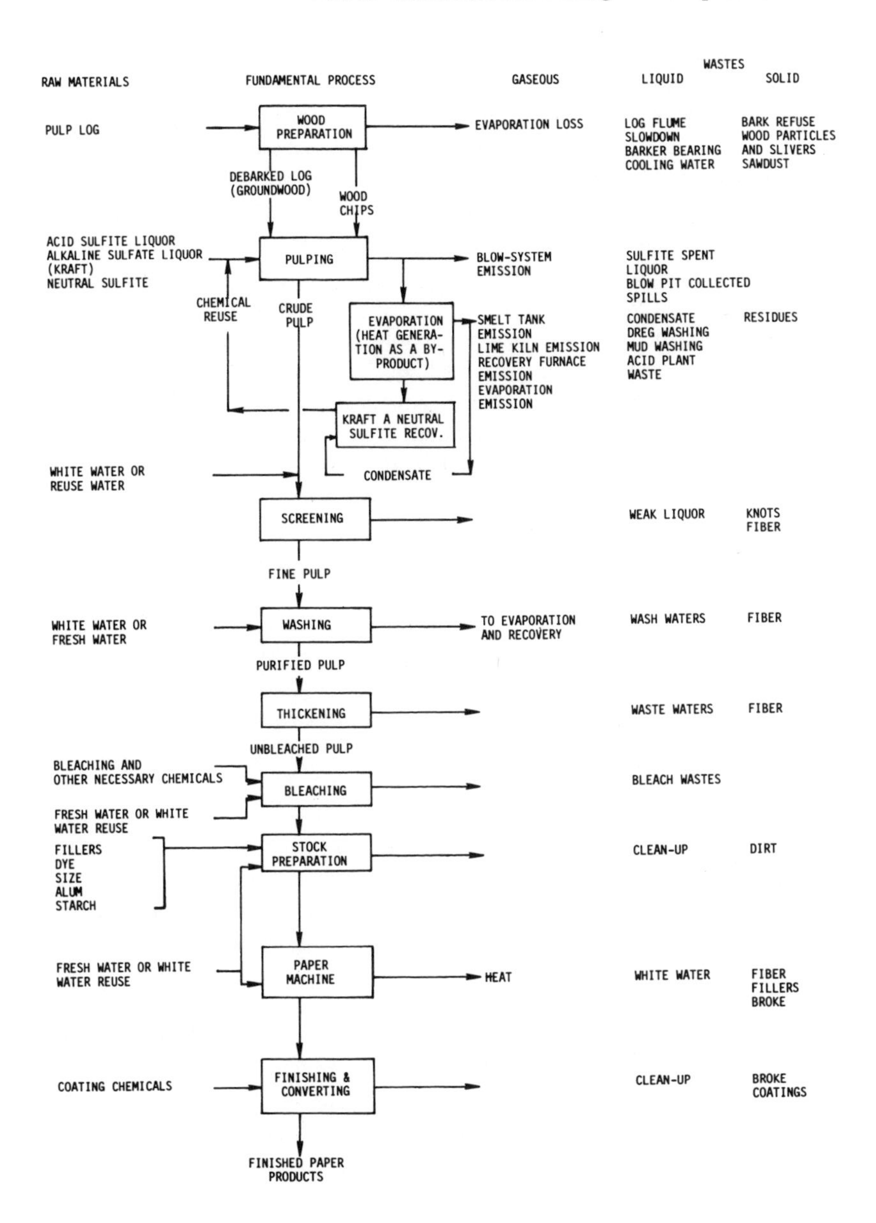

Figure 38: Pulp, paper, and paperboard industry–general flow diagram.

De-inking: Reclamation of waste paper involves removal of ink, fillers, coatings, and other noncellulose materials. De-inking processes produce a wastewater high in BOD_5 and total suspended solids concentrations.

A brief description of specific processes and products associated with each subcategory is presented below.

Integrated Mills

011 Alkaline-Dissolving. At these mills, a highly bleached wood pulp is produced in a full cook process, using a sodium hydroxide and sodium sulfide cooking liquor and a pre-cook operation called "pre-hydrolysis". The principal product is a highly purified, dissolving pulp used mostly for the manufacture of rayon and other products which require the virtual absence of lignin and a very high alpha cellulose content.

012 Alkaline-Market. At mills in this subcategory, a bleached, paper-grade market wood pulp is produced in a full cook process, using a highly alkaline sodium hydroxide cooking liquor. Sodium sulfide is also usually present in varying amounts in the cooking liquor.

013 Alkaline-BCT. At these mills, bleached alkaline pulp is produced and manufactured into paperboard, coarse, and tissue (BCT) grades of paper. Bleached alkaline pulp is produced by a process similar to that used in the Alkaline-Market subcategory.

014 Alkaline-Fine. These mills produce a bleached, alkaline pulp which is manufactured into fine papers, including business, writing, and printing papers.

015 Alkaline-Unbleached. An unbleached wood pulp is produced at these mills, using a full cook process with a highly alkaline sodium hydroxide cooking liquor. Sodium sulfide is also usually present in the cooking liquor in varying amounts. Products are coarse papers, paperboard, and may include market pulp, unbleached kraft specialties, towels, corrugating medium, and tube stock.

016 Semi-Chemical. At semi-chemical mills, a high-yield wood pulp is produced and manufactured into corrugating medium, insulating board, partition board, chip board, tube stock, and specialty boards. A variety of cooking liquors are used to cook the wood chips under pressure. The cooked chips are usually refined before being converted into board or similar products.

017 Alkaline-Unbleached and Semi-Chemical. At mills in this subcategory, high-yield, semi-chemical pulp and unbleached kraft pulp are produced. Cooking liquors from both processes are recovered in the same recovery furnace. Major products in this subcategory include linerboard, corrugating medium, and market pulp.

019 Alkaline-Newsprint. Bleached, alkaline pulp and groundwood pulp are produced in these mills. Newsprint is the principal product.

021 Sulfite-Dissolving. At mills in this subcategory, sulfite pulp and paper or papergrade market pulp are produced. Sulfite wood pulp is produced by a full-cook process, using strong solutions of calcium, magnesium, ammonia or sodium bisulfite, and sulfur dioxide. Purchased groundwood, secondary fibers, or virgin pulp are commonly used in addition to sulfite pulp to produce tissue paper, fine paper, newsprint, market pulp, chip board, glassine, wax paper, and sulfite specialties.

032 Thermo-Mechanical Pulp (TMP). Wood pulp is produced at mills in this subcategory in a process using rapid steaming followed by refining. A cooking liquor, such as sodium sulfite, is added. Principal products are fine paper, newsprint, and tissue papers.
033 Groundwood–CMN. At these mills, groundwood pulp is produced using stone grinders or refiners; no separate steaming vessel is used before the defibration. Purchased fibers are used in addition to groundwood pulp to produce coarse papers, molded fiber products, and newsprint (CMN).
034 Groundwood–Fine. At mills in this subcategory, groundwood pulp is produced using stone grinders or refiners; no separate steaming vessel is used before the defibration. Purchased fibers are used in addition to groundwood pulp to produce fine papers, including business, writing, and printing papers.
Secondary Fiber Mills: No pulp is produced at secondary fiber mills; most of the material furnished is waste paper. Some secondary fiber mills include de-inking to produce a pulp, paper, or paperboard product.
101 De-ink–Fine and Tissue. At mills in this subcategory, a de-ink pulp is produced from waste paper. Principal products made from the de-inked pulp include printing, writing, business, and tissue papers, but may also include products such as wallpaper, converting stock, and wadding.
102 De-ink–Newsprint. Mills in this subcategory produce newsprint from de-ink pulp, derived mostly from over-issue and waste newsprint papers.
111 Wastepaper–Tissue. Mills in this subcategory use paper stock derived from de-inked waste paper. Principal products are facial and toilet paper, paper towels, glassine, paper diapers, and wadding.
112 Wastepaper–Board. Mills in this subcategory use a paper stock derived from waste paper without de-inking. A wide range of products are made, including set-up and folding boxboards, corrugating medium, tube stock, chip board, gypsum liner, and linerboard. Other board products include fiber and partition board, building board, shoe board, bogus, blotting, cover, auto filter, gasket, tag, liner, electrical board, fiber pipe, food board, wrapper, and specialty boards.
113 Wastepaper–Molded Products. At these mills, most of the furnish (paper stock) is obtained from waste paper without de-inking. Principal products are molded items, such as fruit and vegetable packs, and similar throw-away containers and display items.
114 Wastepaper–Construction Products. Included in this subcategory are mills primarily producing saturated and coated building paper and boards. Waste paper is the furnish; no de-inking is employed. The principal products include roofing felt, shingles, and rolled and pre-

pared roofing. Asphalt may be used for saturating, and various mineral coatings may also be used. Some asbestos and nonwood fibers (fiberglass) may also be used. At many mills some groundwood, defibrated pulp, or wood flour may be processed and used in production of the final product.

Nonintegrated Mills: Nonintegrated mills purchase wood pulp or other fiber source(s) to produce paper or paperboard products.

201 Nonintegrated–Fine. These nonintegrated mills produce fine papers from wood pulp or secondary fibers, prepared at another site. No de-inking is employed at the papermill site. Principal products are printing, writing, and business papers, bleached bristols, and rag papers.

202 Nonintegrated–Tissue. Mills in this subcategory produce sanitary or industrial tissue papers from wood pulp or secondary fiber prepared at another site. No pulp is prepared at the papermill site. Principal products are facial and toilet paper, paper towels, glassine, paper diapers, wadding, and wrapping.

204 Nonintegrated–Lightweight. These mills produce lightweight or thin papers from wood pulp or secondary fiber prepared at another site, as well as from nonwood fibers and additives. Principal products are uncoated, thin papers (such as carbonizing, cigarette papers), and some special grades of tissue (such as capacitor, pattern, and interleaf).

205 Nonintegrated–Filter and Nonwoven. Mills in this subcategory produce filter papers and nonwoven items using a furnish of purchased wood pulp, waste paper, and nonwood fibers. Principal products are filter and blotting paper, nonwoven packaging and specialty papers, insulation, and gaskets.

211 Nonintegrated–Paperboard. Mills in this subcategory produce various types of paperboard from purchased wood pulps or secondary fibers. Products include linerboard, folding boxboard, milk cartons, food, chip, stereotype, pressboard, electrical, and other specialty board grades.

Miscellaneous Mill Groupings

Integrated–Miscellaneous. This mill grouping includes three types of miscellaneous mills: (1) mills employing more than one pulping process (exceptions are the Alkaline–Newsprint and Alkaline–Unbleached and Semi-Chemical subcategories); (2) miscellaneous processes not described above (i.e., nonwood pulping, chemi-mechanical, miscellaneous acid, and alkaline pulping mills); and (3) mills producing a wide variety of products not covered above.

Secondary Fiber–Miscellaneous. These mills manufacture products or product mixes not included in the Wastepaper–Tissue, Wastepaper–Board, Wastepaper–Molded Products, and Wastepaper–Construction Products subcategories. Their furnish is more than 50 percent wastepaper without de-inking.

Products in this category may include market pulp from wastepaper and polycoated wastepaper, filters, gaskets, mats, absorbent papers, groundwood specialties, and other grade mixtures. A mill producing less than 50 percent construction paper or any other combination of products, other than secondary fiber subcategory products, would be classified in this grouping.

Nonintegrated–Miscellaneous. This grouping includes any nonintegrated mill not included in the above subcategories. Included are mills making mostly asbestos and synthetic products; paper and paperboard products that are too diverse to be classified; or products with unique process or product specifications, commonly called specialty items.

Wastewater Characteristics

Table 62 shows raw wastewater characteristics for the major subcategories of the pulp, paper, and paperboard industry. Table 63 shows raw wastewater characteristics as a function of product production.

Control and Treatment Technology

In-Plant Control Technology: Woodyard and debarking operations can be modified to achieve zero discharge. A reduction of pulp mill wastewater is possible through the reuse of condensates, reductions in groundwood thickener overflow, and more careful spill collection. Washers and screen rooms will benefit from the use of additional washing units, decker filtrate recycle, and more efficient vibrating screens.

Bleaching operations are sensitive to changes in rinse efficiencies, while evaporation and recovery operations benefit most from condensate recycle, condensor upgrading, spent liquor neutralization, stream segregation, and stringent spill collection.

A number of control measures can be used in the paper mill area. These include: design improvements, the use of high pressure, efficient showers, the recycle of white water and vacuum pump water, and the segregation of cooling streams. Increased overflow storage and equalization can also reduce impacts on downstream treatment units.

Other process controls which improve effluent quality are oxygen bleaching, sequential chlorination, the Rapson-Reeve Closed Cycle Process (used for recovery of sodium chloride), no-sulfur pulping, and displacement bleaching.

End-of-Pipe Treatment Technology: Table 64 shows the removal efficiencies of various treatment technologies based on operating data obtained from pilot or full-scale treatment operations in the pulp, paper and paperboard industry.

Table 62: Pulp, Paper, and Paperboard Industry—Raw Wastewater Characteristics

	Subcategories[a]												
	A												
Wastewater parameter	011	012	013	014	015	016	017	019	021	022	032	033	034
Conventional - mg/l													
BOD_5	295	229	300	266	268-304	399-569	334	225	631	319-336	305	171-210	168-257
Ammonia									105				
Chemical oxygen demand		335	766	576	948	2,410	895			4,794		212	625
Color (platinum cobalt units)		1,680	1,233	850	811	3,915	425			2,013		25	139
Total suspended solids	437	294	294	494	294-379	665-795	420	605	376	170-217	645	492-578	498-788
Inorganics - ug/l													
Chromium	12	85	26	14	29	29				13		13	9
Copper	31	46	22	19	79	38				81		15	28
Cyanide					9	16						9	
Lead	9	17	6	14	95	24				13		13	9
Mercury	<1	<1	<1	<1	<1	<1				<1		<1	<1
Nickel	31	36	16	6	12	10				16		8	5
Zinc	154	138	149	114	143	40				91		483	74
Organic - ug/l													
Abietic acid		583	1,043	191	2,026	124	1,392			135		223	182
Anthracene		<1											
Benzene	<1	<1		<1	3	1				53		9	
Bis (2-ethylhexyl) phthalate	11	3	28	18	21	10				38		8	3
Bromoform													
Butylbenzylphthalate				8	<1								
Chlorobenzene													
Chlorodibromomethane													
Chloroform	1,222	1,550	781	<1	1	1				3,211		<1	99
2-Chlorphenol													
Chrysene													
Dehydroabietic acid		224	739	181	741	168	607			555		427	148
Dichlorobromomethane			4							9			
Dichlorodehydroabietic acid		29	2	4	<1					<1			
Diethyl phthalate	<1		<1			7				<1			
Di-n-butyl phthalate	3	2	<1	2	4	5				<1			<1
Di-n-octyl phthalate													
1,1-Dichloroethane										4			
1,2-Dichloroethane													
2,4-Dichlorophenol	4	1	2							<1			
9,10-Epoxystearic acid							133			49			
Ethylbenzene	14	<1		<1	<1								<1
Isophorone				4									

(continued)

Table 62: (continued)

Wastewater parameter	Subcategories[a] A: 011	012	013	014	015	016	017	019	021	022	032	033	034
Isopimaric acid		58	96	48	325	34	547			62		14	29
Linoleic acid		698	508	94	453	61	441			57		16	168
Linolenic acid		35		10	169					12			125
Methylene chloride	<1	2	2	34	6	58				464			2
Monochlorodehydroabietic acid		50	52	41						82			
Napthalane					2					34			
Oleic acid		298	1,084	175	1,070	115	618			168		74	171
Parachlorometal cresol													
PCB 1242													
PCB 1248													
PCB 1254						<1							
PCB 1260													
Pentachlorophenol		6	3		<1	1				4			3
Phenol	70	55	7	85	230	56				53		16	28
Pimaric acid		78	115	40	323	27	152			8			50
Tetrachloroethylene		<1	<1	<1									<1
Tetrachloroguaiacol		11	5	6						1			
Toluene	1	1	23	4	2	7				15		293	13
1,1,1-Trichloroethane			8			3				414			
Trichloroethylene		<1			5	<1				5			
Trichlorofluoromethane													
3,4,5-Trichloroguaiacol		9	<1	2						<1			
2,4,6-Trichlorophenol	11	8	11							4			
Xylenes					14	<1	11			<1		4	

Wastewater parameter	Subcategories[a] B: 101	102	111	112	113	114	C: 201	202	204	205	211
Conventional - mg/l											
BOD_5	466-599	235	223[c]	319-1,262	521-1,112		176	181	29-57	29	98
Ammonia											
Chemical oxygen demand	1,366	3,733	363	8,833	291	3,487	433	395	313	240	185
Color (platinum cobalt units)	320	88	960	121							
Total suspended solids	1,759-2,012	1,820	714	387-1,359	217	816-1,329	622	531	93-171	146	412

(continued)

Table 62: (continued)

	Subcategories[a]										
	B						C				
Wastewater parameter	101	102	111	112	113	114	201	202	204	205	211
Inorganics - ug/l											
Chromium	22	29	20	170	9	81	5	15	2	6	1,323
Copper	34	76	55	107	16	145	43	45	19	61	128
Cyanide	68		9	74	9	352			9	11	610
Lead	61	163	44	153	22	264	5	11	6	4	6,667
Mercury	<1	<1	<1	<1	<1	<1	<1	<1	<1	<1	<1
Nickel	8	15	21	37	23	40	5	1	1	2	18
Zinc	149	335	492	1,433	392	998	71	92	16	159	1,273
Organic - ug/l											
Abietic acid	636	3,467	54	407	210	4,179	205	53			1,477
Anthracene											
Benzene	2		<1	<1	ND[b]	<1	<1				<1
Bis (2-ethylhexyl) phthalate	7	13	10	23	2	30	1,193	8	5	85	14
Bromoform				40							
Butylbenzylphthalate		5		80		3		797			
Chlorobenzene	14										
Chlorodibromomethane						<1					
Chloroform	1,772	< 1	2	19		3	7	3	27		
2-Chlorphenol	< 1										
Chrysene											
Dehydroabietic acid	2,173	3,700	372	467	453	899	483	213		33	667
Dichlorobromomethane				<1		2					
Dichlorodehydroabietic acid	2										
Diethyl phthalate	1	1	13	79		29		12			4
Di-n-butyl phthalate	5	<1	3	32		16		<1	<1		180
Di-n-octyl phthalate											
1,1-Dichloroethane											
1,2-Dichloroethane	<1					<1					
2,4-Dichlorophenol	2										
9,10-Epoxystearic acid			413	10							
Ethylbenzene	11	2	13			1		13,081		<1	3
Isophorone											
Isopimaric acid	295	510	16	84	48	958	40	37			117
Linoleic acid	153	750	42	207	850	33					
Linolenic acid	40		23								
Methylene chloride	4	<1	87	<1	<1	<1	7		<1		<1
Monochlorodehydroabietic acid	126										
Napthalane	42		26								
Oleic acid	549	1,367	183	290	473	1,307	19	260			260
Parachlorometal cresol											
PCB 1242	1					< 1					

(continued)

Table 62: (continued)

Wastewater parameter	Subcategories[a]										
	B						C				
	101	102	111	112	113	114	201	202	204	205	211
PCB 1248											
PCB 1254			< 1	< 1		< 1				15	
PCB 1260	1										
Pentachlorophenol	18			1,050	2	35					
Phenol	38	1	41	457	8	1,233	94	1	2	65	7
Pimaric acid	69	257	3	41	57	471	12	10			25
Tetrachloroethylene	32		74	< 1		< 1		4			3
Tetrachloroguaiacol	5										
Toluene	25	14	2	4		81	< 1	130	2	2	3
1,1,1-Trichloroethane	7			2		6					
Trichloroethylene	168			1		7					
Trichlorofluoromethane						< 1					
3,4,5-Trichloroguaiacol	3										
2,4,6-Trichlorophenol											
Xylenes	4	46	28	< 1	5		13,547	5			8

Wastewater parameter	Subcategories[a]			
	D			
	Groundwood chemi-mechanical	Nonwood pulping	Secondary fiber-miscellaneous	Nonintegrated miscellaneous
Conventional - mg/l				
BOD_5				
Ammonia				
Chemical oxygen demand	567	1,300		283
Color (platinum cobalt units)				
Total suspended solids				
Inorganics - ug/l				
Chromium	3	5		13
Copper	40	39		46
Cyanide	13	9		9
Lead	2	17		14
Mercury	< 1	< 1		< 1
Nickel	3	5		20
Zinc	403	75		543
Organic - ug/l				
Abietic acid	2,700	82		59
Anthracene				
Benzene				

(continued)

Table 62: (continued)

Wastewater parameter	Subcategories[a]			
	D			
	Groundwood chemi-mechanical	Nonwood pulping	Secondary fiber-miscellaneous	Nonintegrated miscellaneous
Bis (2-ethylhexyl) phthalate	7	8		26
Bromoform				
Butylbenzylphthalate				
Chlorobenzene				
Chlorodibromomethane				
Chloroform		417		3
Chrysene				
Dehydroabietic acid	1,400	249		121
Dichlorobromomethane				
Dichlorodehydroabietic acid		6		
Diethyl phthalate		2		
Di-n-butyl phthalate	3	< 1		
Di-n-octyl phthalate				
1,1-Dichloroethane				
1,2-Dichloroethane				
2,4-Dichlorophenol				
9,10-Epoxystearic acid				
Ethylbenzene	< 1			
Isophorone				
Isopimaric acid	1,020	16		28
Linoleic acid	307	274		9
Linolenic acid				
Methylene chloride	5	< 1		
Monochlorodehydroabietic acid	54	6		
Napthalane				
Oleic acid	1,280	220		
Parachlorometal cresol				
PCB 1242				
PCB 1248				
PCB 1254	< 1			1
PCB 1260				
Pentachlorophenol		12		24
Phenol	31	5		5
Pimaric acid	747	10		11
Tetrachloroethylene				
Tetrachloroguaiacol				
Toluene	3	< 1		< 1
1,1,1-Trichloroethane		33		
Trichloroethylene				
Trichlorofluoromethane				
3,4,5-Trichloroguaiacol				
2,4,6-Trichlorophenol				
Xylenes	57	4		3

[a] See Table 61 for identification of subcategories.

[b] ND indicates concentrations below detectable limits.

Table 63: Pulp, Paper and Paperboard Industry—Raw Wastewater Characteristics—Production-Based Data

Subcategory[a]	Designation	Wastewater flow/unit of product produced[b]	BOD_5[c]	TSS[c]
A. Integrated Mills				
Alkaline - BCT	013	36.5	91.3	83.0
Alkaline - Dissolving	011	47.5	107.6	153.0
Alkaline - Fine	014	26.5	61.0	132.3
Alkaline - Market	012	42.8	83.0	63.0
Alkaline - Newsprint	019	22.5	42.2	113.0
Alkaline - Unbleached	015	11.2	28.3	32.0
Alkaline - Unbleached and Semi-Chemical	017	13.4	37.3	47.0
Groundwood - CMN	033	21.2	37.1	97.0
Groundwood - Fine	034	16.4	35.2	107.0
Semi-Chemical	016	7.8	36.9	43.0
Sulfite - Dissolving	021	61.6	306.0	180.6
Sulfite - Papergrade	022	36.6	97.3	66.0
Thermo-Mechanical Pulp	032	14.4	36.5	77.0
B. Secondary Fiber Mills				
De-Ink - Fine and Tissue	101	19.5	97.4	286.0
De-Ink - Newsprint	102	16.2	31.7	246.0
Wastepaper - Board	112	3.7	12.9	15.0
Wastepaper - Construction Products	114	2.2	11.5	16.0
Wastepaper - Molded Products	113	11.3	11.4	21.0
Wastepaper - Tissue	111	9.4	17.5	54.0
C. Non-Integrated Mills				
Non-Integrated - Filter	205	41.2	10.0	50.0
Non-Integrated - Fine	201	11.6	17.0	60.0
Non-Integrated - Lightweight	204	63.9	30.6	91.0
Non-Integrated - Paperboard	211	24.6	20.0	84.5
Non-Integrated - Tissue	202	17.6	26.5	77.9

[a] Data not currently available for "Miscellaneous Mill Grouping" subcategory.

[b] In units of kilogallons of flow per ton of product produced.

[c] In units of pounds of pollutant per ton of product produced.

Table 64: Pulp, Paper, and Paperboard Industry—Removal Efficiencies of End-of-Pipe Treatment Technologies

Treatment technology	Mill type	Pollutant removal efficiencies - percent			
		BOD_5	Total suspended solids	Color	Resins and fatty acids
Primary					
Screening, sedimentation	Integrated	10-30	5-80		
Dissolved air flotation and mechanical	Non-integrated	10-50			
Secondary and Advanced					
Aerated stabilization basins	Bleached kraft	85			
	Unbleached kraft	87			96
Air activated sludge	Bleached kraft	75-90			84-90
Chemical flocculation and clarification	Bleached kraft	35	76-87	70-90	
Oxygen activated sludge	Sulfite newsprint	90	50		75
	Alkaline unbleached	91	25-70		
Rotating biological contactor	Bleached kraft	80			
	Board mill	77			
	Sulfite mill				
Aerobic contact filter	Sulfite lab	80-88			
Granular filtration	---		45-85		
Activated carbon adsorption	Unbleached kraft	85-99	36-99	100	80
	Chemical bleached Kraft	95		0-82	46-66
Microscreening	Non-integrated	67	97		
Electrochemical treatment	Bleached kraft	69			
	Caustic extraction	89		99	
Ion flotation				95	
Untrafiltration	Bleached kraft and Caustic extraction	30-50		20-50	73-93
Reverse osmosis	Neutral sulfite	90			0-10
Reverse osmosis/freeze concentration	Bleach plant		98		
Amine treatment (organ-ophillic precipitation)	Bleached kraft	10-74	36-78	90-99	

Type of Residue Generated

Table 65 summarizes sludge production data in each subcategory. In terms of residues generated from treatment of the wastewaters, the pulp and paperboard industry ranked third in the U.S., producing 1.93 million dry tons of sludge in 1977 and a projected 2.4 to 3.5 million dry tons in 1987. Further, the nonrecoverable residues generated by air pollution control scrubbers in one segment of this industry — kraft pulping — ranked the industry third in sludge production from air pollution controls, producing 4.7 million dry tons in 1977 and a projected amount of up to 8.8 million dry tons in 1987. The following discussion applies only to wastewater treatment residues; detailed information is not available to allow discussion of air pollution scrubber sludges.

Table 65: Pulp, Paper, and Paperboard Industry–Wastewater Sludge Production Summary [a]

Subcategory	Product production, tons per day	Estimated Solids Production 1,000 pounds per day, dry basis		
		Primary only	Primary plus biological	Chemical clarification
011 Alkaline-Dissolving	1,000	70	103	26
012 Alkaline-Market	600	24	38	10
013 Alkaline-BCT	800	44	63	12
014 Alkaline-Fine	800	56	72	12
015 Alkaline-Unbleached	1,000	19	27	5
016 Semi-Chemical	425	10	15	2
017 Alkaline-Unbleached and Semi-Chemical	1,500	41	59	9
019 Alkaline-Newsprint	1,400	87	112	17
021 Sulfite-Dissolving	600	76	121	18
022 Sulfite-Papergrade	450	21	32	6
032 Thermo-Mechanical Pulp	350	15	20	3
033 Groundwood-CMN	600	28	36	6
034 Groundwood-Fine	500	27	36	5
101 Deink-Fine and Tissue	180	37	44	5
111 Wastepaper-Tissue	45	1.6	2	0.3
112 Wastepaper-Board	160	0.4	0.5	0.1
113 Wastepaper-Molded Products	50	0.3	0.4	0.1
114 Wastepaper-Construction Products	350	0.2	0.3	0.1
201 Nonintegrated-Fine	215	6.4	7.5	1.3
202 Nonintegrated-Tissue	180	4.6	5.5	1.0
204 Nonintegrated-Lightweight	60	1.9	2.3	1.2
205 Nonintegrated-Filter and Nonwoven	20	0.4	0.4	0.4
211 Nonintegrated-Paperboard	40	1.6	1.8	0.4

[a] Applies to model mills.

Sludges from this industry are mostly nontoxic. They are composed principally of cellulose (wood) fibers, chips, knots, and other small particles; of biological sludges; and from inert clay and mineral pigments, coatings, and filler materials. Several organic materials that are toxic to aquatic life are present in the raw wastewaters, but these are effectively removed by biological wastewater treatment, and end up in the sludge. Heavy metals, principally chromium, zinc, and lead are associated only with wastewaters from a few types of mills, especially those using reclaimed fibers such as newspapers. These metals are associated with inks found in the reclaimed fibers, and will end up in the sludges. Sludges may also contain traces of the pulping chemicals, such as sodium, magnesium, calcium, or ammonium sulfide, sulfite, sulfate or hydroxide, and of lignin color-bodies from the original wood.

Residual Management Options

Because of their high fiber content, sludges from this industry are relatively easy to dewater. Primary sludges may have as much as 15 percent solids, and secondary sludges usually average from 2 percent to 8 percent solids. The main options used for disposal of sludges from this industry have been: landfilling, burning as fuel, and incineration. Other disposal techniques which have been used (some on an experimental basis), include: land spreading, both as a semi-solid cake and as a liquid, for agricultural and silvicultural purposes; animal feed supplements, used because of high protein or nitrogen and phosphorus content; and (after composting) for horticultural purposes.

Also closely related to the choice of final disposal techniques is the "ash" content of the sludge. "Ash" is an inert and noncombustible material used in the paper-making process as pigments, coatings, and fillers. Sludges with a high ash content — e.g. from mills making coated paper — are less suitable for burning as fuel or for incineration, and are more likely to be landfilled. Some techniques have been developed for recovering the coating or filler materials, cleaning them, and reusing them in the process. Although these techniques are desirable since they facilitate burning or incineration of the remainder of the sludge, as well as recover a reusable material (coating or filler), they have not yet been widely adopted.

As a last resort, sludges may be sent to a landfill. Disposal by landfilling is a problem in this industry because of: a) the large amount of material; b) its poor bearing capacity after placement; c) and its susceptability to leaching. The latter problem can be addressed with clay (or "ash") blankets above, below and around the landfilled sludge, and with collection and treatment of leachate from the landfill.

In an industry survey of 98 plants, it was found that 70 mills landfilled their sludge, 20 burned it as fuel along with other wastes, six incinerated it, and two used other methods of disposal.

RUBBER PROCESSING

General Industry Description

The rubber processing industry includes the manufacture of tires and inner tubes, synthetic and reclaimed rubber, and molded, extruded, fabricated and latex-based products. Each of the above segments of the industry uses different services and raw materials.

Industrial Categorization

The rubber processing industry is broadly subdivided into three main categories: tire and inner tube industry, synthetic rubber industry, and fabricated and reclaimed rubber industry. For the purposes of raw waste characterization and delineation of pretreatment information, the industry is further subdivided into eleven subcategories, as shown in Table 66.

Table 66: Subcategorization of Rubber Processing Industry

Main category	Subcategory	Designation
Tire and inner tube	Tire and inner tube	A
Synthetic rubber	Emulsion crumb rubber	B
	Solution crumb rubber	C
	Latex rubber	D
Fabricated and reclaimed rubber	Small-sized general molded, extruded, and fabricated rubber plants	E
	Medium-sized general molded, extruded, and fabricated rubber plants	F
	Large-sized general molded, extruded, and fabricated rubber plants	G
	Wet digestion reclaimed rubber	H
	Pan, dry digestion, and mechanical reclaimed rubber	I
	Latex-dipped, latex-extruded, and latex-molded rubber	J
	Latex foam	K

Process Description

Tire and Inner Tube (A): Tire manufacurers produce many types of tires designed for multiple uses. General product categories include passenger, truck, bus, farm tractor and implements, and aircraft tires. Basically, the tire consists of five parts, namely: tread, sidewall, cord, bead, and inner liner. Basic tire ingredients are synthetic rubbers, natural rubber, filler extenders and reinforcers, curing and accelerating agents, antioxidants and pigments. The typical tire manufacturing process consists of the following:

1. Preparation or compounding of the raw materials;
2. Transformation of compounded materials into the five tire components;
3. Building, molding, and curing of the final product.

The flow diagram for a typical tire plant is shown in Figure 39. The Banbury mixer and roller mill are the basic machinery units used in the compounding operation. Fillers, extenders, reinforcing agents, pigments, and antioxidants are added and mixed into the raw rubber stock. Nonreac-

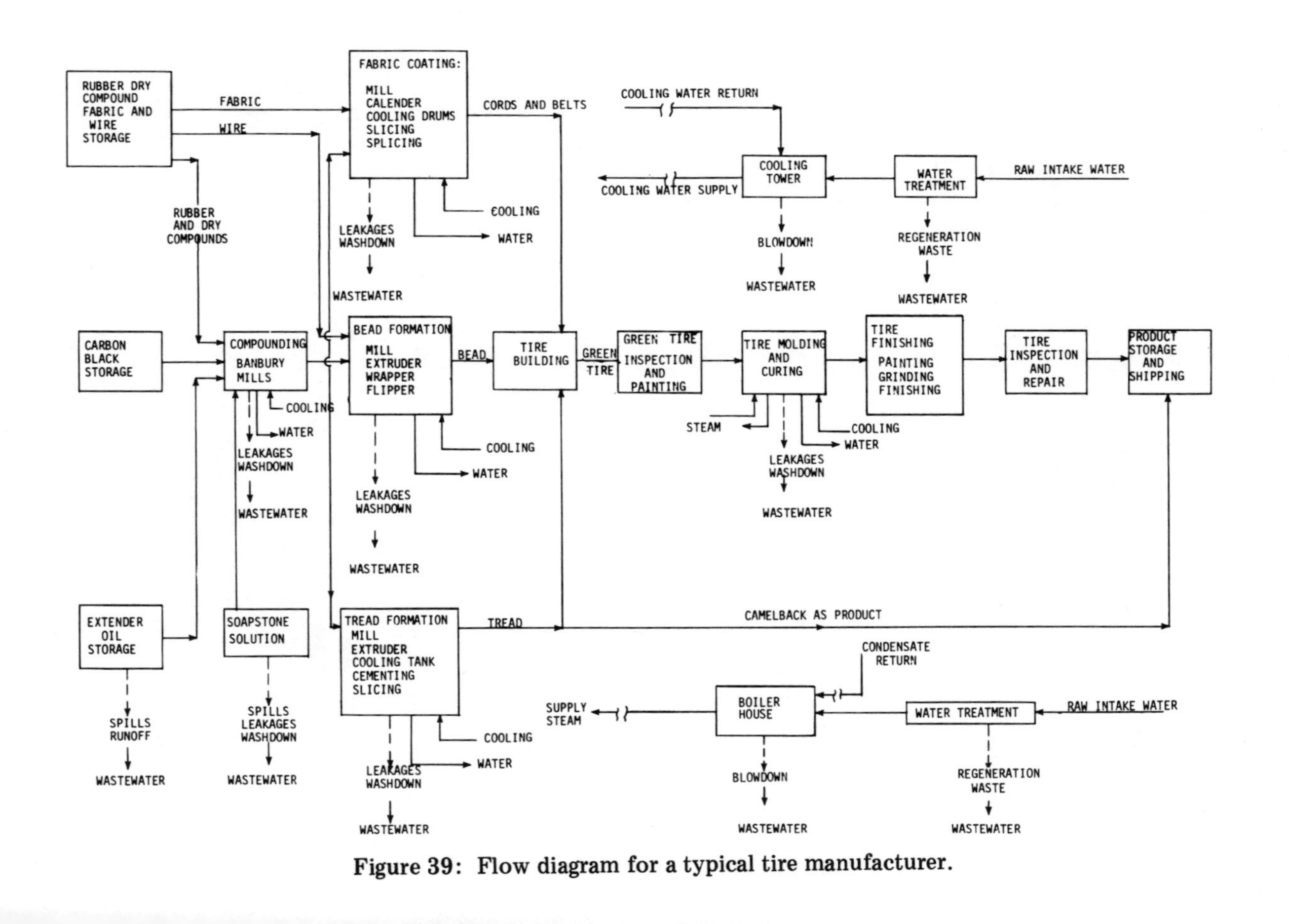

Figure 39: Flow diagram for a typical tire manufacturer.

tive rubber stock, which contains no curing agents, has a long shelf life and may be stored for later use. The reactive rubber stock, which contains curing agents, has a short shelf life and must be compounded and used immediately.

Carbon black and oil are added to the rubber in the compounding operation. After mixing, the compound is formed into sheets in a roller mill, extruded into sheets, or pelletized. Sheeted material is tacky and must be coated with a soapstone solution to prevent sticking.

Sheeted rubber and other raw materials, such as cord and fabric, are then transformed into one of the basic tire components by several parallel processes. The tire is built up as a cylinder on a collapsible, rotating drum. Uncured tires are sprayed with release agents before molding, and then cured in an automatic press. After the molding and curing operation, the tire proceeds to the grinding operation, where the excess rubber which has escaped through the weepholes is ground off. For whitewall tires, additional grinding is required to remove a black protective strip, followed by adding a protective coat of paint. After inspection and final repairs (if needed), the tire is ready to be shipped.

Inner tube manufacture is very similar to tire manufacture, and consists of the same three basic steps: compounding of raw materials, extension of compounded materials, and the building, molding and curing operations which form the final product. One distinction in inner tube manufacturing is the use of large quantities of butyl rubbers. A flow diagram of a typical inner tube plant is shown in Figure 40.

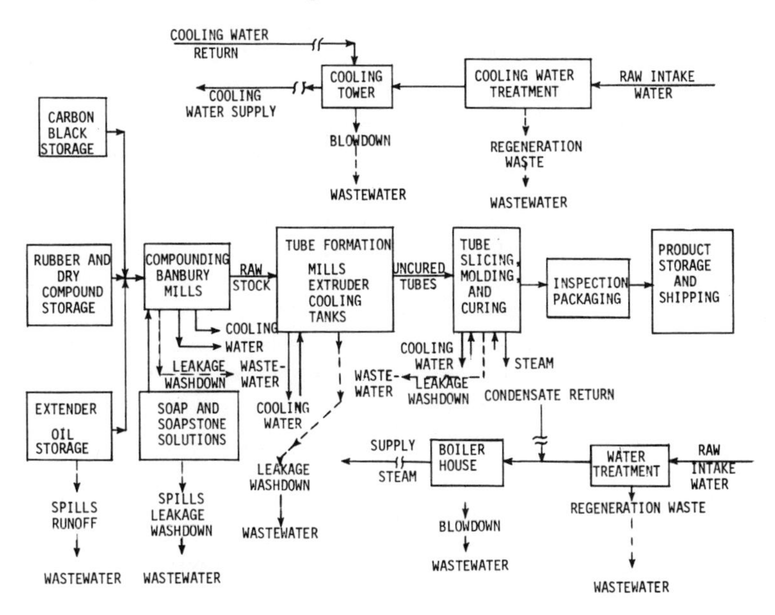

Figure 40: Flow diagram for a typical inner tube manufacturer.

Wastewater contaminants from the tire and inner tube industry are: oils from runoff, roller mills, hydraulic system, and presses; organics and solids from Banbury seals, soapstone dip tank, dipping operation, spray painting operation, and air pollution equipment discharges; and solvent-based cements from the cementing operation.

Emulsion Crumb Rubber (B): Figure 41 shows the process flow diagram for the continuous production of crumb styrene-butadiene rubber (SBR), by the emulsion polymerization process. SBR is the principal synthetic rubber in use today. In this process, styrene and butadiene (monomers) are piped to the plant, and the inhibitors are removed by caustic scrubbers. Soap solution, catalyst, activator, and modifier are added to the monomer mixture prior to entering the polymerization reactors. The reactor system is capable of producing either "cold" (40°F to 45°F, 0 to 15 psig) or "hot" (122°F, 40 to 60 psig) rubber. Product rubber is formed in the emulsion phase of the reaction mixture. The product is a milky white emulsion called latex. Short-stop solution is added to the latex leaving the reactors to stop polymerization when the desired conversion has been achieved. Unreacted monomers are stripped from the latex and recycled to the feed area. An antioxidant is added to the stripped latex in a blend tank to protect the rubber from attack by oxygen and ozone. The latex is now stabilized, and different batches, recipes, or dilutions can be mixed. After coagulation, screening, rinsing, and dewatering, the rubber crumb is finally dried, pressed in bales, and stored until to shipment. Wastewater contaminants from the emulsion polymerization process are: dissolved and separable organics from monomer recovery, crumb dewatering, monomer stripping, and tanks and reactors; uncoagulated latex from tanks, reactors, and monomer stripping; and suspended and dissolved solids from coagulation, crumb dewatering, monomer stripping, and reactors.

Solution Crumb Rubber (C): The production of synthetic rubber by solution polymerization is a stepwise operation, and, in many aspects, is very similar to emulsion polymerization (B). However, solution polymerization requires extremely pure monomers, and the solvent (e.g., hexane) should be completely anhydrous. However, in contrast to emulsion polymerization, where the monomer conversion is approximately 60 percent, solution polymerization achieves conversion levels which are typically in excess of 90 percent. Wastewater contaminants are: dissolved and separable organics from solvent purification and monomer recovery; suspended dissolved solids from crumb dewatering; and highly alkaline caustic scrubber water.

Latex Rubber (D): Latex production follows the same processing steps as emulsion crumb production, with the exception of latex coagulation, crumb rinsing, drying, and bailing. Polymerization is carried out to 98 to 99 percent conversion levels.

As a result, monomer recovery is not economical, and the process is

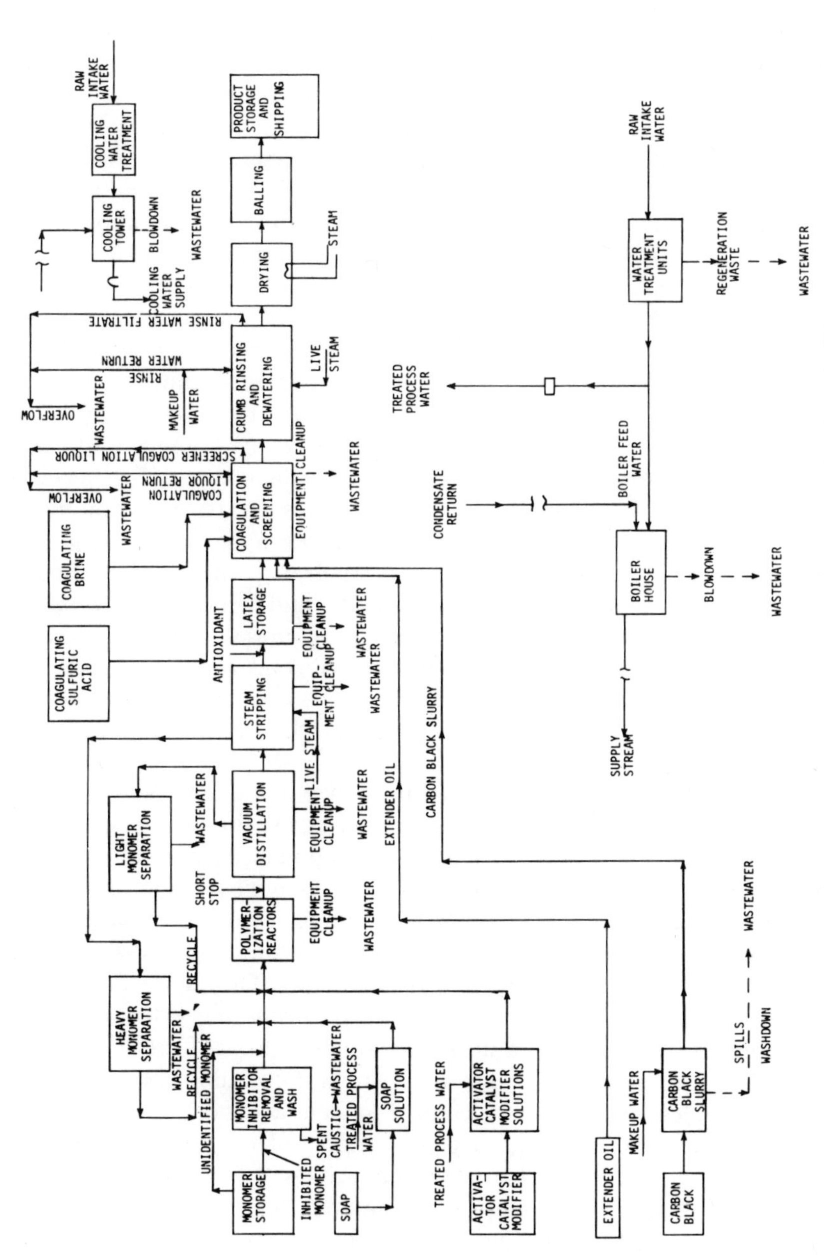

Figure 41: Typical emulsion crumb rubber production facility.

directed towards maximum conversion on a once-through basis. The nature and origins of principal wastewaters are: dissolved and separable organics from excess monomer stripping, reactors and tank cars; suspended and dissolved solids from reactors, strippers, tank cars, and tank trucks; and uncoagulated latex from reactors, tank cars, and tank trucks.

General Molded, Extruded, and Fabricated Rubber Plants (E,F,G): Categories in the general molded rubber product segment include battery parts, seals, packing, heels, shoes, medical supplies, druggist supplies, stationery supplies, etc. General extruded products include belting and sheeting. Products manufactured in the general fabricated segment are rubber hose, footwear, friction tape, fuel tanks, boats, pontoons, life rafts and rubber clothing, and coated fabrics.

During the molding of rubber products, rubber is cured as it is shaped. Curing (often referred to as vulcanization) is an irreversible process during which a rubber compound, through a change in its chemical structure, becomes less plastic and more resistant to swelling by organic liquids. In addition, properties of elasticity over a greater range of temperature are conferred, improved, or extended. The principal methods used for the manufacture of general molded products are compression, transfer, and injection molding processes. All three of these processes may be in use at one plant location. The processes typically consist of compounding of the rubber stock, preparation of mold preforms, molding, and deflashing. The nature and origins of wastewaters are: oils from curing presses, compounding and pickup by storm runoff; solids from soapstone dip tank and wet air pollution equipment discharges; rubber fines from rinse water; and anti-tack agents from cooling water overflow. Figure 42 shows a flow diagram for a typical molding operation.

Manufacture of sheeting and belting serves as a good example of the production methods used to make extruded items. Rubber stock is compounded on a Banbury mixer or compounding mill. After compounding, the rubber is worked on a warm-up mill and fed to the extruder. Extruded rubber is produced in sheets. In some cases, the extruded or calendered rubber is dipped in soapstone slurry for storage. Belting or extruded sheeting is cured using a rotacure or press curing technique. The nature and origins of wastewaters are: oils from machinery, and calendering, extrusion, compounding processes, and from storm runoff in the storage area; solids from the anti-tack agent and tank and wet air pollution equipment discharges; organics and lead from steam vulcanizer condensate; and cooling water.

Hose production provides a good example of the fabricated rubber manufacturing processes. The nature and origins of wastewater contaminants are: oils from machinery, compounding, and storm runoff; solids from soapstone dip tank, ply formation and latex storage; dissolved organics from ply formation, shoe building, and latex storage; and anti-tack agents from cooling water overflow. Figure 43 illustrates a flow diagram for the manufacture of typical hose items.

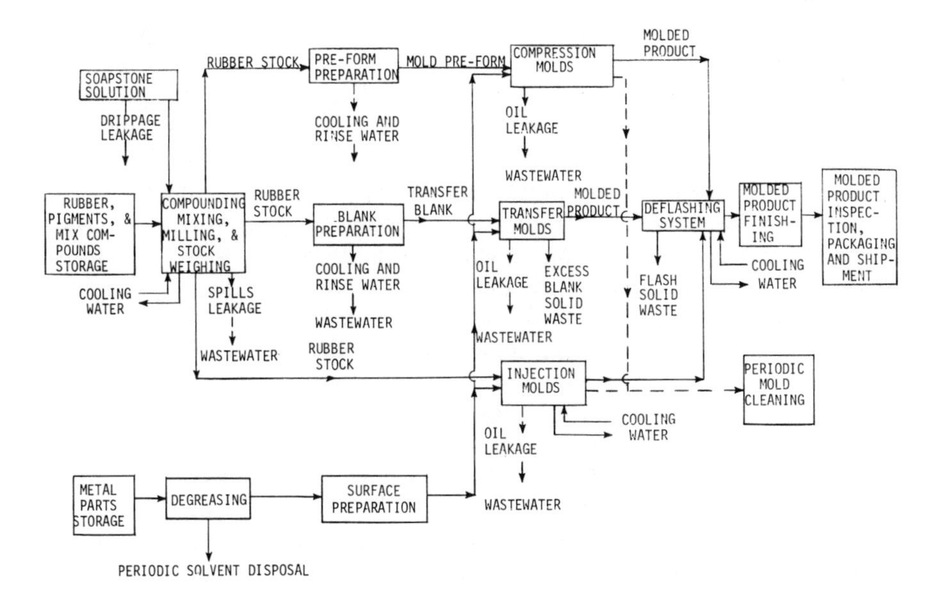

Figure 42: Flow diagram for the production of a typical molded item.

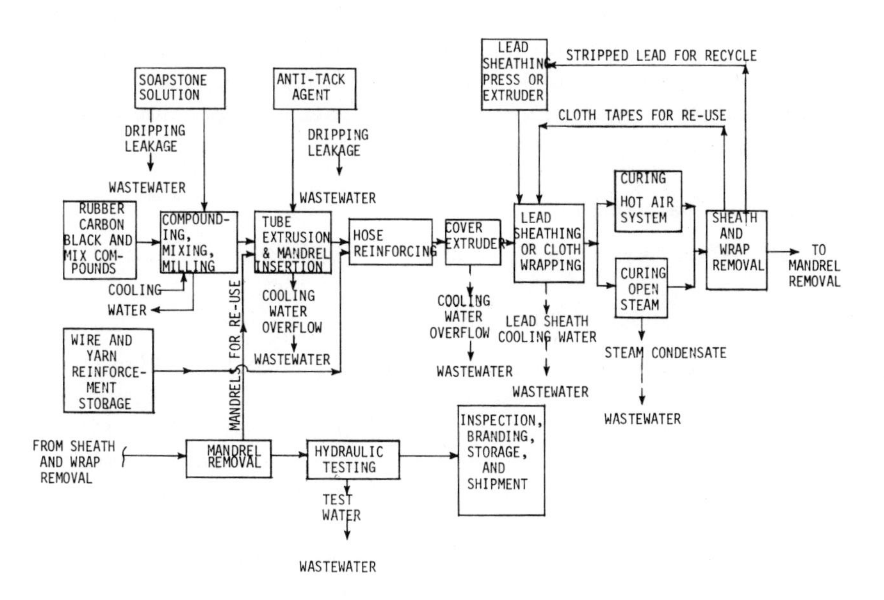

Figure 43: Flow diagram for the production of typical hose items (including reinforced types).

Wet Digestion Reclaimed Rubber (H) and Pan, Dry Digestion and Mechanical Reclaimed Rubber (I): Reclaimed rubber is prepared primarily from scrap tires and scrap inner tubes. Three basic techniques are used to produce reclaimed rubber: the digester process, the pan process, and the mechanical process. The reclaiming process can be divided into three major, broad components, two of which are mostly mechanical and the third predominately chemical. By far the most important source of raw material is tire scrap. The rubber scrap is separated and ground, then given heat treatment for depolymerization, and finally processed for intensive friction milling. All three processes employ similar rubber-scrap separation and size reduction methods. They differ in depolymerization and the final processing steps. The nature and origins of wastewater contaminants are: oil from depolymerization, blowdown tanks, dewatering, dryers and compounding; solids, caustics and organics from depolymerization, defibering, dewatering and soapstone dip tanks; and fibrous material removed from tires. Figure 44 shows a generalized flow diagram for the three processes.

Latex-Dipped, Latex-Extruded and Latex-Molded Rubber (J) and Latex Foam (K): To manufacture sundry rubber goods from latex compounds, it is necessary to convert the compounds into solids of desired form. The latex is compounded with various ingredients such as anti-oxidants. Several manufacturing processes are used for fabricating different types of rubber goods from latex mixtures. Principal wastewater contaminants for the latex-dipped, latex-extruded, and latex-molded rubber are suspended solids, dissolved solids, oil and surfactants. In addition, another important contaminant for the latex foam industry is zinc. The process flow diagrams for a typical latex-based, dipped item and a latex foam item are shown in Figures 45 and 46, respectively.

Wastewater Characteristics

Table 67 shows raw wastewater characteristics from this industry.

Control and Treatment Technology

In-Plant Control Technology: Significant in-plant control of both waste quantity and quality is possible for most subcategories of the rubber processing industry. For tires and inner tubes (A), in-plant control includes the proper handling of soapstone, latex dip, and discharges from air pollution control equipment. A closed-loop recirculation system eliminates the continuous discharge of large quantities of soapstone. Alternatives to recirculation include the use of substitute solutions which require the system to be cleaned on a less frequent basis. A common practice among larger manufacturers is to eliminate the latex-dipping operation from the tire facility. In plants that still dip fabric, an effective control measure is to seal off drains,

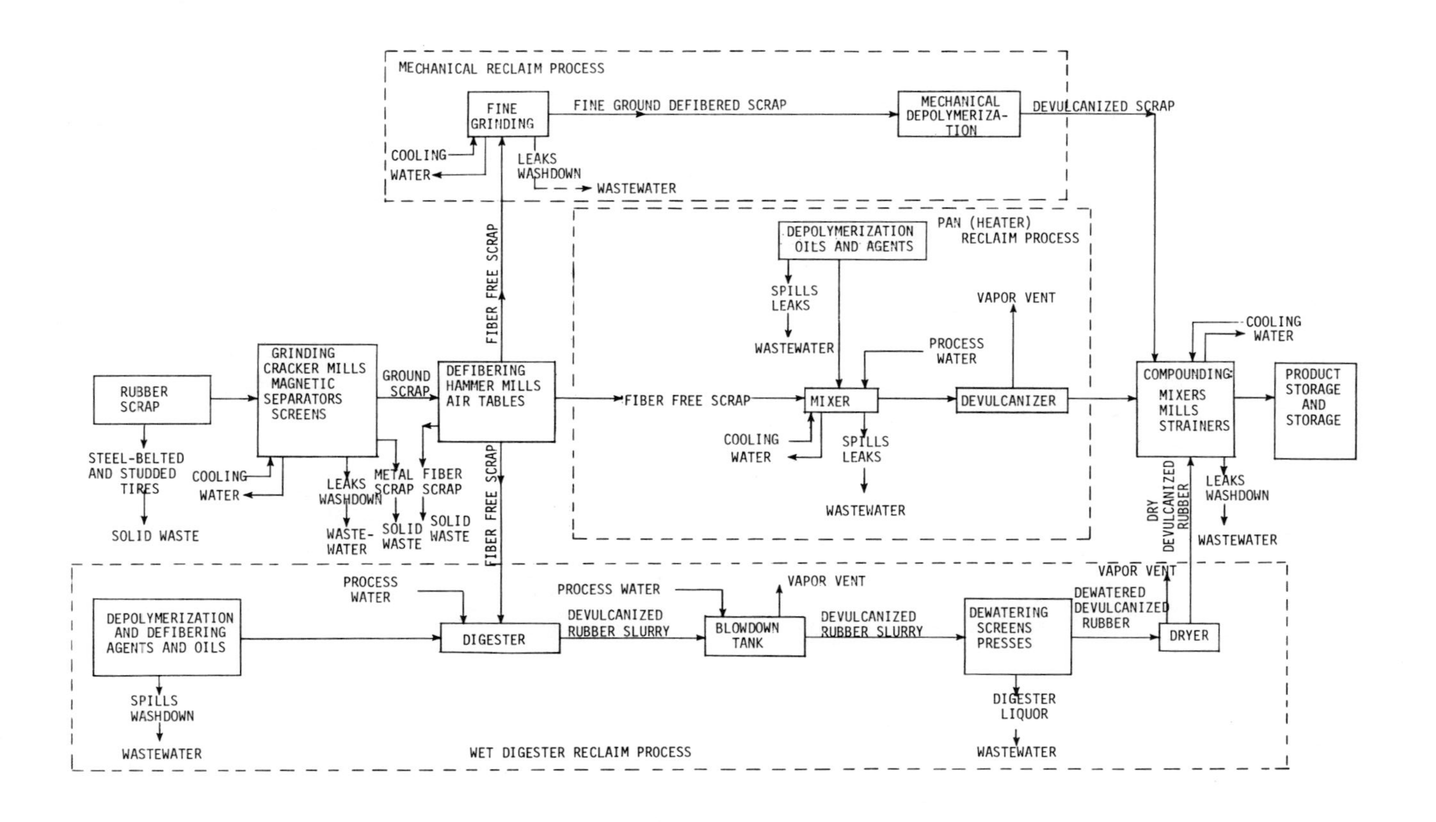

Figure 44: Flow diagram of a typical mechanical, pan (heater), and wet digester reclaim process.

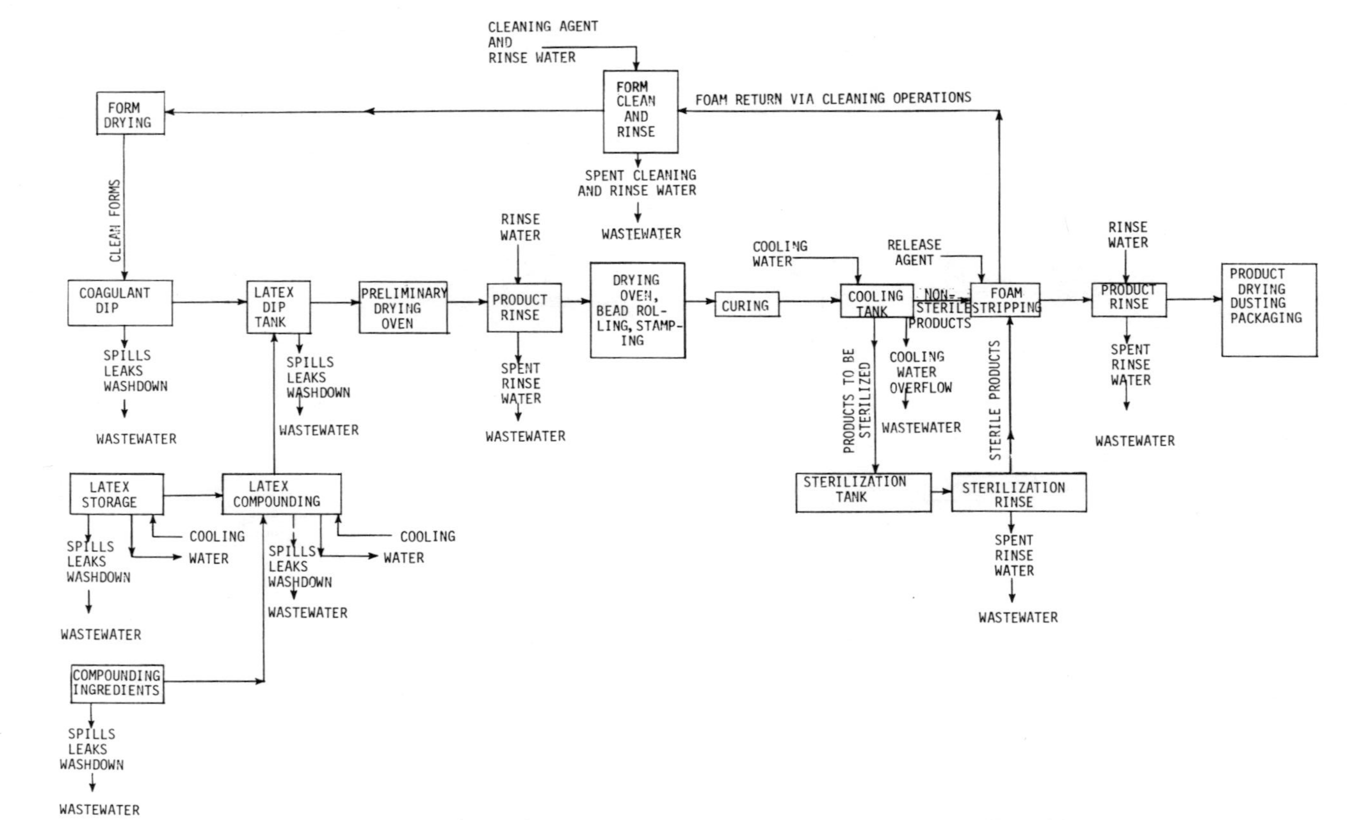

Figure 45: Flow diagram for the production of typical latex-based, dipped items.

supply the area with curbing, and drum the waste solutions for landfill disposal. Solids from the wet scrubber discharge in the tire finishing area can be settled out in a sump. The particulates are large, and with a properly designed separator, the clarified water can be completely reused. Further in-plant measures for the tire and inner tube industry include the control of spills and leakage by providing curbing and oil sumps, the use of dry sweeping equipment for prevention of process-area washdowns from contaminating wastewaters, and the diking of all oil storage areas to prevent contamination of wastewaters by oil spillages.

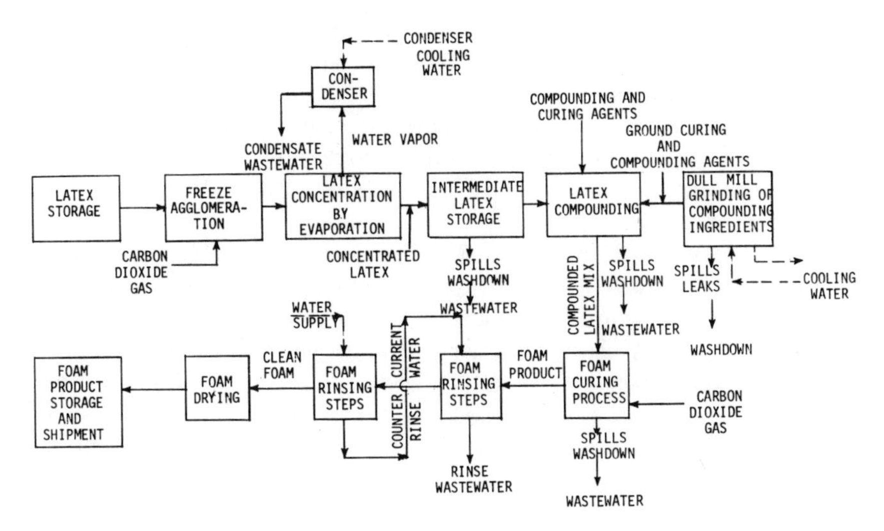

Figure 46: Process flow diagram for latex foam.

Table 67: Rubber Processing Industry—Raw Wastewater Characteristics

Wastewater parameter (mg/l)	Subcategories[a] A	B	C	D	E	F	G	H	I	J	K
BOD	0.2-31	115-183	10-123	377-418	4-24	10-24	6-28	10	7.2	133-152	1,155
COD	0-298	528-1,886	50-1,168	2,440-2,790	3-20	20-50	57-261	52	37.6	176-678	4,285
Lead						0.008	0-8				
Oil	1-96	7-191	8-195	28	1-26	0.8-7	4.6-31	4.7	2.8	6-129	571
Surfactants										1.8-6.4	5.1
TDS	0-757				384-657	607-790	231-3,100	132	104	385-1,146	1,353
TSS	9-1,065	124-770	18-470	450-470	1-9	1-13	10-62	21	16.6	78-3,019	492
Zinc											200

[a] See Table 66 for identification of subcategories.

In-plant control measures for the general molded, extruded, and fabricated rubber subcategories (E,F,G) include proper handling and isolating general spills and leaks of soapstone and other anti-tack agents, as well as latex compounds, solvents and rubber cements, metal preparation wastes, and air pollution control equipment discharges. Contamination by machinery oils, greases and suspended solids can be reduced by blocking of floor drains, removing oil leaks promptly with dry absorbent granules, and curbing the problem area. The spillage of soapstone and other anti-tack solutions can be controlled by similar methods. An effective way to handle latex is through the use of latex drums with plastic liners which can be discarded when the drum is reused. Latex spills around storage and transfer facilities are coagulated with alum and scraped from the ground. Solvents and rubber cements should be mixed and stored in areas without floor drains to control spills and leaks.

If acid pickling is used to prepare metal components, precipitation of metals and pH adjustment should be carried out. Pickling wastes can also be containerized and hauled from the plant, rather than entering the sewage system.

In the wet digestion reclaimed rubber subcategory (H), significant in-plant control measures include the defibering of scrap rubber by mechanical or physical techniques as an alternative to chemical defibering, and the recycling of process oils and digester liquor. Recycling of process oils and the control of vapor condensates and spills and leaks are significant in-plant measures for the pan, dry digestion and mechanical reclaimed rubber subcategory (I). In the latex-based products subcategories (J,K), prevention of latex spills and leaks, and reduction in the volumes of foam rinse waters and cleaning wastes by employing countercurrent rinsing, constitute the most significant in-plant control measures.

End-of-Pipe Treatment Technology: The various wastewater treatment practices for each of the 11 subcategories of the rubber processing industry are summarized in Table 68. The removal efficiencies given reflect raw waste loads of process effluents from each of the subcategories.

Type of Residue Generated

Information on the types and characteristics of residues generated by this industry is insufficient and cannot be discussed at this time.

Residual Management Options

Lack of sufficient information on types and characteristics of residues generated does not permit a discussion of specific management options at this time.

Table 68: Rubber Processing–Wastewater Treatment Practices

Pollutant and method	Subcategory[a,b] - removal efficiencies - percent							
	A	B	C	D	H[c]	I	J	K
Biochemical oxygen demand								
° Sedimentation and holding lagoon, recirculation of soapstone	76-83		99					79
° Coagulation, clarification		82	81	51				
° Coagulation, clarification, activated sludge		81	86					
° Coagulation, clarification, zinc precipitation, and clarification								65
Chemical oxygen demand								
° Sedimentation and holding lagoon, recirculation of soapstone	62-87		78					32
° Coagulation, clarification		72-74	74	80				
° Coagulation, clarification, activated sludge			52	76				
° Oil separator and holding lagoon					44-63	94	9	
Oil and grease								
° Sedimentation and recirculation of soapstone	96		91					
° Gravity separator	60				95	99.8	93	28
° Coagulation, clarification		99	99	52				
° Coagulation, clarification, activated sludge			88					
Suspended solids								
° Coagulation, clarification	89		81				73	
° Equalization, activated sludge			83	62				
° Clarification, stabilization lagoon			89-94					
° Gravity separator	80				72	95	82	
° Coagulation, clarification, zinc precipitation, and clarification								90

[a] See Table 66 for identifcation of subcategories.
[b] Data for categories E, F, and G were not available.
[c] Values indicated reflect reductions due to recycle as well as wastewater treatment.

SOAP AND DETERGENT MANUFACTURING

General Industry Description

The Soap and Detergent industry produces liquid and solid cleaning agents for domestic and industrial use, including laundry, dishwashing, bar soaps, specialty cleaners, and industrial cleaning products. Discharges from this industry are generally nontoxic and readily responsive to treatment, except in the industrial surfactant subcategory. More than 95 percent of plant effluents are treated in municipal treatment plants.

The industry is broadly divided into two categories: soap manufacture, which is based on processing of natural fat; and detergent manufacture, which is based on the processing of petrochemicals.

Industrial Categorization

A useful categorization system for the purposes of raw waste characterization and the establishment of pretreatment information from this industry is given in Table 69.

Table 69: Soap and Detergent

Main category	Subcategory	Designation
Soap manufacture	Batch kettle and continuous	A
	Fatty acid manufacture by fat splitting	B
	Soap from fatty acid neutralization	C
	Glycerine recovery	
	Glycerine concentration	D
	Glycerine distillation	E
	Soap flakes and powders	F
	Bar soaps	G
	Liquid soap	H
Detergent manufacture	Oleum sulfonation and sulfation (batch and continuous)	I
	Air - SO_3 sulfation and sulfonation (batch and continuous)	J
	SO_3 solvent and vacuum sulfonation	K
	Sulfamic acid sulfation	L
	Chlorosulfonic acid sulfation	M
	Neutralization of sulfuric acid, esters, and sulfonic acids	N
	Spray dried detergents	O
	Liquid detergent manufacture	P
	Detergent manufacturing by dry blending	Q
	Drum dried detergents	R
	Detergent bars and cakes	S

Process Description

A flow diagram for the entire industry is given in Figure 47.

Soap Manufacturing and Processing: Soap manufacturing consists of two major operations: the production of neat soap (65 to 70 percent hot soap solution), and the preparation and packaging of finished products into flakes and powders (F), bar soaps (G) and liquid soaps (H). Many producers of neat soap also recover glycerine as a by-product for subsequent concentration (D) and distillation (E).

Neat soap is generally produced in either the batch kettle process (A), or the fatty acid neutralization process, which is preceded by the fat splitting process (B,C). Descriptions of processes for the production of neat soap will follow.

Batch Kettle and Continuous (A) — The production of neat soap by this process consists of receiving and storing raw materials, fat refining and bleaching, and soap boiling.

Major wastewater sources are washouts of both storage and refining tanks, as well as leaks and spills of fats and oils around these tanks. These streams are usually skimmed for fat recovery prior to discharge to the sewer.

The fat refining and bleaching operation removes impurities which would cause color imperfections and odor problems in the finished soap. Wastewaters have a high soap concentration and contain treatment chemicals, fatty impurities, emulsified fats, and sulfuric acid solutions of fatty acids. Where steam is used for heating, the condensate may contain fatty acids of low molecular weight, which are highly odorous, partially soluble materials.

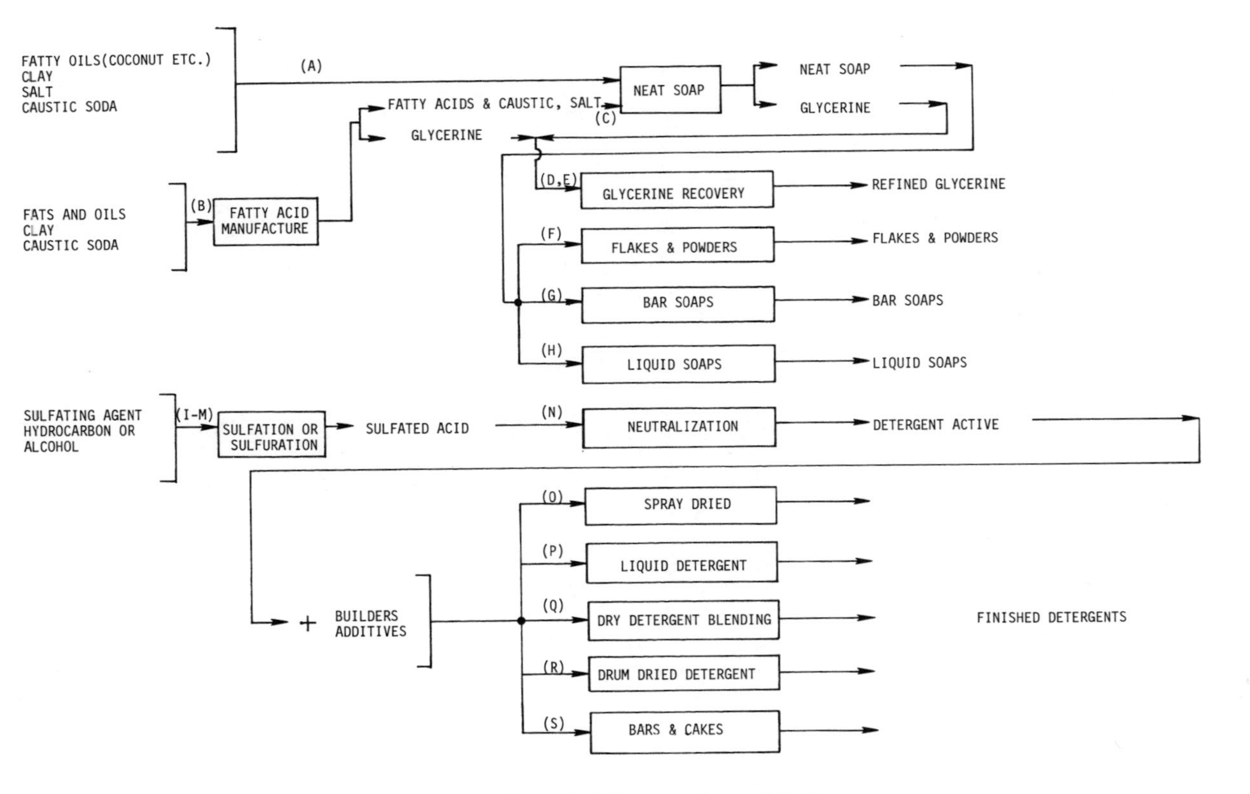

Figure 47: Soap and detergent manufacture.

The soap boiling process produces two concentrated waste streams: sewer lyes, which result from the reclaiming of scrap soap, and the brine from Nigre processing. Both of these are low volume wastes, having high pH's and BOD's as big as 45,000 mg/l.

Fatty Acid Manufacturing by Fat Splitting (B) and Fatty Acid Neutralization (C) — Soap produced by the neutralization process is a two step process: fatty acid manufacturing or fat splitting, and fatty acid neutralization.

Fat Splitting (B) — Manufacturing of fatty acid from fat is called "fat splitting". Washouts from storage, transfer, and pretreatment stages are the same as those for process (A). Process condensate and barometric condensate from fat splitting will be contaminated with fatty acids and glycerine streams. These waste streams are settled and skimmed to recover the insoluble fatty acids which are processed for sale. Water will typically circulate through a cooling tower and be reused. Occasional purges of part of this stream result in high concentrations of BOD, and some grease and oil being released to the sewer.

In the fatty acid distillation process, wastewater is generated as a result of an acidification process which breaks the emulsion. This wastewater is neutralized and sent to the sewer. Wastewater will contain salt from neutralization, zinc, and alkaline earth metal salts from the fat splitting catalysts, and emulsified fatty acids, and fatty acid polymers.

Fatty Acid Neutralization (C) — Soap-making by fatty acid neutralization is a faster process than the kettle boil process, and generates less wastewater effluent.

In fatty acid neutralization, a caustic is used to convert, by neutralization, fatty acids to soap. Sodium carbonate is often used in place of caustic. When soaps that are liquid at room temperature are desired, more soluble potassium soaps are made by substituting potassium hydroxide for the sodium hydroxide (lye). This process is relatively simple, and high purity raw materials are converted to soap with essentially no wastewater by-products. The only wastewater of consequence is sewer lye from reclaiming of scrap. Sewer lyes contain excess caustic soda and salt, added to grain out the soap, and some dirt and paper which is not removed in the strainer.

Glycerine Recovery Process (D,E) — A process flow diagram for the glycerine recovery process shows the glycerine by-products from kettle boiling (A) and fat splitting (B). The process consists of three steps: (1) pretreatment to remove impurities; (2) concentration of glycerine by evaporation; and (3) distillation to a finished product at 98 percent purity.

There are three wastewater streams of consequence which are generated by this process: two barometric condensates–one from evaporation, one from distillation; plus the glycerine foots, or still bottoms. Contaminants from the condensates are essentially glycerine and a small amount of entrained salt. In the distillation process, glycerine foots or still bottoms leave a glassy, dark brown, amorphous solid, which is rich in salt and is disposed into the wastewater stream. Also present in the wastewater stream will be glycerine, glycerine polymers, and organics — BOD_5, COD, and dissolved solids. Sodium chloride will cause increases in dissolved solids levels. Little or no suspended solids, oil and grease, or pH problems should result.

Glycerine can also be purified by the use of ion exchange resins to remove the sodium chloride salt, followed by evaporation of the water. This process puts additional salts into the wastewater, but results in less organic contamination.

Soap Flakes and Powders (F) — Neat soap may or may not be blended with other products before flaking and powdering. After thorough mixing, the finished formulation is run through various mechanical operations to produce flakes and powders. Since all of the evaporated moisture goes to the atmosphere, there is no wastewater effluent.

Some operations will include a scrap reboil to recover reclaimed soap. Scrap reboil is salted out for soap recovery, and the salt water is recycled. After frequent recycling, the water becomes so contaminated that it must be discharged.

Occasional washdown of the crutcher may be needed. The tower is usually cleaned using a dry process. There is also some gland water which passes over the pump shaft, picking up any minor leaks. These operations contribute very little to the total effluent loading.

There are a number of possible effluents from this process. However, surveys of the industry have shown that operating plants generally recycle wastewater or use dry clean-up processes.

Bar Soaps (G) — The procedures for bar soap manufacture will differ significantly from plant to plant, and will depend upon product specifications.

The amount of water used in bar soap manufacturing also varies greatly. In many cases, the entire bar soap processing operation is done without generating a single wastewater stream. Equipment is usually cleaned using dry processes, with no washups. In other cases, one or more wastewater stream will be generated, due to housekeeping requirements associated with the particular bar soap processes.

Major wastewater streams in bar soap manufacturing are from filter backwash, scrubber waters or condensate from a vacuum drier, and from equipment washdown. The main contaminant in all these streams is soap.

Liquid Soap (H) — In the making of liquid soap, neat soap is blended in a mixing tank with other ingredients, such as alcohols or glycols, or with pine oil and kerosene for a finished product with greater solvency and versatility. Before being drummed, the final blended product is frequently filtered to achieve a sparkling clarity. In making liquid soap, water is used to wash out the filter press and other equipment. There are very few effluent leaks. Spills can be recycled or disposed as a dry waste. Washout between batches is usually unnecessary or can be recycled indefinitely.

Detergent Manufacturing and Processing: Detergents can be formulated with a variety of organic and inorganic chemicals depending upon the cleaning characteristics desired. There are four main groups of detergents: Anionics, Cationics, Nonionics, and Amphoterics.

Anionics comprise the most prevalent group of detergents on the market. They are usually produced from the sodium salts of an organic sulfate or sulfonate, of animal or petroleum origin.

Cationic detergents are known as "inverted soaps" and are produced in quite small volumes. They are relatively expensive and somewhat harsh on the skin. However, they make excellent bacteriostats and fabric softeners.

Nonionic detergents are an increasingly popular active ingredient for automatic washing machine formulations. The products are effective in hard water and produce very low levels of foam. They are made by the addition of ethylene oxide to an alcohol.

Amphoterics are those detergents which can be either anionic or cationic, depending upon the pH of the system wherein they work. They account for only a small portion of the detergent market.

A finished, packaged detergent customarily consists of two main components–the active ingredient (surfactant) and the builder. The surfactant acts as the cleaning agent, while the builder performs such functions as buffering the pH, soil dispersion, and anti-soil redeposition. The process descriptions below include the manufacture of the surfactant as well as the preparation of the finished detergent.

Oleum Sulfonation and Sulfation (I) — One of the most important active ingredients of detergents is the sulfate or sulfonate compounds made via the oleum route. In most cases, the sulfonation/sulfation process is carried out continuously in a reactor where the oleum (a solution of sulfur trioxide in sulfuric acid) is brought into intimate contact with the hydrocarbon or alcohol. Reaction occurs rapidly in this process. The stream is then mixed with water, where the surfactant separates out and is sent to a settler. Spent acid is drawn off and usually reprocessed elsewhere, and sulfonated/sulfated material is sent to be neutralized.

This process is normally operated continuously and performs indefinitely, without need for periodic cleanout. A stream of water is generally passed over the pump shafts to pick up leaks, as well as to cool the pumps. Wastewater flow from this source is small but continual.

Air-SO_3 Sulfonation and Sulfation (J) — This process for surfactant manufacturing has many advantages, and is used extensively. SO_3 sulfation produces no water in the reaction process. Because of this reaction's particular tendency to char the product, the reactor system must be thoroughly cleaned on a regular basis. In addition, there are usually several airborne sulfonic acid streams which must be scrubbed, resulting in wastewater discharges to the sewer during sulfation.

SO_3 Solvent and Vacuum Sulfonation (K) — Undiluted SO_3 and organic reactant are fed into the vacuum reactor through a mixing nozzle. This system produces a high quality product. Offsetting this advantage is the high operating cost of maintaining the vacuum. With the exception of occasional washout discharges, the process is essentially free of wastewater generation.

Sulfamic Acid Sulfation (L) — Sulfamic acid is a mild sulfating agent and is used only in certain products of very specialized quality, because of the high reagent price. Washouts are the only wastewater effluents from this process.

Chlorosulfonic Acid Sulfation (M) — For products requiring high quality sulfates, chlorosulfonic acid is an excellent agent. It is a corrosive agent and generates hydrochloric acid as a by-product. The effluent washouts are minimal.

Neutralization of Sulfuric Acid Esters and Sulfonic Acids (N) — This step is essential in the manufacturing of active ingredients in detergents. During this step, sulfonic acids or sulfuric acid esters (products produced by processes I - M) are converted into neutral surfactants. This process is a potential source of some oil and grease. Occasional leaks and spills around the pump and valves are the only expected source of wastewater contamination.

Spray Dried Detergents (O) — In this segment of the processing, neutralized sulfonates and/or sulfates are first blended with builders and additives in the crutcher. The slurry is then pumped to the top of a spray tower (15 to 20 feet in diameter by 150 to 200 feet high), where nozzles spray out detergent slurry. A large volume of hot air enters the bottom of the tower and rises to meet the falling detergent. The design preparation of this step will determine the detergent particle's shape, size, and density, which in turn determines its solubility rate in the washing process.

Air coming from the tower will be carrying dust particles which must be scrubbed, thus generating a wastewater stream. Spray towers are periodically shut down and cleaned. Tower walls are scraped and thoroughly washed down. This final step is mandatory since the manufacturers must be very careful to avoid contamination of subsequent formulations.

Wastewater streams are rather numerous. Washouts of equipment from the crutchers down the line to the spray tower itself constitute many effluent streams. One high volume wastewater flow occurs from the air scrubber, which cleans and cools the hot gases rising from this tower.

All of the plants recycle some of the wastewater produced, with some plants recycling all of the flows generated.

Due to increasingly stringent air quality requirements, it is expected that fewer plants will be able to maintain a complete recycle system of all water flows in the spray tower area.

Liquid Detergents (P) — Active ingredients in detergents are pumped into mixing tanks, where they are blended with numerous other ingredients, ranging from perfumes to dyes. From here, the fully formulated liquid detergent is sent to the filling line for filling, capping, labeling, etc. Whenever the filling line is to change to a different product, the filling system must be thoroughly cleaned out to avoid product contamination.

Dry Detergent Blending (Q) — In this process, fully dried surfactant materials are blended with additives in dry mixers. In normal operations, many succeeding batches of detergent can be mixed in the same equipment without the need for anything but dry cleaning. However, when a change in formulation occurs, the equipment must be completely washed down. A modest amount of wastewater is thus generated.

Drum Dried Detergents (R) — This process is one method of converting a slurry to a powder, and should be essentially free of wastewater discharges other than occasional washdown effluent.

Detergent Bars and Cakes (S) — Detergent bars are either 100 percent synthetic detergent or a blend of detergent and soap. They are blended in essentially the same manner as that used for conventional soap. Fairly frequent cleanups generate a wastewater stream.

Wastewater Characteristics

Table 70 shows raw wastewater characteristics. Table 71 shows raw wastewater characteristics as a function of production. Most plants contain two or more of the subcategories shown on the table; their wastewaters will therefore be a composite of these individual unit processes.

Control and Treatment Technology

In-Plant Control Technology.: Significant in-plant control of both wastewater quantity and quality is possible, particularly in the soap manufacturing subcategories, where maximum flows may be 100 times minimum flow. Considerably less in-plant water conservation is possible in the detergent industry, where flows per unit of product are smaller.

The largest in-plant modification that could be made involves changing or replacing the barometric condensers (applicable to processes A,B,D,E). The quantity of wastes discharged from these processes could be significantly reduced by recycling the barometric cooling water through fat skimmers, recovering valuable fats and oils, and then through the cooling towers. Replacing barometric condensers with surface condensers has been used in several plants to reduce both the waste flow and the quantity of organics wasted.

Significant reduction of water usage is possible in the manufacture of

Table 70: Soap and Detergent—Raw Wastewater Characteristics

Wastewater parameter[b]	Subcategories[a] A	B	C	D	E	F	G	H	I	J
BOD	3,600	60-3,600	400				1,600-3,000		75-2,000	380-520
Boron										
Chlorides	20,000-47,000									
COD	4,267	115-6,000	1,000						220-6,000	920-1,589
Nickel		Present								
Oil and grease	250	13-760	200						100-3,000	
pH (units)	5-13.5	High	High	Neutral	Neutral	Neutral	Neutral	Neutral	1-2	2-7
Surfactant									250-7,000	
TSS	1,600-6,420	115-6,000	775						100-3,000	
Zinc		Present								

Wastewater parameter[b]	Subcategories[a] K	L	M	N	O	P	Q	R	S
BOD				8.5-6,000	48-19,000	65-3,400			
Boron				Present	Present	Present	Present	Present	Present
Chlorides									
COD				245-21,000	150-60,000	640-11,000			
Nickel									
Oil and grease									
pH (units)	Low	Low	Low	Low					
Surfactant							60-2,000		
TSS				100-3,000					
Zinc									

[a] See Table 69 for identification of subcategories.

[b] Measurements are in mg/l, unless otherwise noted.

Table 71: Soap and Detergent—Raw Wastewater Characteristics—Production-Based

Wastewater parameter	Subcategory[a,b,c] A	B	C	D	E	F	G	H	I	J	K	L	M	N	O	P	Q	R	S
BOD	6	12	0.1	15	5	0.1	3.4	0.1	0.2	3	3	3	3	0.10	0.1-0.8	2-5	0.1	0.1	7
Chloride													5						
COD	10	22	0.25	30	10	0.3	5.7	0.3	0.6	9	9	9	9	0.3	0.3-25	4-7	0.5	0.3	22
Oil and Grease	0.9	2.5	0.05	1	1	0.1	0.4	0.1	0.3	0.5	0.5	0.5	0.5	0.1	0.3			0.1	0.2
TSS	4	22	0.2	2	2	0.1	5.8	0.1	0.3	0.3	0.3	0.3	0.3	0.3	0.1-1.0		0.1	0.1	2
Surfactant									0.7	3	3	3	3	0.2	0.2-1.5	1.3-3.3		0.1	5

[a] Data based on batch flow conditions.

[b] See Table 69 for description of each industrial subcategory.

[c] Units are in kilograms per 1,000 kilograms of product produced.

liquid detergents (P) by the installation of water recycle piping and tankage, and by the use of air rather than water to blow down filling lines.

In the production of bar soaps (G), the volume of discharge and the level of contamination can be reduced materially by installation of an atmospheric flash evaporator before the vacuum drier step.

Pollutant carry-over from distillation columns such as those used in glycerine concentration (D) or fatty acid separation (B) can be reduced by the addition of two trays.

Treatment Technology: The industry routinely institutes a broad range of pretreatment processes to control its effluent. Treatment methods which are used are shown in Table 72. Also shown in this table are the anticipated removal efficiencies of each treatment process for the various pollutants generated. A composite flow sheet showing a complete treatment system for the soap and detergent industry is shown in Figure 48. As a minimum, even small plants with batch operations should employ equalization to smooth out peak discharges. In addition, larger plants with an integrated product line may require both suspended solids and organics removal. The bulk of the large solid material present in the industry's wastewater can be removed by coagulation and sedimentation. Fine solid material can be removed by sand or mixed-bed filtration applied as a tertiary step after biological oxidation.

Table 72: Treatment Methods Used in the Soap and Detergent Industry

Pollutant and method	Percentage of pollutant removed
Oil and grease	
° API type separation	Up to 90 percent of free oils and greases. Variable on emulsified oil.
° Carbon adsorpton	Up to 95 percent of both free and and emulsified oils.
° Flotation	Without addition of alum or iron, 70-80 percent removal of both free and emulsified oil. With the addition of alum or iron, 90 percent.
° Mixed media filtration	Up to 95 percent of free oils. Efficiency in removing emulsified oils unknown.
° Coagulation (with either iron, alum, or other material) plus clarification	Up to 95 percent of free oil. Up to 90 percent of emulsified oil.
Suspended solids	
° Mixed media filtration	70-80 percent.
° Coagulation-clarification	50-80 percent.
Residual suspended solids	
° Sand or mixed media filtration	50-95 percent.
Dissolved solids	
° Ion exchange, reverse osmosis	Up to 90 percent.
Biochemical and chemical oxygen demand	
° Biological treatment	60-95 percent.
° Carbon adsorption	Up to 90 percent.

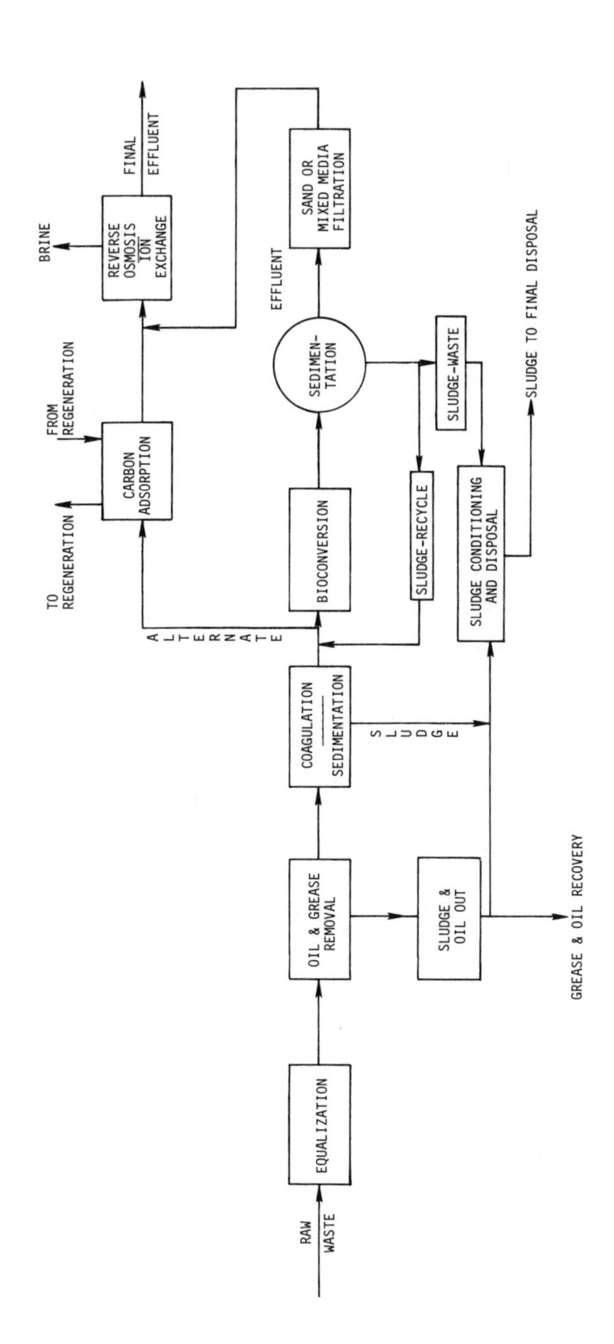

Figure 48: Composite waste treatment flow sheet–soap and detergent industry.

Organics removal is typically provided by either of several forms of biological oxidation, or by powdered or granulated activated carbon adsorption. A few plants employ reverse osmosis or ion exchange as a tertiary step for the removal of individual dissolved pollutants or total dissolved solids.

Type of Residue Generated

Information on the types and characteristics of residues generated by this industry is insufficient and cannot be discussed at this time.

Residual Management Options

Lack of sufficient information on types and characteristics of residues generated does not permit a discussion of specific management options at this time.

STEAM ELECTRIC

General Industry Description

Steam electric power plants are involved in the generation, transmission, and distribution of electric energy. Fuel (coal, oil, natural gas, uranium) is used to generate heat, which converts water to pressurized and superheated steam. The steam transfers energy to rotating blades of a turbine which, in turn, drives an electric generator or alternator to convert the imparted mechanical energy into electrical energy. Steam exiting the turbine is condensed to water, where it is recycled and begins the production cycle again. Liberated heat is transferred to a cooling medium (usually water), which is either discharged or recirculated. Unlike other industries, the product, electricity, cannot be economically stored, and therefore, the industry must be ready to produce enough electricity to meet consumer demand at any time.

Industrial Categorization

For the purpose of establishing wastewater effluent limitation guidelines, the Steam Electric Utility industry has been subcategorized as shown in Table 73. This subcategorization is based on the wastes generated by various operations.

Process Description

Wastewaters are produced from a number of continuous or intermittent steam electric power plant operations. Figure 49 shows a process flow diagram for a fossil-fueled steam electric power plant and indicates the wastewater sources.

Table 73: Steam Electric Utility Industry Subcategorization

Category	Designation
Condenser cooling system	I
Once-through	IA
Recirculating	IB
Water treatment	II
Clarification	IIA
Softening	IIB
Ion exchange	IIC
Evaporator	IID
Filtration	IIE
Other treatment	IIF
Maintenance cleaning	IV
Boiler or steam generator tubes	IVA
Boiler fireside	IVB
Air preheater	IVC
Miscellaneous small equipment	IVD
Stack	IVE
Cooling tower basin	IVF
Ash handling	V
Oil-fired plants	VA
fly ash	VA1
bottom ash	VA2
Coal-fired plants	VB
fly ash	VB1
bottom ash	VB2
Drainage	VI
Coal pile	VIA
Contaminated floor and yard drains	VIB
Ash pile	VIC
Air pollution (SO_2) control devices	VII
Miscellaneous waste streams	VIII
Sanitary wastes	VIIIA
Plant laboratory and sampling systems	VIIIB
Intake screen backwash	VIIIC
Closed cooling water systems	VIIID
Low-level radioactive wastes	VIIIE
Construction activity	VIIIF

Condenser Cooling System (I): The efficiency of a power plant depends largely on the cleanliness of its heat transfer surfaces. Internal cleaning of this equipment is usually done by the use of chemicals, thus allowing scale, corrosion deposits and biological growth to be removed.

Once-Through Cooling Water (IA) — Approximately 67 percent of the steam electric power plants use once-through cooling water for condensation. Condenser cooling water is obtained from oceans, rivers, estuaries, or groundwater, and is discharged after its passage through the condenser system. Chlorine or hypochlorite is usually added to once-through condenser cooling systems to minimize the biofouling of heat transfer surfaces.

Wastewater composition varies with source water characteristics and condenser tube materials. Tube construction has major bearing on: (1) corrosion products formed, and (2) the amount of biocide needed to control tube biofouling. Copper alloy tubes, for example, have biocidal activity of their own and require less chlorine.

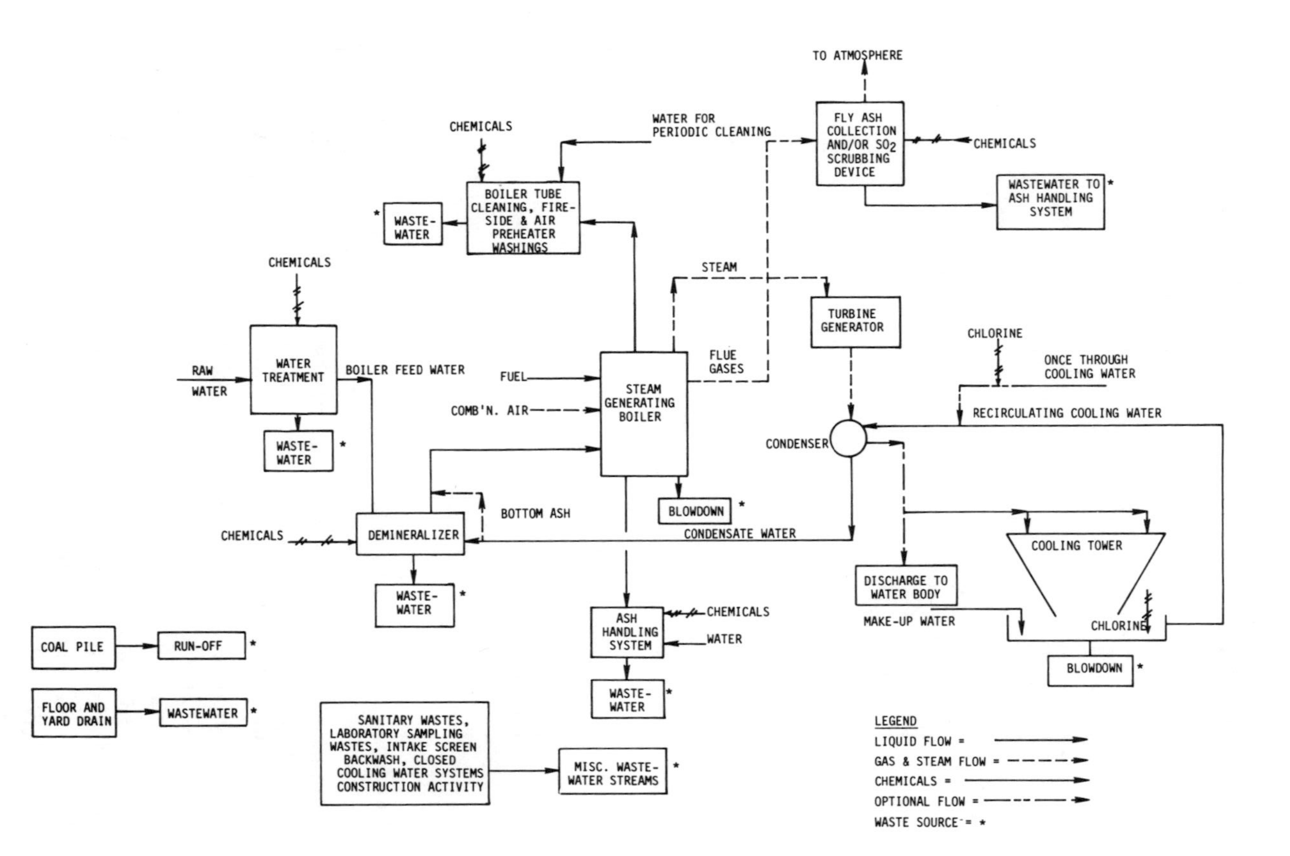

Figure 49: Process flow diagram for fossil-fueled steam electric power plant.

Some facilities use mechanical cleaning, resulting in possible increased metal content of wastewaters. Heat treatment has been used in saltwater coolant systems to remove marine encrustations from intake and discharge conduits.

Recirculating (IB) — As more steam electric power plants are built and ready sources of low cost once-through cooling water become more scarce, there is an increasing trend toward reuse or recycling of cooling water. The choice of the recycling or recirculating cooling system waters, by either cooling pond or cooling tower catchments, has a significant effect on the pollutants produced, the additives needed, and the composition of the blowdown streams. Cooling tower blowdown may contain various corrosion and deposit inhibitors, dispersants, biofouling control agents, solvents or carriers, corrosion products, and extracted wood preservatives.

There are three major types of cooling systems: cooling ponds and canals, mechanical-draft evaporative cooling towers, and natural-draft evaporative cooling towers. Each of these are used to remove heat from the recirculated water; in addition, each system has a number of adaptions.

Cooling ponds are generally most appropriate in relatively dry climates and in locations where large adjacent land areas are available. In some cases where land area is not readily available, spray facilities have been installed to reduce the required pond size. One important advantage to these systems is that, in many cases, chemical treatment of recirculating water can be avoided.

Mechanical-draft evaporative cooling towers are by far the most popular method used for cooling large steam electric power plants. The mechanical-draft tower uses fans to move air past the droplets or films of water to be cooled. Evaporation of water into the air stream provides the primary mechanism for cooling influent water. Advantages include smaller land requirements than for cooling ponds, and lower capital costs than for natural-draft towers.

Like the mechanical-draft towers, the wet natural-draft towers rely on water evaporation for cooling effect. Instead of fans, the tower is designed so that air will naturally flow from the bottom to the top of the tower as a result of: (1) density differences between ambient air and moist air inside the tower, and (2) the "chimney" effect of the tower's tall structure. Natural draft towers are often selected over mechanical draft towers in areas where low wet-bulb temperatures and higher humidity prevail.

Water Treatment (II): Boiler feedwater is treated to remove suspended and dissolved solids and prevent scale formation. Processes include clarification, softening, ion exchange, evaporator, filtration, and others.

Clarification (IIA) — Suspended solids are coagulated using alum, polyelectrolytes, etc., and allowed to settle. The clarified water is drawn off and usually filtered to remove the last traces of turbidity. Sludge is discharged to solid waste piles.

Softening (IIB) — Chemical precipitation, using calcium hydroxide or sodium carbonate, is practiced to reduce hardness and alkalinity of process waters. Clarification is generally followed by filters. Sludge is discharged to solid waste piles.

Ion Exchange (IIC) — When water is passed through a selected ion exchange resin, a high quality feedwater can be produced. The resin removes dissolved mineral salts through chemical reaction at active sites on the resin surface. When the resin is exhausted, it ceases to function and must be regenerated. Regeneration involves an initial backwash, contact with regenerating chemicals, and a final rinse. The rinse water and spent regenerating chemicals contain the associated contaminants from the waste stream.

Evaporation (IID) — Evaporation is a process used to purify boiler feedwater by vaporization, subsequent condensation, and external collection. In this process, a portion of the boiling water is drawn off as blowdown. As water evaporates from the pool, the raw water salts in the pool become concentrated. If allowed to concentrate too much, the salts will scale the heating surfaces, and heat transfer rates diminish. To prevent scaling, a portion of the pool water is drawn off as blowdown. Chemical composition is similar to that of the raw water feed, except that it is several times concentrated.

Filtration (IIE) — Water is filtered through an underdrained bed of granular material, usually sand, anthracite coal or other filter media, to remove the last traces of turbidity. Periodically, filters are backwashed, generating a wastewater stream containing suspended solids.

Other Treatment (IIF) — Reverse osmosis utilizes a semipermeable membrane which selectively permits the passage of water through the membrane, but rejects the passage of brine. This brine constitutes the wastewater stream from this process.

Boiler or Steam Generator (III): Approximately 70 percent of all power plants generate boiler blowdown. Power plant boilers are either a once-through or drum-type design. Once-through designs are used almost exclusively in high pressure, super-critical boilers; no wastewater streams are directly associated with their operation. Drum-type boilers, on the other hand, operate at sub-critical conditions, where steam generated in the drum-type units is in equilibrium with boiler water. Boiler water impurities are, therefore, concentrated in the liquid phase.

The concentration of impurities in drum-type boilers must not exceed certain limitations imposed by boiler operating conditions. Boiler blowdown, therefore, serves to maintain specified limitations for dissolved and suspended solids. The sources of impurities in the blowdown are the intake water, internal corrosion of the boiler, and chemicals added to the boiler system. A number of chemicals are added to the boiler feedwater to control scale formation, corrosion, pH, and solids deposition.

Boiler blowdown is usually of high quality, and may be even higher quality than the intake water. This water is usually suitable for internal reuse in the power plant, for example as cooling water makeup.

Maintenance Cleaning (IV)

Boiler or Steam Generator Tubes (IVA) — Cleaning is designed to remove scale and corrosion products that accumulate on the boiler's steam side and on the water side of the steam condenser. There are various chemicals used in boiler cleaning operations, each with different characteristics and applications. They include alkaline degreasers, ammoniated citric acid, ammoniated EDTA, ammoniacal sodium bromate, acids with additives. Waste constituents contain mostly boiler-dissolved metals, some oils and organics, and maintenance chemicals. Wastewaters have extreme pH levels, high dissolved solids and BOD_5 contents, and high chemical oxygen demand concentrations.

Boiler Fireside (IVB) — Boiler firesides collect fuel ash, corrosion products and airborne dust, and are commonly washed by spraying high pressure water against boiler tubes while they are still hot. Waste effluents from this washing operation contain an assortment of dissolved and suspended solids. Acid wastes are common for boilers fired with fuels which are high in sulfur content.

Air Preheater (IVC) — Air preheaters are used to heat ambient air prior to combustion, thus economizing thermal energy use. Two types are used: tubular or regenerative. In either case, part of the sensible heat of the combustion flue gases is transferred to the incoming fresh air. Both are periodically washed to remove soot and fly ash. Washing typically occurs once-a-month, however, industry has reported variations ranging from 4 to 180 washings per year.

Fossil fuels with high sulfur content will produce sulfur oxides, which will absorb on air preheater deposits. Water washing of these deposits will produce an acidic effluent. Alkaline reagents are often added to wash water to neutralize acidity, prevent corrosion of metallic surfaces, and maintain an alkaline pH.

Preheater wash water has a high solids content (both suspended and dissolved). Solids primarily include sulfates and heavy metals such as copper, iron, nickel and chromium.

Miscellaneous Small Equipment (IVD) — At infrequent intervals, other plant components such as condensate coolers, hydrogen coolers, air compressor coolers, stator oil coolers, etc. are cleaned chemically. Inhibited hydrochloric acid is commonly used. Detergents and wetting agents are also added when necessary. The waste volume from these washings is smaller than that resulting from other types of chemical cleanings. Pollutant parameters are extreme pH levels, total suspended solids, metals and oil.

Stack (IVE) — Depending upon the fossil fuel used, power plant stacks may have deposits of fly ash and soot. These deposits can be highly acidic as a result of sulfur oxides in the flue gases. If a wet scrubber is used to clean the flue gas, process or equipment upsets can result in additional scaling on the stack interior. Normally, high-pressure water is used to clean the deposits on stack walls. These wastewaters may contain total suspended solids, high or low pH values, various metals, and oil.

Cooling Tower Basin (IVF) — Depending upon the quality of the makeup water used in the cooling tower, carbonates can be deposited in the tower basin. Similarly, some algae growth may occur on basin walls. Debris carried in atmosphere may also collect in the basin. Consequently, water is used for periodic basin washings. The primary pollutant in the wastewater is total suspended solids.

Ash Handling (V): Steam-electric power plants which are fueled by oil or coal produce ash as a combustion waste by-product. The total ash content consists of bottom ash and fly ash. Bottom ash is the residue which accumulates in the furnace bottom, in either a dry or molten state; and fly ash consists of fine particles trapped by dust collectors, usually electrostatic precipitators for the flue gas stream.

Ash-handling is the conveyance of the accumulated waste products to a disposal system. The method of conveyance may be either wet (sluicing) or dry (pneumatic). Only wet handling methods will be discussed. Chemical characteristics of ash-handling wastewater depend upon the type of fuel burned, usually oil or coal.

Oil-Fired Plants (VA) — The quantity of ash produced in an oil-fired plant is very small. However, the settling characteristics of oil ash are not as favorable as those of coal ash. Most oil ash deposits are partially soluble and can be removed by water washing. Oil ash can contain significant amounts of metals.

Coal-Fired Plants (VB) — Two types of ash are generated in coal-fired power plants: fly ash and bottom ash.

Fly ash may be conveyed dry or as a slurry to storage and disposal. The slurry may be separately discharged to a fill, or may be combined with the bottom ash handling system. However, due to its extremely low settling rate, fly ash should not be pumped into a pond from which water is to be reclaimed, unless the pond is greatly enlarged to provide long retention times.

Bottom ash in a dry or molten state is handled by sluicing or jetting, and is discharged either directly to fill, to dewatering bins for removal by truck or railroad car, to an ash pond, or to a sedimentation basin. A water recovery system permits recirculation and reuse of conveying water. Wastewaters contain large quantities of solids.

Drainage (VI)

Coal Pile (VIA) — For coal-fired generating plants, outside storage of coal at or near the site is necessary to assure continuous plant operation. Normally, a 90-day supply is maintained. Contact of open piles with air and moisture results in oxidation of metal sulfides to sulfuric acid. When rain falls on these piles, the acid is washed out and eventually winds up in coal pile runoff. Storage piles are sometimes sprayed with a tar to seal their outer surface and prevent rain water infiltration.

Water pollution associated with coal pile runoff is due to the chemical pollutants and suspended solids usually transported in coal pile drainage. Drainage quality and quantity is variable, depending on the meteorological condition, area of pile and type of coal used.

Coal pile runoff is commonly characterized as having a low pH (high acidity), and a high concentration of total dissolved solids including — iron, magnesium and sulfate. Undesirable concentrations of aluminum, sodium, manganese, and other metals may also be present.

Contaminated Floor and Yard Drains (VIB) — When floors are washed, floor drains collect dust, fly ash, coal dust and floor scrubbing detergent. This waste stream also contains lubricating oil or other oils, which are washed away during equipment cleaning, oil from leakage of pump seals, etc., and oil collected from spillage around storage tank area.

Air Pollution (SO_2) Control Devices (VII): SO_2 is absorbed from stack gases using a flue gas desulfurization system (FGD), which scrubs the exiting gases with an alkaline slurry. This may be preceded by partial removal of fly ash from the stack gases. FGD processes may be nonregenerable or regenerable. Nonregenerable processes include lime, limestone, lime/limestone combinations, and double alkaline systems. These processes involve the precipitation of $CaSO_3$ and $CaSO_4$ solids, and generate large amounts of sludges. After settling, the supernatant may be recycled back into the scrubbing process. Regenerable processes include magnesium oxide and Wellman-Lord. There is no wastewater or sludge stream associated with regenerable processes.

Miscellaneous Waste Streams (VIII): Miscellaneous waste streams include sanitary wastes, plant laboratory and sampling systems, intake screen backwash, closed cooling water systems, low-level radioactive wastes, and dusts and other wastes from construction activity.

Wastewater Characteristics

Table 74 shows raw wastewater characteristics. Characteristics of the individual wastewater streams vary depending upon fuel composition, type of boiler, construction materials, type of scrubber (for air pollution wastewaters), feedwater composition, additives, and cleaning chemicals used. Wastewaters are produced relatively continuously from the following sources (where applicable): cooling water systems, ash handling systems, wet scrubber air pollution systems, and boiler blowdowns. Wastewaters are produced intermittently by water treatment operations (ion exchange, filtration, clarification, evaporation), maintenance cleaning, and rainfall runoff. Important pollutants include residual chlorine, metallic wastes (particularly copper, iron and zinc), dissolved and suspended solids, extreme pH levels, and some organics.

Table 74: Steam Electric Utility Industry—Raw Wastewater Characteristics

Wastewater Parameters	Subcategories[a]					
	1A	1B	IIC	IID	III	IV A
Conventional - mg/l						
COD						24.9-9,900
Iron				0.54	0.53	0-10,800
Nitrogen (ammonia)						140-5,200
Nitrogen (nitrate)						0.01-0.5
Nitrogen (organic)						10-870
Oil and grease			0.13-22	2.1	1.7	5-41
pH (units)			1.7-10.6			0.5-10.5
Phosphorus (total)					17	1.2-300
TDS			1,800-9,600			340-74,000
TOC						90-4,600
TSS			3-305	28.4	66	8-2,300
Inorganics - ug/l						
Antimony		1.5				
Arsenic		1-35				5-307,000
Asbestos (fibers/l)		$(0.13\text{-}160)10^6$				
Beryllium		3.4				10
Cadmium		5-100				1-51
Chromium		6-300				0-27
Copper		10-3,800		390	140	2,000-11,700,000
Cyanide		30-230				
Lead		1-300				10-5,000
Mercury		0.24-50				0-15,000
Nickel		3-200				0-500,000
Selenium		3-32				2-24,000
Silver		2-40				10-70
Thallium		4-8				
Zinc		17-730				60-840,000
Organics - ug/l						
Bis (2-ethylhexyl) phthalate	12-49	15-157				
Bromoform	97-580	4-154				
Chlorodibromomethane		3-59				
Chloroform	20	5-25.5				
1,3-Dichlorobenzene	12	20-26				
1,1-Dichloroethylene	16					
Di-n-butyl phthalate	12-46	18-31				
Diethyl phthalate	11	10				
Methylene chloride	480-9,400	9,400				
Pentachlorophenol		13				
Phenol	10	20				20-70
Tetrachloroethylene	78	1				
1,2-Trans-dichloroethylene	11					
1,1,1-Trichloroethane	12	45				
2,4,6-Trichlorophenol		35				
Toluene		1-14				

(continued)

Table 74: (continued)

Wastewater Parameters	Subcategories[a]						
	IV B	IV C	V B1	V B2	V B (1+2)	VI	VII
Conventional - mg/l							
COD		50-70					1-390
Iron	1,000-41,000	0.05-8,250			0.23-2.3	23-1,800	0.02-8.1
Nitrogen (ammonia)					0.04-0.64		
Nitrogen (total)							0.001-0.002
Oil and grease		0.25-8.5					
pH (units)		3.2-3.5			6.3-11.5	2.3-3.1	3-12.7
Phosphate (PO_4)					0.01-0.12		0.03-0.41
TDS		600-750			151-452	2,300-16,000	2,800-92,500
TOC	5,000						
TSS	250-50,000	30-10,211			12-140	8-2,300	
Inorganics - ug/l							
Antimony					2	4-1,800	90-2,300
Beryllium					2-10	10-70	2-180
Chromium	1.5	1,000-1,450	7-70	4	3-200	5-20	10-500
Copper	2,000-250,000	100-6,000	16	35-80	2-100	100-1,400	2-560
Cyanide					8-18	30-200	
Lead			111	5	10-25	10	10-520
Mercury			23		0.2-38	0.2-1,000	0.1-70
Nickel	70,000	18,000-25,000	7-21	30	0.8-900	150-4,500	50-1,500
Selenium					1-65	1-30	0.6-2,700
Silver			4	3	5-10		5-600
Thallium					9-200		
Zinc	4,000	1,000-1,500	1,000-1,400	70-100	250	1,100-16,600	10-590
Organics - ug/l							
Acenaphthylene					16		
Bis (2-ethylhexyl) phthalate			17-25	309	3-33	12	
Butyl benzyl phthalate					10		
Chloroform						17	
2-Chloronaphthalene					52		
1,3-Dichlorobenzene			41	65	35-64	20	
2,4-Dichlorophenol				83			
2,4-Dinitrophenol					50		
Di-n-butyl phthalate			6-46	26-46	8-46	32	
Methylene chloride			140-9,400	9,400	140-9,000	3,900	
Pentachlorophenol				51			
Phenol					4-6		
1,2-Trans-dichloroethylene						53	
1,1,1-Trichloroethane					27		

[a] See Table 73 for identification of subcategories.

Control and Treatment Technology

In-Plant Control Technology

Process Changes — Once-through cooling water systems may be modified to a recirculating system, resulting in a substantial reduction in the amount of wastewater discharged. Substantial reductions can also be achieved by preparing boiler feedwater in a combined system of hot lime softening and ion exchange processes. Because of the reduced dissolved solids load from the ion exchange process, the system would require less frequent regeneration. Substitution of reverse osmosis or electrodialysis for ion exchange may reduce discharges.

Coal pile runoff contamination can be reduced by spraying coal piles with tar or covering them with plastic sheeting to seal the surface against water infiltration. Groundwater contamination is minimized by placing an impervious liner below coal pile.

Substitution of Process Ingredients — A large number of biocides are available as a substitution for chlorine. Cooling tower blowdown can be reduced by increasing the concentration cycle. This can be achieved by lime softening of cooling water makeup, by substituting scale-forming ions with more soluble ions, or by using sequestering agents such as polyesters and phosphonates to prevent deposition of precipitates. The dispersing agents, however, become pollutants in blowdown water. Installation of plastic, concrete, or ceramic system components which are more corrosion and erosion resistant would reduce water treatment requirements. Synthetic organic corrosion inhibitors, which are more compatible with biological treatment systems, can be substituted for chromate-based compounds.

Water Conservation and Reuse — Once-through cooling and cooling tower blowdown water can be used to sluice ash from boilers of coal-fired plants or as makeup water for SO_2 scrubbing solutions. Wastewater from ion exchange regeneration can be neutralized, treated in settling chambers, and used as makeup to recirculating cooling water, ash transport water, or for flue gas scrubbing. Boiler blowdown, a high quality water, can be reused for any process in the plant.

Sluice water can be completely recirculated from bottom ash and combined ash ponds. Chemical addition is not normally required for pH adjustment or scale prohibition. Recirculation of fly ash sluice water requires softening for a closed-loop fly ash system.

Metal cleaning wastes, after treatment, can be used for ash transport or scrubber supply.

Recovery and Reuse of Process Chemicals — Nearly all chromates used as corrosion inhibitors can be recovered using ion exchange systems. Regenerant streams from these beds, high in chromate concentration, can be returned to the cooling system. Lime sludge can be used in the air scrubber system.

Control of Specific Waste Constituents — The reduction of chlorine in once-through cooling systems may be accomplished by two approaches: reducing the chlorine dose, and dechlorination using SO_2 or sodium sulfate. The chlorine dose can be reduced by developing and implementing a chlorine minimization program. This involves a preliminary study of the raw water source, the type of microorganisms present, and the determination of the minimum concentration of chlorine required for biofouling control. Dechlorination can be achieved by adding SO_2 (or its derivatives) at the discharge channel. The control of chlorine concentration in the blowdown of recirculating cooling waters is achieved by one of the previously mentioned approaches, or by blowdown water retention, or substitution of chlorine with other available biocides. Since residual chlorine dissipates during the chlorination cycle, the system may be bled during periods when the chlorine content is at a minimum. Blowdown water retention for reuse or disposal by evaporation can control residual chlorine in the discharge stream.

Good Housekeeping Practices — Precautionary measures against spills and leaks and good maintenance procedures will reduce water consumption and improve water quality.

End-of-Pipe Treatment Technology: Wastewater effluents are usually treated by one or more of three processes: pH control, removal of dissolved materials, and separation of phases. With the increased use of organic materials for cleaning solutions, the use of incineration of boiler chemical cleaning solutions in power plant boilers has become more prevalent. However, emissions of metals into the atmosphere may be a problem. Some utilities employ ash ponds for the treatment of boiler chemical cleaning wastes. Many ash ponds are naturally alkaline, and thus have good potential for metal hydroxide precipitation. Indeed, fly ash has been used in water treatment to increase the rate of floc growth. A number of treatment schemes employing physical/chemical processes have been tested, designed and implemented for the treatment of boiler chemical cleaning wastes. The basic mechanism behind these treatment schemes is neutralization with caustic or lime followed by precipitation of the metal hydroxide compounds, clarification, and pH adjustment. Additional processes used alone or in combination include oxidation, sulfide addition, filtration and carbon adsorption.

Coal pile and other runoff can be collected, neutralized to precipitate dissolved metal ions, and followed by sedimentation to remove suspended solids. In certain cases, flocculant addition would be required. Where applicable, coal pile runoff could be evaporated in ponds provided with an impervious liner to prevent groundwater contamination.

Bench-scale test data on the effectiveness of several treatment technologies for cooling tower blowdown and ash pond overflow streams at three power plant sites are shown in Tables 75 and 76. Reverse osmosis is effective for concentrating copper, nickel, zinc, lead, silver, chromium, and arsenic.

Table 75: Steam Electric Utility Industry—Removal Efficiencies of Pollutants in Cooling Tower Blowdown at Three Plants—Bench Scale Results

Wastewater pollutant	Percent removed by reverse osmosis			Percent removed by chemical precipitation to pH 11.5			Percent removed by chemical precipitation to pH 11.5 plus Fe^{+2}			Percent removed by by activated carbon		
	A	B	C	A	B	C	A	B	C	A	B	C
Inorganics												
Arsenic	-[a]	75		86	25	-	86	25				
Antimony	60	-		40	43			-				
Beryllium			85			76			85			
Cadmium	-	-	37		-	38		11	37			
Chromium		60	94		-	76		40	95			
Copper	82	79	92	73	62	89	86	91	96			
Cyanide	-	80	-									
Lead	-	0	96	-		96		0				
Mercury	-		60		-	60		0	60			
Nickel	83	50	10	-	52	43	50	0	10			
Selenium	50											
Silver	-	14	92	-	-	44	-	43	93			
Thallium			50			88			88			
Vanadium	8	-	-	-	78	45	-	56	-			
Zinc	99	92	97	82	92	97	95	92	97			
Organics												
Acenaphthene		10	41									
Benzene	33		33							33		33
1,2-Benzanthracene		63	41							8	69	
3,4-Benzofluoranthene										53		
Chloroform	58		58							58		58
Diethylphthalate			63									
Dimethylphthalate		84	56									
Di-n-butylphthalate		-								-	-	
Fluoranthene										60		
Pyrene			71							69		
Tetrachloroethylene	35	35										
Toluene	15									86		

[a] "-" implies effluent higher than influent.

Table 76: Steam Electric Utility Industry–Removal Efficiencies of Pollutants in Ash Pond Effluents at Three Plants–Bench Scale Results

Wastewater pollutant	Percent removed by reverse osmosis			Percent removed by chemical precipitation to pH 11.5			Percent removed by chemical precipitation to pH 11.5 plus Fe^{+2}			Percent removed by by activated carbon		
	A	B	C	A	B	C	A	B	C	A	B	C
Inorganics												
Arsenic		89	99		89	99		67	99			
Antimony	50[a]	86	50	14	-	20	-	-	30			
Beryllium	-			80			80					
Cadmium		35		50	0		50	-				
Chromium	75	67		50	-		50	33				
Copper	89	29	65	71	29	54	80	50	31			
Cyanide	82		23									
Lead	67	25			25			25				
Mercury	-											
Nickel	89	9	40	95	-	12	95	-	20			
Selenium	67	75	85		0	-		12	24			
Silver	64	60	0	9	20	-	9	20	-			
Thallium	-		89						22			
Vanadium	81	82	32	37	0	39	37	-	39			
Zinc	82	57	82	90	5	82	92	14	82			
Organics												
Acenaphthene		0										
Benzene	30	0	0							50		
1,2-Benzanthracene	-	-										
Diethylphthalate	80									80		
Dimethylphthalate		-									35	
Di-n-butylphthalate	38									38		
Toluene	20											

a "-" implies effluent higher than influent.

Also, chemical precipitation with lime is effective in decreasing copper, zinc, arsenic, chromium and lead. A combination of lime and ferrous sulfate results in higher removal efficiencies. In most cases, the organics in the two waste streams were at such low concentrations that the removal efficiencies could not be determined.

Table 77 shows laboratory treatment results of boiler cleaning wastes using dilution and pH adjustment to 9.5. Metals were precipitated and filtered.

Table 77: Steam Electric Utility Industry–Treatment of Boiler Cleaning Wastewater (IVA) Summary of Jar Tests[a]

Dissolved metals	Concentration before treatment, mg/l	Concentration after treatment - mg/l, Dilution factor prior to treatment: 20:1	10:1	5:1	None
Copper	306	0.03	0.34	0.32	0.35
Iron	5,140	0.14	0.31	0.60	0.52
Manganese	41	0.01	0.01	0.14	0.12
Nickel	375	0.04	0.13	0.31	2.9
Vanadium	0.8	0.1	0.1	0.1	0.5
Zinc	335	0.02	0.045	0.2	0.74

[a] Treatment includes dilution of boiler cleaning wastes (HCL) with copper complex cleaning agent, pH adjustment to 9.5.

Type of Residue Generated

As prevously indicated, numerous wastewater streams are generated in steam electric power plants. However, in terms of mass and characteristics of residues generated, there are three principal waste streams: ash transport, flue gas scrubbing, and cooling tower blowdown. Thus, these three will be discussed first; the other flows will be discussed later.

Ash from boilers comes off as fly ash (the ash entrained in the flue gases) and bottom ash or slag (the material which accumulates in the bottom of the combustion chamber). The amount of ash and the proportion coming off as fly ash and bottom ash or slag, are determined by the type of fuel — gas, oil or coal — and the characteristics of the fuel from the specific source used. Also, in the case of coal, ash production depends on: preparation, cleaning, and sizing procedures; design of the furnace — pulverized feed, cyclone or stoker feed; and methods of ash collection, handling, and disposal.

Burning of natural gas produces no ash. Burning of oil produces small amounts of fine ash, which may contain significant amounts of metals, and may be partially soluble. Burning of coal produces significant amounts of ash; the volume of this residue is typically from 3 to 30 percent of the volume of coal burned. The range of noncombustible minor and "trace" elements and components that form the bulk of the ash varies so much that few generalizations are possible. Table 78 gives the ranges for various elements. Several heavy metals are conspicuous on the list, especially Cr, Ni, Cu, and Pb.

Table 78: Range of Trace Metals in U.S. Coals[a]

Element	Range of concentration, mg/l
Antimony	0.2-9
Arsenic	0.5-106
Beryllium	0-31
Boron	1.2-356
Bromine	4-52
Cadmium	0.1-65
Chromium	0-610
Cobalt	0-43
Copper	1.8-185
Fluorine	10-295
Gallium	0-61
Germanium	0-819
Lanthanum	0-98
Lead	4-218
Manganese	6-181
Mercury	0.01-1.6
Molybdenum	0-73
Nickel	0.4-104
Phosphorus	5-1,430
Scandium	10-100
Selenium	0.4-8
Tin	0-51
Uranium	< 10-1,000
Vanadium	0-1,281
Yttrium	< 0.1-59
Zirconium	8-133

[a] Source: Los Alamos Scientific Laboratory. Environmental Contamination from Trace Elements in Coal Preparation Wastes. Springfield, VA: National Technical Information Service, August 1976. PB 267 339.

Pulverized feed is used for larger furnaces. If the melting point of the bottom "ash" is low enough, a slag is produced which can be tapped and drawn from the furnace in its molten state. Otherwise, a solid, granular bottom ash forms, which can be removed either dry or sluiced away with water. Coals which form bottom slags in such furnaces tend to produce about 50 percent fly ash and 50 percent bottom slag. Coals which do not form a slag tend to produce about 80 percent fly ash and 20 percent bottom ash by volume.

Cyclone furnaces use a coarse size of coal, generally of the type which forms bottom slags. Typically, such furnaces give 15 to 20 percent fly ash and 80 to 85 percent bottom slag per unit of coal burned.

Stoker feed furnaces use an even coarser coal feed, but are practical only for small size units. Coal feed and bottom ash removal are done mechanically. Different types of stokers can produce fly ashes in the range of 10 to 55 percent of the coal burned.

Fly ash is generally captured in electrostatic precipitators, flue gas scrubbers, or other flue gas handling equipment. Removal of such fly ash, like the removal of bottom ash from the furnace, can be done either dry, or by sluicing with water. Only sluicing will produce a wastewater and a wastewater treatment residue. Heavy metals are a likely component of the ash residues. The fine particle size and the surface properties of fly ash cause suspended particles to have poor settling properties.

As SO_2 removal becomes a more widespread practice at power plants, flue gas scrubber sludges are increasingly being generated. The sludge is the product of the reaction of the sulfur dioxide with lime or other alkalis, plus some fly ash and other materials which pass through the dust removal system. Scrubber sludge is mostly a mixture of calcium sulfite and sulfate. Particle size and surface properties are quite different for sulfite and sulfate salts, with sulfate having much better settling and dewatering characteristics. Sulfite can be readily oxidized in air to produce sulfate, at some additional expense. Because of the entrained fly ash, scrubber sludge can also be "contaminated" with heavy metals.

Cooling tower blowdown is used to control the levels of total dissolved solids. Evaporation of water from the tower leaves behind those solids originally in the water. The concentration of these scale-forming solids tends to build up if not controlled. Because certain compounds are added to the cooling tower water to inhibit corrosion and biological growths in the tower, the inhibitors—frequently chromate salts—are also present in the blowdown. The most common method of chromate treatment is to reduce the chromic ion with SO_2 or a sulfite salt, followed by precipitation as a metal hydroxide at high pH. The resulting sludge is relatively easy to dewater if lime has been used as the precipitant, but, because it has a high chromium content, it is likely to be designated a hazardous waste.

Analytic data for the other sludges generated by this industry is not available. However, some of these other sludge streams can also cause problems, even though they are generated in smaller quantities. Maintenance cleaning sludges (IV) can be expected to contain heavy metals. Water treatment sludges (II) consist of clarifier sediments and filter washes. Alum sludge is a bulky gelatinous substance, composed of aluminum hydroxide, inorganic particles, color colloids, microorganisms, and other organic matter. The major constituent from lime soda softening is calcium carbonate. Other constituents may include metal hydroxides, organic, and inorganic matter.

Many small wastewater streams are simply added to, and treated with, the major ash sluicing flows. Thus, the concentrated brines from water treatment by ion exchange, evaporation, or reverse osmosis are diluted out and dissipated in much larger flows. Also, periodic cleaning of boiler tubes—either "steam" or "fire" side—, air preheaters, stacks, or the cooling tower basin, as well as drainage from the coal piles, floor drains, or yard drains, may add small increments of pollutants to the main flows. Incremental increases in the sludge, and its heavy metals content, metal corrosion products, scale, soot, fly ash, and coal dust may result.

Based on a 1978 survey, an estimated 68.1 million dry tons of fly ash, bottom ash, and slag were produced by this industry. Approximately 16.4 million dry tons were used and 51.7 million dry tons were disposed of in ash disposal ponds and/or landfills. Also, another survey estimated that 43.8 million dry tons of air pollution (flue gas scrubber) sludge was generated by this industry in 1977, and that 216.7 million dry tons would be generated in 1987, making this industry by far the largest generator of air pollution control sludges.

Residual Management Options

Ash transport and disposal can be done by either "wet" or "dry" methods. The wet method involves sluicing the ash as a water slurry to a large settling and storage pond. The dry method involves transport as an air-borne suspension to storage silos and bulk transfer in trucks to landfills. Because the dry method is not a wastewater problem, only the wet method is of interest here. A 1978 survey showed that the wet method is used in 51 percent of the cases; however, a growing trend toward using the dry method is developing. Limitations on the quality of water discharged from ash ponds, more stringent topographic and site requirements for ponds than landfills, and the less efficient use of land area when used for ponds, have caused this trend. Because of the sheer volume of material and land area required, siting and site restoration are major considerations. Soil permeability, seepage, potential groundwater contamination, and uncertain possibilities for revegetation are other major factors.

There are also limited uses of fly ash as pozzolanic additives in Portland cement concrete. Use of fly ash as structural fill is governed by its tendency to behave like fine-grained soils such as wet silt or clay.

Flue gas scrubber sludges composed of $CaSO_3/CaSO_4$ mixtures can be disposed of "wet" in ponds, or "dry" by dewatering to cake form and placing in landfills. Sulfite impairs settling and dewatering characteristics and bearing capacity, and makes land containing sludge ponds permanently unusable. Oxidation of the sulfite to sulfate by aeration improves these characteristics, but this method is still experimental. The presence of some fly ash, either picked up in the scrubber or deliberately added afterward, also improves sludges' physical properties.

Gravity thickening, vacuum filtration, and centrifugation are the most commonly used methods for dewatering such sludges. Table 79 shows the solids concentrations achievable by these methods. In an April 1978 survey of 30 existing and 39 planned scrubber installations, 19 existing installations and 26 planned ones used gravity thickeners, 6 and 13 used vaccum filters and 1 and 3 used centrifuges.

Table 79: Typical Solids Concentrations of FGD Sludges

Sludge source	Sulfite sludges, percent solids	Sulfate sludges, percent solids
Scrubber bleed	5-15	5-15
Gravity thickened or settling ponds	20-45	60-65
Vacuum filtration or centrifuging following primary clarification	50-60	80-85

Another alternative to enhance the disposal of scrubber sludges is fixation. This involves mixing with proprietary additives or various bulk materials such as fly ash, lime, or blast-furnace slag. Fixation is more important for scrubber sludges that are to be disposed of "dry", that is, in landfills. Landfilled material must have a certain minimum shear strength in order to not flow away on sites with a slight slope. The intent of fixation is to insure such minimum strength. Sludge disposed of "wet" however, is retained behind dikes as if it were a liquid, and shear strength is unimportant.

Considerations of seepage, site reclamation, etc., also affect scrubber sludge ponds and landfills in the same way they affect ash ponds and landfills.

Flue gas scrubber sludges that are low in $CaSO_3$ and fly ash have also been used in Japan and Europe as gypsum wallboard. Other such uses of this sludge by-product are under intensive study. Certain alternate scrubbing techniques that produce less, or no, waste sludge are also being investigated.

Both ash and flue gas scrubber sludges can have significant toxic heavy metals contents. Cooling tower blowdown residues, especially those with high chromate contents, can be greatly decreased by substitution of other organic-type inhibitors. Alternately, one can mix the blowdown flow with the water intended for makeup, and then treat the mixture by lime precipitation/softening. The softened water has sufficiently low scale-forming characteristic, making it a potential candidate for makeup water. The chromate inhibitor from the blowdown, being soluble, remains in the supernatant/overflow and returns to the cooling tower. With this system, chromium loss in the wastewater and sludge is drastically reduced, and the softened sludge has a low enough chromium content that it may not be considered "hazardous".

TEXTILE MILLS

General Industry Description

The textile industry involves the manufacture of fabrics from cotton, wool, and synthetics. Synthesis and spinning of synthetic fibers is not included in this industry, but rather is covered under synthetic organic chemicals.

The textile industry may be broadly divided into two groups. Facilities in the first group are primarily engaged in receiving and preparing fibers; transforming these materials into yarn, thread or webbing; converting the yarn and web into fabric or related products; and finishing these materials at various stages of production. All of the above processes are included in the Textile Mills Category. The second group consists of facilities principally engaged in receiving woven or knitted fabric for cutting, sewing and packaging. In general, all processing in this category is dry, and little or no discharge results; therefore, this group is excluded from the Textile Mills Category.

Basic raw materials in this industry are wool, cotton, and man-made fibers. Of the three major textiles, wool represents the smallest market, and synthetic textiles the largest. The natural fibers are supplied in staple form (staple being short fibers). The man-made fibers are supplied in either staple or continuous filament form. In either case, the fiber is spun into yarn, which is simply a number of filaments twisted together. The yarn is woven or knitted into a fabric, and the fabric is then dyed and treated to impart such characteristics as shrink resistance, crease resistance, fireproofing, etc. The finished fabric is delivered (directly or through converters, jobbers, and wholesalers) to the manufacturer of textile products.

In a USEPA survey of 1,973 textile mills, 245 mills were direct dischargers, 974 were indirect dischargers, 196 were zero dischargers, and 558 had an undefined type of discharge.

Industrial Categorization

A useful categorization for the purposes of raw waste characterization and organization of pretreatment information is shown in Table 80.

Table 80: Textile Industry Subcategories

Subcategory	Designation
Wool scouring	A
Wool finishing	B
Low water use processing	C
Woven fabric finishing	D
Knit fabric finishing	E
Carpet finishing	F
Stock and yard finishing	G
Non-woven manufacturing	H
Felted fabric processing	I

Process Description

A general diagram of the textile industry is shown in Figure 50.

Wool Scouring (A): Wool is a natural fiber from sheep, and contains many impurities which must be removed before further use. Wool scouring is the process that converts the raw wool into cleaned wool yarns. Two methods of wool scouring, solvent and detergent scouring, are practiced. In the United States, the latter is used almost exclusively. Figure 51 shows a typical wool scouring process flow diagram.

In detergent scouring, the fleece is carried through a series of scouring bowls. Detergent is added to emulsify greases and oils. The scour liquor which results contains significant quantities of oil and grease. It also contains materials derived from sheep urine, feces, blood, tars, branding

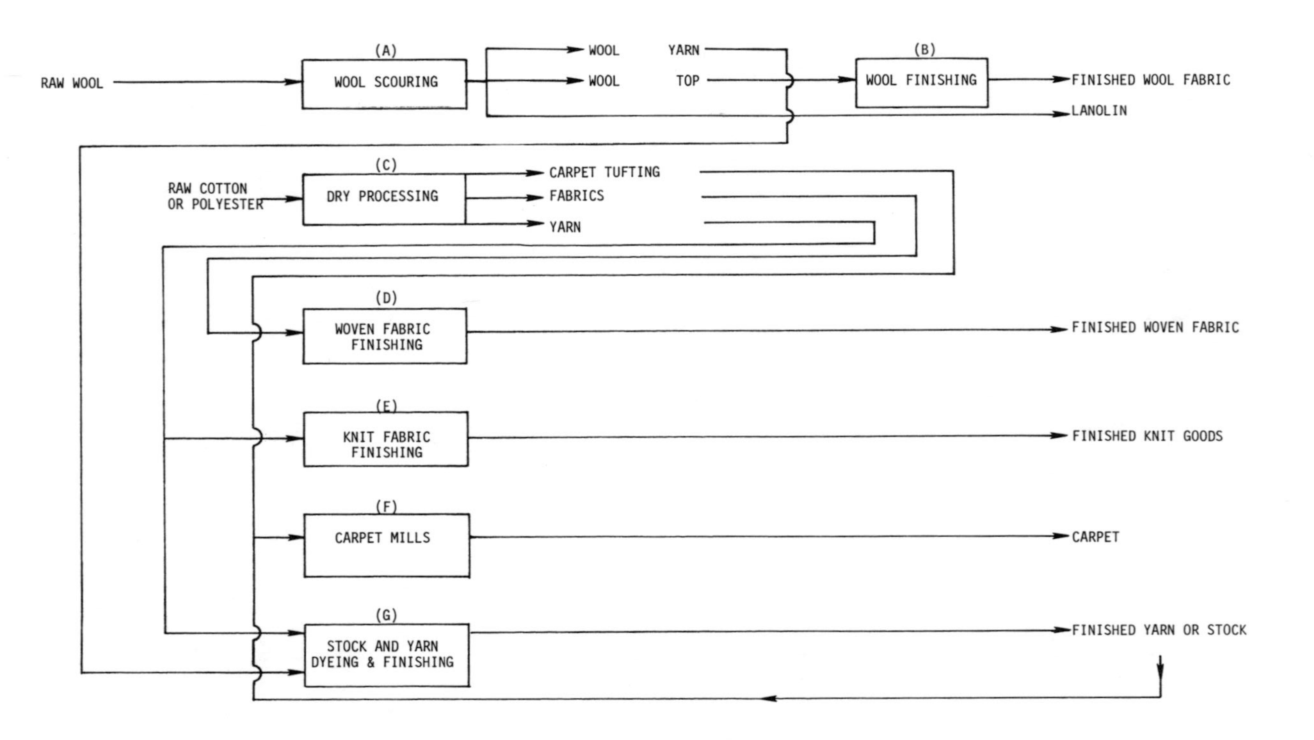

Figure 50: Process flow diagram–textile manufacturing industry.

fluids, and insecticides, as well as grit. A grease reduction step is necessary to reduce pollution, and due to the commercial value of wool grease, grease recovery is currently practiced by most scouring operations. Two methods may be used: centrifuging and acid cracking. Wastewaters may, however, contain significant quantities of oil and grease, even after in-process recovery.

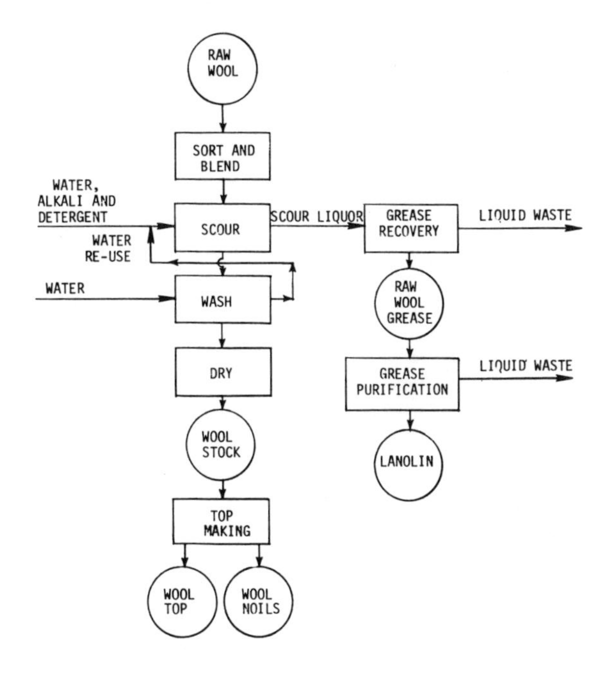

Figure 51: Typical wool scouring process flow diagram.

Wool Finishing (B): This process converts wool fibers into finished wool fabric, with washing, drying, weaving, and knitting as the intermediate steps. Wool finishing has a higher water usage rate than any other fiber finishing category. Scouring is the term applied to the washing of the fabric by the use of detergents, wetting agents, emulsifiers, alkali, ammonia, or other washing agents. The purpose of this scour is to remove oils, grease, dirt, etc. Figure 52 shows a typical wool finishing process flow diagram.

Scours may be either heavy or light, depending on fabric type and acquired impurities. In particular, heavy-weight, closely woven fabrics with a high percentage of recycled wool require very heavy detergents, long wash times, and extensive rinsing to clean the goods.

With either type of scour, large amounts of water are used for rinsing. Heavy scour produces higher organic and hydraulic loadings than light scour.

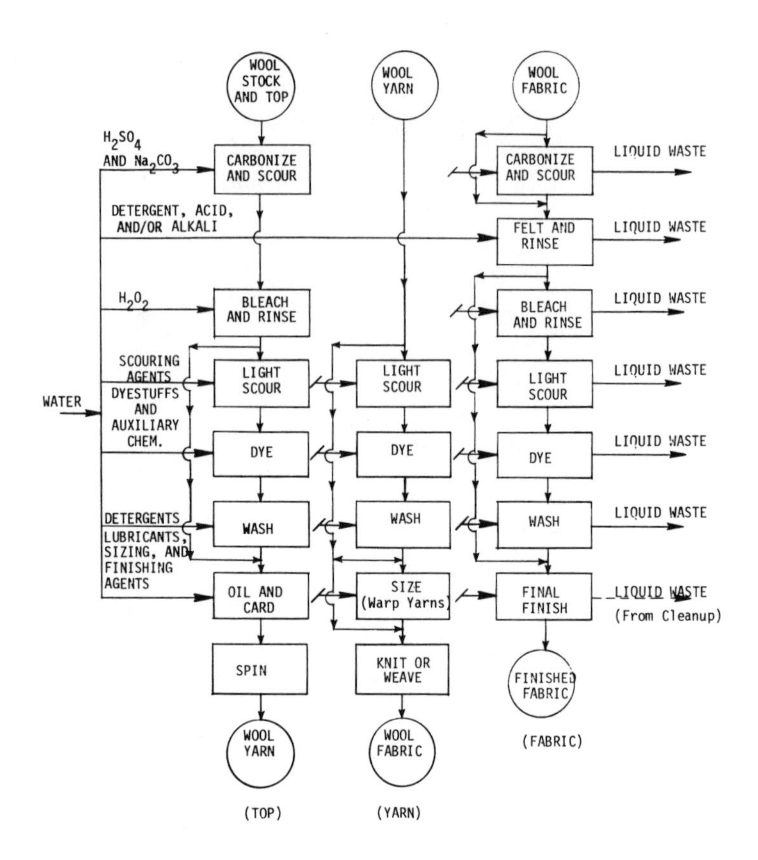

Figure 52: Typical wool finishing process flow diagram.

Carbonizing consists of soaking the fabric in sulfuric acid and baking the wool to oxidize any contaminants. This is followed by preliminary rinsing, neutralization, and final rinsing. The acid bath is discharged when it is too contaminated for further use, which is about once every two days.

Fulling is usually used on 100 percent woolen fabrics to stabilize the dimensions of the wool and to impart thickness and a substantial feel to the fabric. There are two common methods of fulling; alkali and acid. In both methods, the fabric is first impregnated with fulling solution and then is mechanically worked in the presence of heat, and sometimes pressure. Following this work, the fabric is rinsed extensively.

Bleaching is a common preparation for dyeing operations. Hydrogen peroxide is the most frequently used bleaching agent, although hypochlorite, peracetic acid, chlorine dioxide, sodium perborate, or a reducing agent may be used.

The more commonly used dyes for wool or wool blends are acid dyes or metallized dyes, and a small amount of chrome may be expected in the effluent. In the dyeing process, generally 90 percent or more of the dye is exhausted, and the dye bath is discharged to the sewer. Since dyes are very expensive, effort is made to assure as high an exhaustion level as possible.

After the fabric is dyed and rinsed, finishing agents may be applied, such as mothproofing, soil repellents, and fire retardents. Any of the finishing chemicals can appear in the wastewater when equipment is washed.

Low Water Use Processing(C): Dry processing textile operations include products and processes which by themselves do not generate large wastewater discharges. Such operations include yarn manufacturing, yarn texturizing, unfinished fabric manufacture, fabric coating, fabric laminating, tire cord and fabric dipping, carpet tufting, and carpet backing.

Two typical flow diagrams are shown in Figure 53. Generally, the only liquid waste from these processes originates in washouts from the slashing or sizing operation. Prior to being woven, the yarns are coated with a sizing material to give the yarn lubrication and strength. Cottons are generally coated with starch, and synthetics with polyvinyl alcohol. Wool and wool blends are seldom sized. Wastewater is also generated in water jet looms, but these are not commonly used in the United States. The wastewater generally represents a low percentage of the total plant flow.

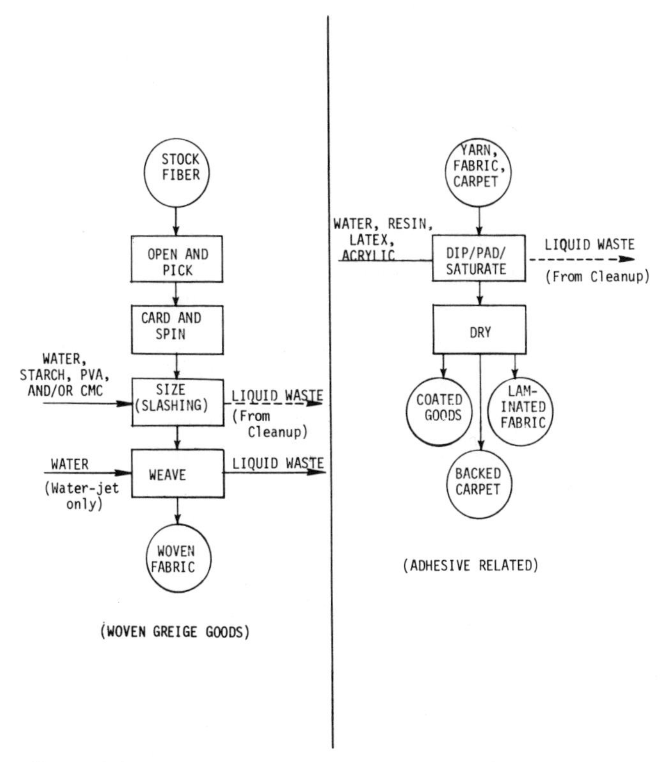

Figure 53: Typical low water use processing flow diagrams.

Woven Fabric Finishing(D): This subcategory covers facilities that primarily finish fabrics, a majority of which are woven. Due to the large number of processes and the resulting larger number of combinations of processes possible in a given plant, it is important that this subcategory be further subdivided as follows:

- Simple Processing - This Woven Fabric Finishing subdivision covers facilities that perform fiber preparation, desizing, scouring, functional finishing, and/or one of the following processes applied to more than five percent of total production: bleaching, dyeing, or printing. This subdivision includes all Woven Fabric Finishing mills that do not qualify under either the Complex Processing or Complex Processing Plus Desizing subdivisions.
- Complex Processing - This Woven Fabric Finishing subdivision covers facilities that perform fiber preparation, desizing of less than 50 percent of their total production, scouring, mercerizing, functional finishing, and more than one of the following, each applied to more than five percent of total production: bleaching, dyeing, and printing.
- Complex Processing Plus Desizing - This Woven Fabric Finishing subdivision covers facilities that perform fiber preparation, desizing of greater than 50 percent of their total production, scouring, mercerizing, functional finishing, and more than one of the following, each applied to more than five percent of total production: bleaching, dyeing, and printing.

A typical flow diagram for Woven Fabric Finishing is shown in Figure 54.

Desizing removes the sizing compounds applied to the yarns in the slashing operation. It consists of solubilizing the size with mineral acid or enzymes (starch size only) and thoroughly washing the fabric. Desizing generates starch solids or polyvinyl alcohol, fat or wax, dissolved solids, suspended solids, and some oil or grease. The pH may be neutral or very low, depending on the desizing method.

Scouring cotton to remove impurities generates a strongly alkaline wastewater. This effluent contains significant levels of dissolved solids, oil and grease, and a modest amount of suspended solids. Scouring of synthetic woven goods generates a low level of dissolved solids from surfactant, soda ash, or sodium phosphate.

Mercerization swells the cotton fibers, since alkali is absorbed into the fibers to provide increased tensile strength and abrasion resistance. The fabric is fed through a series of alkali baths and then washed to remove the caustic. Mercerization wastes predominantly consist of the alkali used in the process. The waste stream also contains high levels of dissolved solids.

In most mills, caustic soda is recovered and concentrated for reuse, thus saving chemicals and avoiding a sizeable waste load. Bleaching, with either hydrogen peroxide (H_2O_2) or sodium chlorite (NaOCl), and subsequent washing contributes very small waste loads, most of which are dissolved solids.

Dyeing is the most complex of all textile finishing processes. Various chemicals may be used to help deposit the dye, or to develop the color. Dye loadings vary widely, depending upon the weight of fabrics being treated and the depth of color desired. Dyed goods are generally washed and rinsed to remove excess dye and chemicals from the cloth. Dyeing processes contribute substantially to textile wastes. Color is an obvious waste. A high level of dissolved solids is expected. Suspended solids levels should be low. Carriers, which are essential for dyeing polyester, have high BOD. Plants using sulfur dyes will contain sulfides in the raw waste. Due to the need for auxiliary chemicals in most dyeing operations, a number of compounds may appear in dye process wastewater.

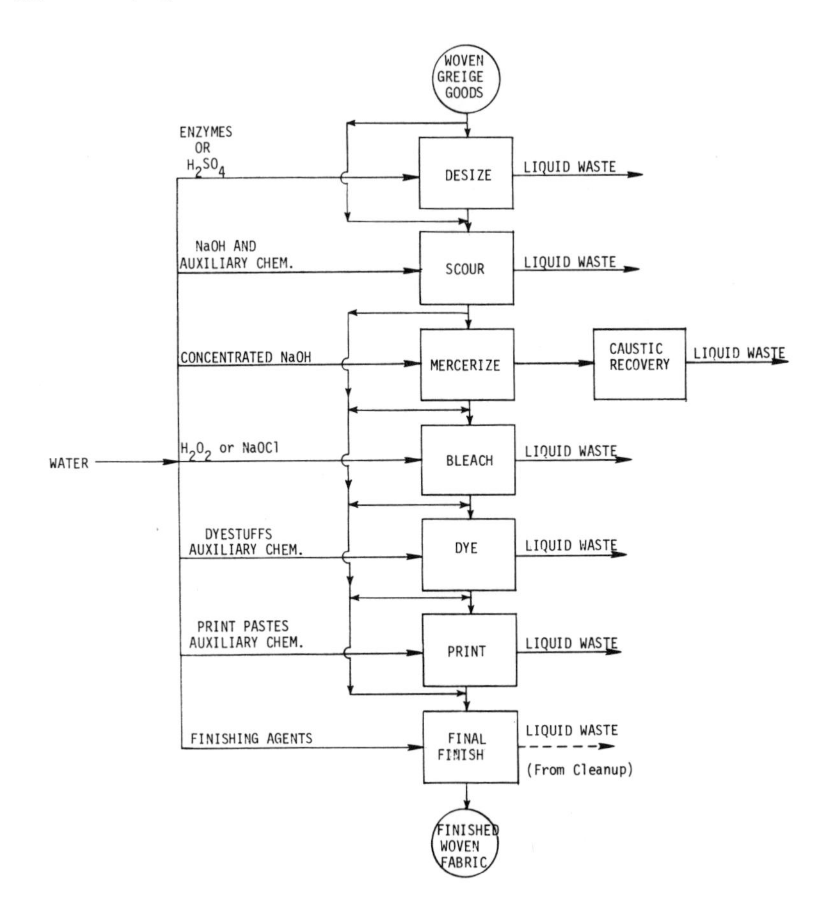

Figure 54: Typical woven fabric finishing process flow diagram.

Printing involves application of dyes or pigments onto the fabric in the form of a pattern. Chemicals and thickeners are used, depending upon the dye type and the fibers used. Printing wastes will contribute to BOD. Much of the wastewater comes from the cleaning of tanks and equipment. These relatively concentrated wastes may justify segregated treatment.

Special finishes such as resin treatment, waterproofing, flameproofing, and soil release endow the fabric with a particular property desired by consumers. The range of chemicals resulting from these processes is very broad. However, the amount of wastes are generally small, since the chemicals are applied with little water use.

Knit Fabric Finishing (E): Plants manufacturing knit fabrics are the source of finished knit piece or yard goods for the apparel, industrial, and household goods trades, and also serve to augment supplies of fabric to underwear and outerwear manufacturers.

A process flow diagram is shown in Figure 55.

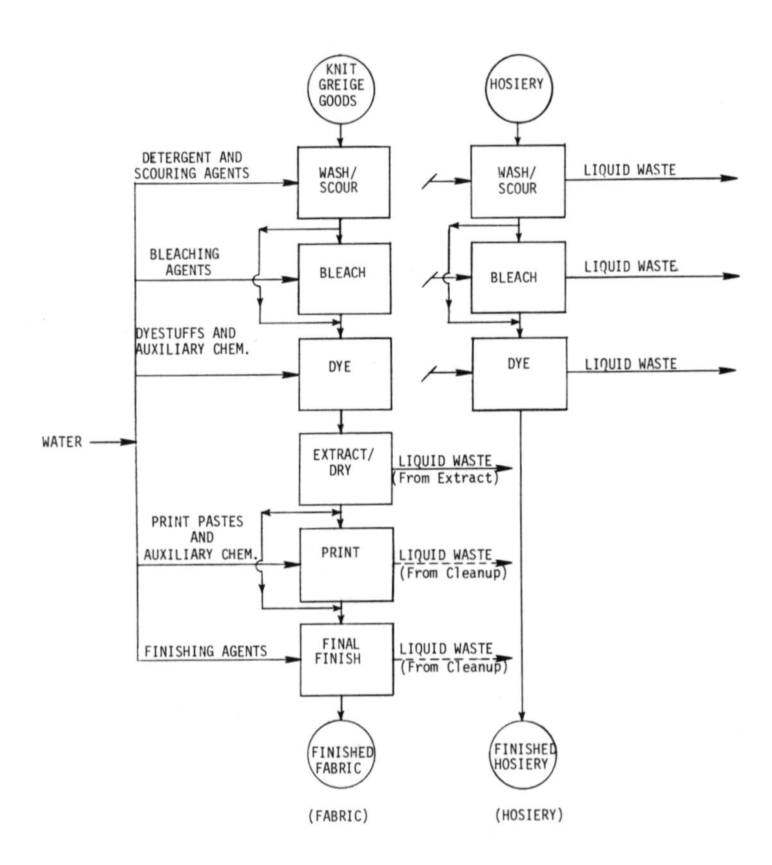

Figure 55: Typical knit fabric finishing process flow diagram.

As with woven fabric finishing, the number of processes performed at a single mill may vary considerably. In addition, certain processes, such as hosiery manufacturing are distinct in terms of manufacturing and raw wastewater characteristics. Consequently, internal subdivisions are required for this subcategory:

- Simple Processing - This Knit Fabric Finishing subdivision covers facilities that perform fiber preparation, scouring, functional finishing, and/or one of the following processes applied to more than five percent of total production: bleaching, dyeing, or printing. This subdivision includes all Knit Fabric Finishing mills that do not qualify under either the Complex Processing or Hosiery Products subdivisions.
- Complex Processing - This Knit Fabric Finishing subdivision covers facilities that perform fiber preparation, scouring, functional finishing, and/or more than one of the following processes, each applied to more than five percent of total production: bleaching, dyeing, or printing.
- Hosiery Products - This Knit Fabric Finishing subdivision covers facilities that are engaged primarily in dyeing or finishing hosiery of any type. As compared to other Knit Fabric Finishing facilities, Hosiery Finishing mills are generally much smaller (in terms of wet production), more frequently employ batch processing, and more often consist of only one major wet processing operation. All of these factors contribute to their lower water use and much smaller average wastewater discharge.

Fabrics may be knitted from dyed or undyed yarns. Fabrics knitted from dyed yarn are scoured or dry cleaned to remove knitting oils before dyeing and/or printing. The types of dyestuffs, auxiliaries, and conditions employed for dyeing knit goods are essentially the same as for woven goods (D). The main differences between knit (E) and woven (D) fabric processing operations are that knit yarns are treated with lubricants, rather than with the starch or polymeric sizes used for woven goods yarns, and that mercerizing operations are not employed with knit goods. Otherwise, waste characteristics are similar to those of woven fabrics (D).

Carpet Finishing: Carpet industry wastewaters are very similar to those from Category D. When polyester is dyed, the carriers present a problem. Although steps are being taken to produce polyester fiber that can be dyed without carriers, disposal of carriers still remains a problem. The pH of carpet wastes is usually close to neutral. Hot dye wastes sometime present a problem to biological treatment systems. The color problem is similar to that of other finishing categories. Where carpets are printed or dyed continuously, as in fabric printing, the thickeners present a high BOD load.

Carpet yarn is generally dyed in another mill and then brought to the carpet mill. The yarn is tufted onto a backing in a dry operation. Washing to remove residual dye, acid, thickeners, and other additives follows. Because of the lower degree of processing required, the wastewaters generated are typically dilute. A process flow diagram is shown in Figure 56.

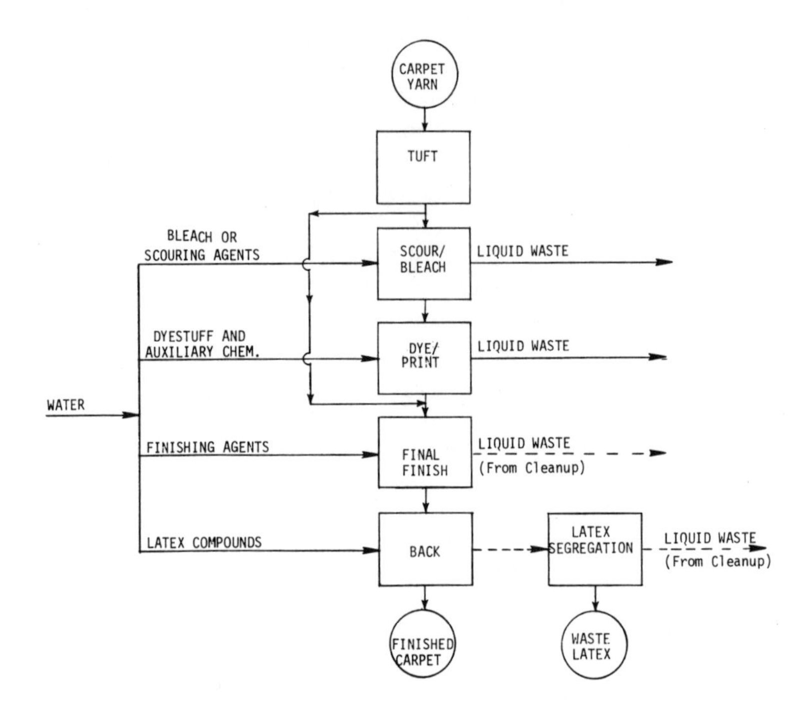

Figure 56: Typical carpet manufacturing process flow diagram.

Stock and Yarn Finishing: In this subcategory, crude yarn is obtained from a spinning facility. Yarn may be natural, synthetic, or blended. Wet processes used by yarn mills include scouring, bleaching, mercerizing, dyeing, and finishing. Wastes generated will depend upon whether natural fibers, blends, or synthetics are produced. A process flow diagram is shown in Figure 57.

When synthetics are handled, only light scouring and bleaching is required, and wastes contain low levels of BOD and dissolved solids. Dyeing contributes a stronger waste, due to carriers found in polyester, and to some acetic acid. These wastes also contain some color.

Scouring, bleaching, and mercerizing of cotton generates BOD and color from fiber impurities, and a higher level of dissolved solids from the mercerizing processes.

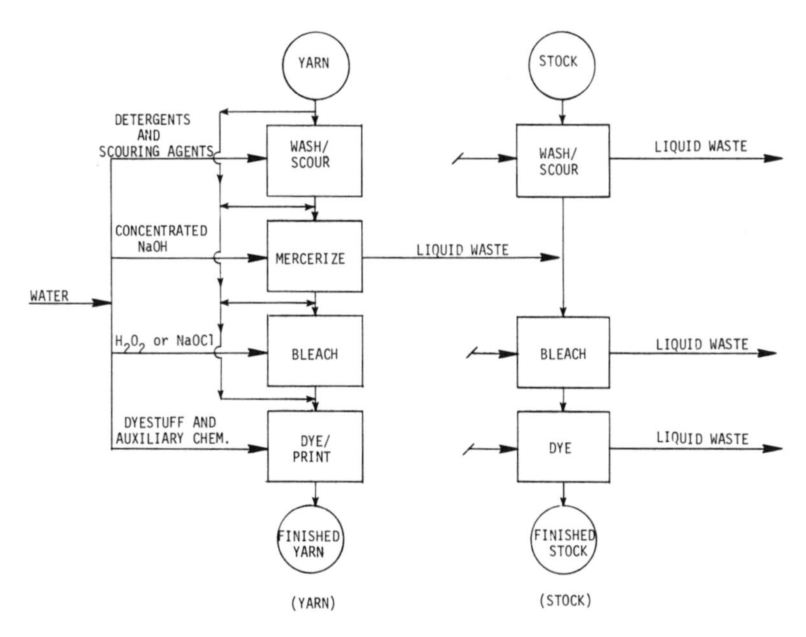

Figure 57: Typical stock and yarn finishing process flow diagram.

Non-Woven Manufacturing: This subcategory covers facilities that primarily manufacture non-woven textile products of wool, cotton, or synthetics, by themselves or as blends, by mechanical, thermal, and/or adhesive bonding procedures. Non-woven products produced by fulling and felting processes are covered under Felted Fabric Processing. A flow diagram for non-woven operations is shown in Figure 58.

A typical operation consists of web formation, bonding, and finishing. The nature of non-woven manufacturing is such that a typical facility has relatively small hydraulic and pollutant contaminants. At a few facilities, special manufacturing operations or activities common to other subcategories might be performed, resulting in higher water use; however, this is the exception rather than the rule.

Felted Fabric Processing: This subcategory covers facilities which primarily manufacture non-woven products, using fulling and felting operations as a means of achieving fiber bonding. Fabrics are then dyed and finished, as in several other categories.

Wool, rayon, and blends of wool, rayon, and polyester are typically used to process felts. Felting is accomplished by subjecting the web or mat to moisture, chemicals (detergents), and mechanical action. Wastewater is generated during rinsing steps which are required to prevent rancidity and spoilage of the fibers; by dyeing operations; and, to a lesser degree, during finishing processes. A flow diagram of a typical felting operation is shown in Figure 59.

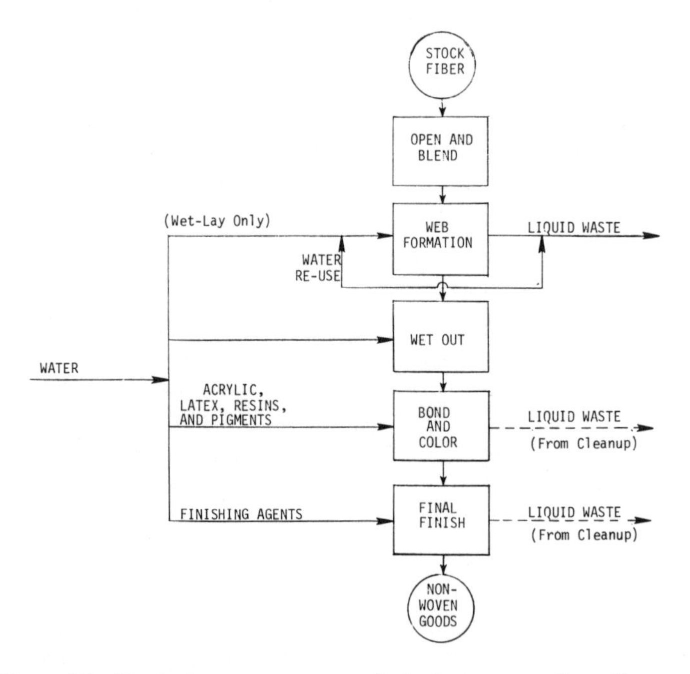

Figure 58: Typical nonwoven manufacturing process flow diagram.

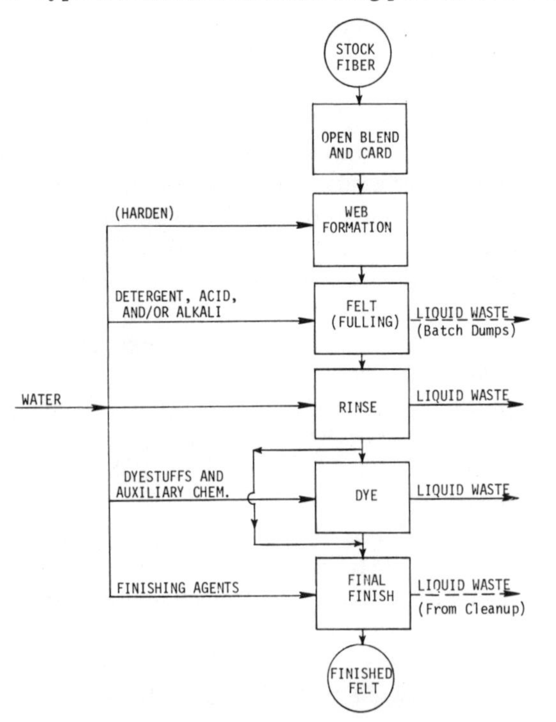

Figure 59: Typical felted fabric processing flow diagram.

Wastewater Characteristics

Table 81 shows raw wastewater characteristics. Table 82 shows raw wastewater characteristics as a functon of production. Textile wastes are generally colored, highly alkaline, and high in BOD, suspended solids, coliform, and temperature. Some dyes which are used are water soluble, others are not.

Biodegradability is highly variable. Metals, such as copper and chromium, are used in some dyeing operations in the industry. Small amounts of zinc and magnesium salts may enter the waste stream from processes that produce durable press goods. Plants using sulfur dyes will discharge sulfides. Desizing operations may contribute high organic and solids loads. Wool processing will also contribute high solids loads, and even if grease recovery is practiced, high grease and oil content may be expected.

Control and Treatment Technology

In-Plant Control Technology

Water Reuse — Reusing the same water in more than one process reduces the hydraulic loading to treatment units. Reuse of cooling water, and reuse of process water in a second, unrelated process are the two major reuse measures commonly practiced.

Cooling water that does not come in contact with fabric or chemicals can often be collected and reused directly. Examples include condenser cooling water, water from water-cooled bearings, heat exchanger water, and water recovered from such equipment as cooling rolls, yarn driers, pressure dyeing machines, and air compressors.

In many operations, reuse of process wastewater is also possible. This may include: reuse of wash water from bleaching in caustic washing, scour makeup and rinse water; reuse of scouring rinses for desizing or washing printing equipment; reuse of mercerizing wash water to prepare scour, chlorine bleach, and wetting out baths; and similar activities.

Water Reduction — Reduced process water usage results in lower hydraulic loading to treatment units. This reduction can be realized by employing more efficient rinsing schemes, conservation measures, or process modifications. Rinse water is a major source of hydraulic load for the textile industry, and therefore a promising candidate for water reduction techniques. An effective reduction method is the countercurrent flow system, in which reused water flows through the process in the direction opposite that of the fabric.

Chemical Substitution — In processes where certain chemicals present a particular problem due to toxicity or pollutant strength, it is sometimes possible to replace the chemical with one which is more amenable to treatment.

Table 81: Textile Mills—Raw Wastewater Characteristics

Wastewater parameter	Subcategory[a]												
	A	B	C	D[b]			E[b]			F	G	H	I
				a	b	c	a	b	d				
Conventional - mg/l													
Biochemical oxygen demand	2,300	170	290	270	350	420	210	270	320	440	180	180	200
Chemical oxygen demand	7,000	600	690	900	1,100	1,240	870	790	1,370	1,190	680	2,360	550
Color (APHA unit)	2,200	1,500	10	80	1,400	1,900	400	750	450	490	570	90	200
Oil and grease	600		80	70	50	70	80	50	100	20	20	60	30
Sulfide	500	3,500		70	110	170	50	150	560	175	200		120
Total suspended solids	3,300	60	180	60	110	150	50	60	80	70	40	80	120
Inorganics													
Antimony	X[c]	X	X			X	X	X	X	X	X		
Arsenic	X				X	X		X	X	X	X		
Beryllium	X			X		X				X			
Cadmium	X	X	X	X	X	X	X	X		X	X	X	
Chromium	X	X	X	X	X	X	X	X	X	X	X	X	
Copper	X	X	X	X	X	X	X	X	X	X	X	X	X
Cyanide	X		X	X		X	X		X	X		X	
Lead	X		X	X	X	X	X	X		X	X		
Mercury	X	X		X	X	X				X	X		
Nickel	X	X	X	X	X	X	X	X		X	X		
Selenium			X	X		X	X		X	X	X		X
Silver	X		X	X	X	X	X	X	X	X	X	X	
Thallium	X			X		X				X			
Zinc	X	X	X	X	X	X	X	X	X	X	X	X	X
Organics													
Acenaphthene				X			X			X			
Acrylonitrile									X				
Anthracene										X			
Benzene	X			X	X	X			X	X		X	
3,4-Benzofluoranthene										X			
11,12-Benzofluoranthene										X			
Bis (2-ethylhexyl) phthalate	X	X	X	X	X	X	X		X	X	X	X	X
Butyl benzyl phthalate												X	
Chlorobenzene	X				X		X	X		X			
Chloroform	X		X	X	X	X	X	X	X	X	X	X	

(continued)

Table 81: (continued)

Wastewater parameter	Subcategory[a]												
	A	B	C	D[b]			E[b]			F	G	H	I
				a	b	c	a	b	d				
2-chloronaphthalene										X			
2-chlorophenol					X								
1,2-Dichlorobenzene	X	X				X	X	X			X		
1,4-Dichlorobenzene		X					X						
Dichlorobromomethane										X			
1,1-Dichloroethane	X												
1,2-Dichloroethane				X									
1,1-Dichloroethylene	X												
2,4-Dichlorophenol						X				X			
1,2-Dichloropropane						X							
Diethyl phthalate	X	X			X	X	X				X		
Dimethyl phthalate				X	X		X				X		
Di-n-butyl phthalate	X		X	X	X	X	X						
2,6-Dinitrotoluene											X		
1,2-Diphenylhydrazine										X			
Ethylbenzene	X	X		X	X	X	X	X		X	X	X	
Fluorene							X						
Hexachlorobenzene	X			X									
Methyl chloride										X			
Methylene chloride				X		X				X			
Naphthalene		X		X		X	X	X	X	X	X	X	
N-nitrosodiphenylamine					X	X			X				
Parachlorometa cresol						X							
Pentachlorophenol		X		X	X	X	X					X	
Phenol (4 APP)	X	X	X	X	X	X	X	X	X	X	X		X
Pyrene				X									
Tetrachloroethylene	X			X				X	X			X	
Toluene	X			X				X	X				X
1,2,4-Trichlorobenzene	X			X		X	X	X					
1,1,1-Trichloroethane	X			X			X						
Trichloroethylene	X	X		X	X	X	X		X	X	X	X	X
2,4,6-Trichlorophenol					X	X			X	X			
Vinyl chloride												X	

[a] See Table 80 for identification of subcategories.

[b] a-simple processing; b-complex processing; c-complex processing and desizing; d-hosiery.

[c] "X" denotes pollutant is of concern for the subcategory indicated.

Table 82: Textile Mills—Raw Wastewater Characteristics—Production-Based Data

	Wastewater Parameter (kg/1,000 kg of product produced)		
Industrial Category	BOD_5	COD	TSS
Wool Scouring	41.8	128.9	43.1
Wool Finishing	59.8	204.8	17.2
Low Water Use	2.3	14.5	1.6
Woven Fabric Finishing			
Simple Processing	22.6	92.4	8.0
Complex Processing	32.7	110.6	9.6
Complex Processing plus Desizing	45.1	122.6	14.8
Knit Fabric Finishing			
Simple Processing	23.4	81.1	6.6
Complex Processing	23.4	115.4	6.6
Hosiery	26.4	89.4	6.7
Carpet Finishing	25.6	82.3	4.7
Stock and Yarn Finishing	20.7	62.7	4.6
Non-Woven Manufacturing	6.7	38.4	2.2
Felted Fabric	70.2	186.0	64.1

Several substitutions practiced in the textile industry include: substituting biodegradable, low-foam detergents for "hard" detergents; switching from chromate oxidizer to peroxides or iodates in dye operations; using sulfuric acid instead of soap in wool fulling; and the use of both mineral acids and oils in place of acetic acid and olive oil in various dyeing and cording operations. All of these result in lower pollutant discharge levels. Before making a substitution, it is essential that a careful analysis be carried out, to insure that a given substitution does not create additional problems while alleviating the difficulties caused by the original chemical.

Material Reclamation — Material reclamation is primarily implemented as a means to reduce operating costs, rather than pollutant loads; hence it is more widely practiced than many other forms of in-plant control. Two types of reclamation currently practiced are caustic recovery and size recovery. Other reclamation schemes, including latex recovery by carpet manufacturers, dye liquor reuse, and solvent recovery by printing mills are being investigated. Increased costs for both chemicals and wastewater treatment are expected to accelerate the use of material reclamation.

Process Changes — Process changes and new process technologies hold great promise for reducing hydraulic and pollutant loads from textile mills. Technological advances in fibers, process chemicals and other raw materials, and fiber process equipment, are constantly being made, and in general these changes have resulted in lower hydraulic and conventional pollutant loadings.

One of the better known and more promising process changes is solvent processing, in which solvents are substituted for water in high water use operations. Problems with process compatibility and solvent recovery have limited the adoption of this system at the present, but future developments may allow more widespread use.

Modifications of existing processes and flow procedures can reduce wastewater generation. These include switching from batch to continuous operation, using standing baths and rinses instead of running ones, and the use of rope washers instead of open-width washers.

End-of-Pipe Treatment Technology: At present, most textile mills employing treatment processes utilize one-stream, nonspecific treatment systems. As treatment costs and removal requirements increase, it will become increasingly attractive to segregate waste streams by a particular type of pollutant load. This allows treatment of a particular pollutant or group of pollutants in as concentrated a form as possible. It is expected that new plant designs will facilitate waste stream segregation.

About 60 percent of textile mills presently discharging to POTWs utilize no pretreatment; about 30 percent employ preliminary treatment (screening, neutralization and equalization); and about 10 percent use secondary treatment.

Secondary treatment used by the textile industry may include chemical, physical, and biological processes.

- Chemical processes used by the textile industry include coagulation and oxidation. Coagulation is used almost exclusively as a process on segregated, high solid streams due to associated high energy and chemical costs. Oxidation may be used to remove color, ammonia, organics, and to disinfect. A number of oxidants are presently used, but ozone is viewed as having the greatest potential for expanded use in the textile industry. Most oxidation is currently done for disinfection and color removal.
- Physical processes, with the exception of clarification, have not found broad acceptance in the textile industry. Filtration is sometimes used for solids removal, and dissolved air flotation has seen limited use in the removal of latex and print paste. Other physical systems, such as ultra- and hyperfiltration, show promise for textile waste treatment, but remain untried.
- Biological processes are the most commonly used in textile mills. The most widely used biological system is aerated lagoons; stabilization ponds and activated sludge systems also are common.

The effectiveness of various treatment technologies in reducing toxic pollutants from textile industry wastewaters is shown in Table 83. These results originate from field sampling and laboratory analysis of textile industry wastewaters, as well as from historical data on removal efficiencies.

Table 83: Textile Mills—Toxic Pollutant Treatment Effectiveness—Reported Values[a]

Wastewater parameter[c]	Chemical Treatment[b]							
	Subcategory A		Subcategory B		Subcategory D		Subcategory E	
	Influent conc.	Effluent conc.	Influent conc.	Effluent conc.	Influent conc.	Effluent conc.	Influent conc.	Effluent conc.
Inorganics								
Antimony			22	23	22-600	10-123	622-684	620-687
Arsenic	38-83	39-43	60	62				
Cadmium	0-130	0-250						
Chromium			116-206	30-47	T[d]-100	T-17	27-360	14-280
Copper	120-320	110-590	4-23	16-130	T-57	T-12	30-110	0-92
Lead	3,500-0		22-30	< 22-30	< 22-34	22-66	20-48	23-53
Mercury							1.8	1.7
Nickel	2,000-0		36-76	36-57	66-190	43-77	0-<10	< 10-44
Selenium							20-62	0-110
Silver	0-500	0-1,300			T-80	T-72	13	9.4-19
Thallium			< 50	< 50				
Zinc	400-1,500	190-460	639-6,400	347-5,730	155-5,200	145-195	47-220	110-190
Organics								
Acrylonitrile								
Benzene					0-15	0-3		
Bis (2-ethylhexyl) phthalate	14-42	23-106	32-760	T-44	T-100	T-34	15	7-45
Chloroform					0-210	9-73		
Cyanide	200-260	0-240			< 10	< 10	0	20
Di-n-butyl phthalate								
1,2-Dichlorobenzene			20			13		
Ethylbenzene								
Methyl chloride								
Methylene chloride			46	28				
N-nitrosodi-n-propylamine								
Naphthalene								
4-Nitrophenol								
Parachlorometa cresol								
Pentachlorophenol								
Phenol						0-T	T-18	
Tetrachloroethylene							17	
Toluene								
1,2,4-Trichlorobenzene			1,580	154				

(continued)

Table 83: (continued)

Wastewater parameter[c]	Chemical Treatment[b]							
	Subcategory A		Subcategory B		Subcategory D		Subcategory E	
	Influent conc.	Effluent conc.	Influent conc.	Effluent conc.	Influent conc.	Effluent conc.	Influent conc.	Effluent conc.
1,1,1-Trichloroethane								
Trichloroethylene								
Trichlorofluoromethane	T-540	T-1,200						
2,4,6-Trichlorophenol								

Wastewater parameter[c]	Physical Treatment[e]									
	Subcategory A		Subcategory B		Subcategory D		Subcategory E		Subcategory G	
	Influent conc.	Effluent conc.	Influent conc.	Effluent conc.	Influent conc.	Effluent conc.	Influent conc.	Effluent conc.	Influent conc.	Effluent conc.
Inorganics										
Antimony			23	12	123	136	<10-888	<10-888	141-177	150-162
Arsenic	39	83	62	102			<10	11	T[d]	T
Cadmium			T	105						
Chromium			41-206	41-101	17-58	14-110	T-98	T-<32	68-91	12-57
Copper	110	120	T-16	105-130	11-323	25-81	T-323	T-41	110-132	20-84
Lead			< 22-30	22-116	14-66	0-64	9-82	10-85	<22-35	<22
Mercury										
Nickel			<36-57	<36-73	28-72	<32-67	0-187	0-188	<36	42-50
Selenium							<41	<102		
Silver			172	158	25-72	28-77	T-73	T-68	T	11-15
Thallium			<50	<50	T-<14	<13-14	<50	<50	0-<50	0-<50
Zinc	190	400	639-5,730	371-5,800	25-195	0-280	34-5,160	40-204	228-283	130-436
Organics										
Acrylonitrile									0-<100	0-<100
Benzene					5-18	5-12	0-15	0-144		
Bis (2-ethylhexyl) phthalate	23	14	44-760	14-80	0-570	0-45	0-109	0-200	76-340	80-170
Chloroform							0-1,020	0-790		
Cyanide	240	260					0-<10			
Di-n-butyl phthalate									0-T	0-T
1,2-Dichlorobenzene					13	T			0-T	T
Ethylbenzene					0-460	0-160				
Methyl chloride					<5-26	<5-30				

(continued)

Table 83: (continued)

Wastewater parameter[c]	Physical Treatment[e]									
	Subcategory A		Subcategory B		Subcategory D		Subcategory E		Subcategory G	
	Influent conc.	Effluent conc.	Influent conc.	Effluent conc.	Influent conc.	Effluent conc.	Influent conc.	Effluent conc.	Influent conc.	Effluent conc.
Methylene chloride			46	47						
N-nitrosodi-n-propylamine							0-T	0-26		
Naphthalene					250	0	0	0	0-13	T
4-Nitrophenol					< 10	< 10				
Parachlorometa cresol									0-T	0-T
Pentachlorophenol				10	0-37	0-30				
Phenol	17	17			24-94	16-20	0-669	0-2,110		
Tetrachloroethylene							17	17	T	T-9
Toluene			14	12	0-320	0-132			T-38	T
1,2,4-Trichlorobenzene							0-29	0-9	19-43	T-21
1,1,1-Trichloroethane					T	99				
Trichloroethylene										
Trichlorofluoromethane										
2,4,6-Trichlorophenol										T

Wastewater parameter[c]	Biological Treatment[f]			
	Subcategory G		Subcategory I	
	Influent conc.	Effluent conc.	Influent conc.	Effluent conc.
Inorganics				
Antimony				
Arsenic				
Cadmium				
Chromium			35	
Copper				18
Lead	36			
Mercury				
Nickel				
Selenium			32	18
Silver				
Thallium				
Zinc	865	123	45	101

(continued)

Table 83: (continued)

Wastewater parameter[c]	Biological Treatment[f]			
	Subcategory G		Subcategory I	
	Influent conc.	Effluent conc.	Influent conc.	Effluent conc.
Organics				
Acrylonitrile				
Benzene				
Bis (2-ethylhexyl) phthalate	48	11	18	
Chloroform				
Cyanide				
Di-n-butyl phthalate				
1,2-Dichloro-benzene				
Ethylbenzene				
Methyl Chloride				
Methylene chloride				
N-nitrosodi-n-propylamine				
Naphthalene			56	
4-Nitrophenol				
Parachlorometa cresol				
Pentachlorophenol				
Phenol				
Tetrachloro-ethylene	48			
Toluene				
1,2,4-Trichloro-benzene				
1,1,1-Trichloro-ethane				
Trichloroethylene				
Trichloro-fluromethane				
2,4,6-Trichloro-phenol				

[a] Both individual values and ranges provided above are based on average, mean, minimum and maximum values reported in individual plant samplings, as well as in historical data.

[b] Coagulation, oxidation.

[c] Values are given in ug/l.

[d] Trace.

[e] Filtration, dissolved air flotation.

[f] Aerated lagoons, activated sludge, biological contactors and stabilization ponds.

Type of Residue Generated

Sludges produced by both direct and indirect dischargers in this industry totalled 330,000 dry tons in 1977; production is projected to be at least 380,000 dry tons in 1987. For indirect dischargers alone, sludge generated by pretreatment is expected to be about 80,000 dry tons in 1984.

Heavy metals, particularly chromium, copper and zinc are used in several manufacturing processes in this industry and, where used, will end up in the sludge. In most cases, there are alternate chemicals that may be used instead of the ones containing heavy metals. Small, but significant amounts of toxic organics, such as trichlorobenzene, polyvinyl chloride, and perchlorethylene, may also be found in raw wastewaters; however, for the most part these are removed during treatment, and are not in the sludge stream.

Residual Management Options

Residues from this industry are typically disposed in general municipal or private landfills. Sludges are frequently stored and/or thickened in lagoons or ponds at the textile plant site, usually in the same lagoons used as wastewater treatment units. Other residue management techniques include: open dumping, segregated landfills, incineration, wet oxidation, reclamation, and landspreading, either with or without dewatering.

Open dumping is rarely an acceptable method, and reclamation is applicable only in special cases. The economics of the other methods are unclear, although segregated landfills and incineration are normally quite expensive. Incineration is expensive partly because it is fuel-intensive; it can also produce an ash with significant toxic heavy metal contents. Landspreading may be attractive where the specific sludge can be shown to have some nutrient values and low toxic heavy metals and toxic organic contents. Land-spreading is still an experimental technique, however, and has shown mixed results with sludges from this industry.

There is the possibility that some textile industry sludges will be designated as hazardous because of the presence of toxic heavy metals or organics. Landfilling in "approved" landfills will then possibly be the only acceptable disposal technique.

TIMBER PRODUCTS PROCESSING

General Industry Description

The industry segments included in this category are wood preserving, insulation board production, and wet process hardboard production. There are over 400 wood preserving plants in the United States. Plants are concentrated in the Southeast, from east Texas to Maryland, and along the Northern Pacific coast. These areas correspond to the natural ranges of the southern pineland Douglas fir, and western red cedar, respectively. The major chemicals used are wood preservatives and fire retardants. Based on 1975 production levels, more than 55 percent of the timber processed was treated by creosote and creosote solutions, more than 25 percent was treated by pentachlorophenol, and about 15 percent was treated by waterborne inorganic salts.

Insulation board is a "noncompressed" fiberboard, and is constructed from lignocellulosic fibers. The principal types of insulation board include: building board, insulating roof deck, roof insulation, ceiling tile, lay-in panels, sheathing, and sound-deadening insulation board.

Hardboard is a "compressed" fiberboard, with a density greater than 0.5 gm/cm , and a thickness ranging between 2 to 13 mm. Production of hardboard by the wet process method is usually accomplished by thermomechanical fiberization of the wood. Chemical additives help the overall strength and uniformity of the product. The end uses of hardboard include paneling, siding, cabinets, furniture, etc. There are 16 wet-process hardboard plants in the United States, representing an annual production in excess of 1.5 million tons per year.

Industrial Categorization

A useful categorization for the purposes of raw waste characterization and organization of pretreatment information is shown in Table 84.

Table 84: Timber Products

Subcategory	Designation
Wood preserving - water-borne or non-pressure	A
Wood preserving steam	B
Wood preserving boulton	C
Insulation board	D
Wet process hardboard	E

Process Description

In wood preserving processes, wood products are injected with chemicals that provide fungistic and insecticidal properties, or impart fire resistance. The most common preservatives are creosote, pentachlorophenol, and various formulations of water soluble inorganic chemicals. Fire retardants are formulations of salts, including borates, phosphates, and ammonium compounds.

The wood preserving process consists of two basic steps: (1) conditioning to reduce natural moisture content and increase the permeability of the wood; and (2) impregnating the wood with preservative. The conditioning step may be performed by one of several methods, including: (1) seasoning or drying wood in large, open yards; (2) kiln drying; (3) steaming the wood at elevated pressure in a retort, followed by application of a vacuum; (4) heating the stock in a preservative bath under reduced pressure in a retort (Boulton process); or (5) vapor drying, heating of the unseasoned wood in a solvent to prepare it for preservative treatment. All of these conditioning methods have as their objective the reduction of moisture content of the unseasoned stock, to a point where the required amount of preservative can be retained in the wood.

Specific process descriptions are described below.

Wood Preserving—Water Borne or Nonpressure (A): This subcategory includes all pressure processes employing waterborne inorganic salts and fluorchromium-arsenic phenol (FCAP) solution. All nonpressure processes are also included. Wood stock to be treated in nonpressure processes is normally conditioned by air seasoning or kiln drying. Widespread use of low cost control and treatment technologies results in more than 85 percent of plants in this subcategory having no wastewater discharges.

Wood Preserving—Steam (B): Conventional steam conditioning (open steaming) is a process in which unseasoned or partially seasoned stock is subjected to direct steam impingement at an elevated pressure in a retort. The maximum permissible temperature is set by American Wood Preservers Association standards at 118°C, and the duration of the steaming cycle is limited by these standards to no more than 20 hours. Steam condensate that collects in the retort exits through traps, and is piped to oil-water separators for removal of free oils. Removal of emulsified oils requires further treatment. Figure 60 shows a diagram of a typical open steaming wood preserving plant.

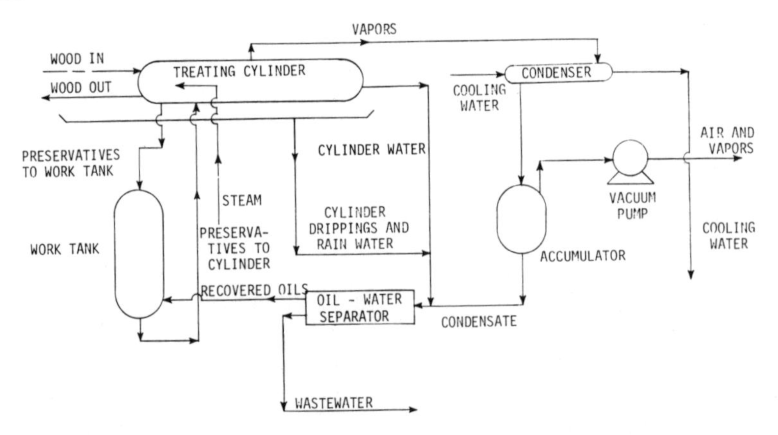

Figure 60: Open steaming process wood treating plant.

In closed steaming, a widely used variation of conventional steam conditioning, steam needed for conditioning is generated in situ by covering the coils in the retort with water from a reservoir and heating the water by passing process steam through the coils. The water is returned to the reservoir after oil separation and reused during the next steaming cycle. There is a slight increase in volume of water in the storage tank after each cycle because of the water removed from the wood. A small blowdown from the storage tank is necessary to remove this excess water and also to control the level of wood sugars in the water.

Modified closed steaming is a variation of the steam conditioning process, in which steam condensate is allowed to accumulate in the retort during the steaming operation until it covers the heating coils. At that point, direct steaming is discontinued and the remaining steam required for the cycle is generated within the retort by using the heating coils.

The vapor-drying process consists essentially of exposing wood in a closed vessel to vapors from any one of many organic chemicals that are immiscible with water and have a narrow boiling range. Selected derivatives of petroleum and coal tar, such as xylol, high-flash naphtha, and Stoddard solvent, are preferred. Numerous other chemicals, including chemical blends, can also be employed as drying agents in the process. Chemicals with initial boiling points of from 212°F to 400°F may be used. Vapors for drying are generated by boiling the chemical in an evaporator or in a cylinder. The vapors are sent to the retort containing the wood, where they condense on the wood, give up their latent heat of varporization, and cause the water in the wood to vaporize. The water vapor thus produced, along with excess organic vapor, exits the vessel to a condenser and then to a gravity-type separator. The water layer is discharged from the separator and the organic chemical is returned to the evaporator or cylinder for reuse.

Wood Preserving—Boulton (C): Wood preserving by the Boulton process is primarily used for hardwood and Western species. Preconditioning is accomplished in the Boulton process by heating the stock in a preservative bath under reduced retort pressure in the retort. The preservative serves as a heat transfer medium. After the cylinder temperature has been raised to operating temperature, a vacuum is drawn and water removed in vapor form from the wood passes through a condenser to an oil-water separator, where low-boiling fractions of the preservative are removed. The Boulton cycle may have a duration of 48 hours or longer for large poles and piling. Figure 61 illustrates the Boulton process.

Insulation Board: Either wood or mineral wool may be used as the raw material for insulation board production. Plants using mineral wool are not covered by this document.

Preliminary Operations — Plants processing wood may receive raw materials in the form of roundwood, fractionated wood, and/or whole tree chips. Fractionated wood can be in the form of chips, sawdust, or planer shavings. Figure 62 shows an illustration of a representative insulation board process.

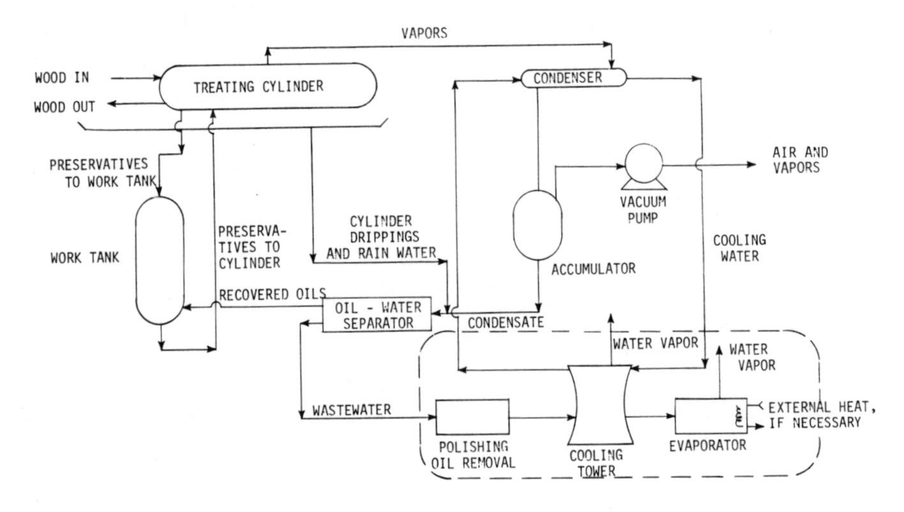

Figure 61: Boulton wood treatment plant.

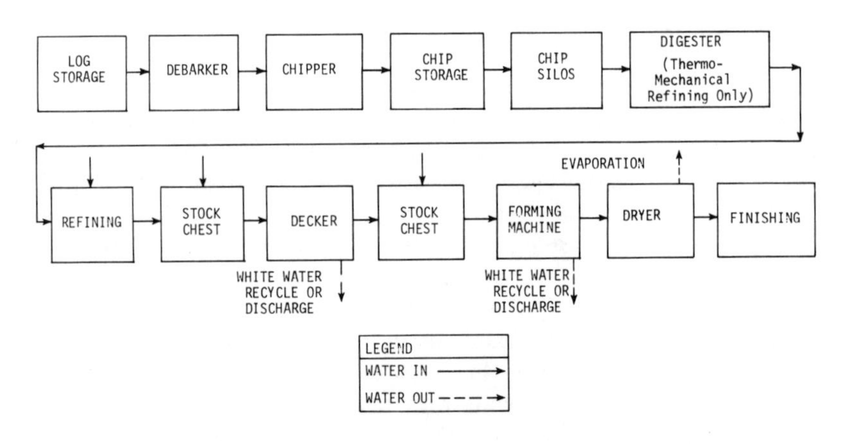

Figure 62: Diagram of a typical insulation board process.

When roundwood is used as a raw material, it is usually debarked by drum or ring barkers before use. The barked wood is then chipped. Chips may pass through a device used to remove metal, grit, dirt, and other trash. This may be either a wet or dry process.

Pulp preparation is usually accomplished by mechanical or thermo-mechanical refining.

Mechanical refining basically consists of two discs between which the chips or wood residues are passed. Disc refiners produce fibers that may pass through a 30 or 40 mesh screen.

Thermo-mechanical refining is similar to mechanical refining except that the feed material is subjected to a steam pressure of 4 to 15 atmospheres for a period of time from 1 to 45 minutes before it enters the refinery. In some cases, the pressure continues through the actual refining process.

Pre-steaming softens the feed material, and thus makes refining easier and reduces energy requirements. The longer the pre-steaming and the higher the pressure, the softer the wood becomes. Heat plasticizes portions of the hemicellulose and lining components of wood which bind the fibers together and produces a longer and stronger fiber.

Subsequent to the refining of the wood, the fibers produced are dispersed in water to achieve consistencies amenable to screening. Screening is done primarily to remove coarse fiber bundles, knots, and slivers. After screening, the fibers produced by any method be sent to a decker or washer.

Decker Operations — Deckers are rotating, wire-covered cylinders, usually recirculated into the system. By dewatering the pulp at this point it can be mixed to the desired consistency with greater accuracy. Washing of the pulp is sometimes desirable in order to remove dissolved solids and soluble organics which may result in surface flaws in the board. High concentrations of these substances tend to stay in the board, and during the drying stages migrate to the surface.

After the washing or deckering operation, the pulp is reslurried in stages. During various dilution stages, additives are used in the pulp suspension to give the board desired properties, such as strength, dimensional stability, and water absorption resistance. After passing through a series of storage and consistency controls, the pulp may pass through a pump-through refiner to disperse agglomerated fiber clumps and to shorten the fiber bundles. The fibrous slurry is then pumped into a forming machine which removes water from the pump suspension and forms a mat.

Forming Operations — The two most common types of forming machines are Fourdrinier and cylinder machines. Both machines produce a mat that leaves the roller press with a moisture content of about 40 to 45 percent and the ability to support its own weight over short spans. The wet mat is then trimmed to width and cut to length by a traveling saw.

Drying and Finishing Operations — After being cut to desired lengths, the mats are dried to a moisture content of five percent or less. Most dryers now in use are gas or oil-fired tunnels, with hot air circulated throughout. Dryers usually have eight or ten decks with various zones of heat, which controls the rate of drying and reduces the danger of fire. These heat zones allow for higher temperatures when the board is "wet" (where the mat first enters), and lower temperatures when the mat is almost dry. The dried board then goes through various finishing operations such as painting, asphalt coating, and embossing. Those operations which manufacture decorative products will usually have finishing operations which use water-base paints containing such chemicals as various inorganic pigments, i.e., clays, talc, carbonates, and certain amounts of binders such as starch, protein, PVA, PVAC, acrylics, urea formaldehyde resin, and melamine formaldehyde resins.

Wet Process Hardboard (E): The basic raw material used in the manufacture of hardboard is wood. Wood receipts may vary in form from unbarked long and short logs to chips. Chip receipts may be from whole tree chipping, forest residue, sawmill waste, plywood trim, and sawdust. Logs are chipped either with the bark attached or after removal of the bark

Fiber Preparation — Before refining or defibering, the chips are pretreated with steam in a pressure vessel or digester. Steaming of the chips under pressure softens the lignin material that binds the individual fibers together and reduces the power consumption required for mechanical defibering. The predominant method used for fiber preparation consists of a combination of thermal and mechanical pulping. This involves a preliminary treatment of the raw chips with a steam and pressure prior to mechanical pulping of the softened chips. The thermo-mechanical process may take place with a digester-refiner as one unit, or in separate units. Primary, secondary, and tickler refiners may be used, depending on the type of pulp required. The pulp becomes stronger with more refining, but its drainage characteristics are reduced. Refining or defibering equipment is of the disc type, in which one disc or both may rotate; the unit may be of pressurized or gravity type.

Stock Washings and Deckers — A washer is used to remove soluble materials. A decker, which is a screen used to separate fibers from the main body of water, also removes some solubles from the fiber bundles. Effluent from a stock washer has a high concentration of soluble organics which are usually mixed into the white water system and must be discharged for treatment, or be recycled within the washing system.

Forming — Most wet process mills form their product on a Fourdrinier-type machine similar to that used in producing paper. Diluted stock is pumped to the headbox where the consistency is controlled (usually with white water) to an average of 1.5 to 1.7 percent, while the stock is fed to the traveling wire of the Fourdrinier. As the stock travels with the wire, water is drained away. At first the water drains by gravity, but as the stock and wire continue, a series of suction boxes remove additional water. As the water is removed, the stock is felted together into a continuous fibrous sheet called a "wet mat." At the end of the forming machine the wet mat leaves the traveling wire and is picked up by another moving screen that carries the mat through one or more roll presses. This step not only removes more water but also compacts and solidifies the mat to a level at which it can support its own weight over short spans. As the wet mat leaves the prepress section, it is cut, on the fly, into lengths as required for the board being produced. The water drained from the mat as it travels across the forming machine is collected in a pit under the machine or in a chest. This "white water" contains a certain amount of wood fibers, dissolved wood chemicals, and dissolved additive chemicals.

Pressing — After forming to the desired thickness, the fibers in the mat are welded together into a grainless board by a hardboard press. The hydraulically-operated press is capable of simultaneously pressing 8 to 26 boards. Press plates may be heated with steam or with a heat transfer medium at temperatures up to 230°C. Unit pressures on the board of up to 68 atm are achieved by the press. Squeezing water from the wet mat removes some of the dissolved solids. The water generated during press squeeze-out has a high organic content and is evaporated or drained away for treatment. From the press the hardboard may be conveyed to an oven, kiln, or humidifier.

Oil Tempering Baking — After pressing, hardboard may receive a special treatment called tempering. This consits of treating the sheets with various drying oils (usually vegetable oils) either by pan-dipping or with roll coaters. In some cases hardboard is passed through a series of pressure rolls, which increase the absorption of oils and remove any excess. The oil is stabilized by baking the sheet from one to four hours at temperatures of 150°C to 177°C. Tempering increases the hardness, strength, and water resistance of the board.

Humidification — Sheets of hardboard are hot and dry as they are discharged from the press or the tempering baking oven. To stabilize the board so as to prevent warping and dimensional changes, it is subjected to a humidification chamber in which the sheets are retained, until the proper moisture content, usually 4.5 to 5 percent, is reached. Figure 63 depicts a flow diagram for this process. Figure 64 depicts a two-side smooth wet process.

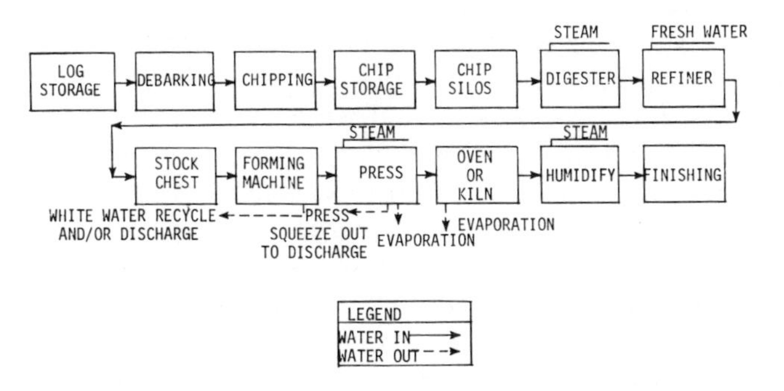

Figure 63: Flow diagram of a typical wet process hardboard mill S1S hardboard production line.

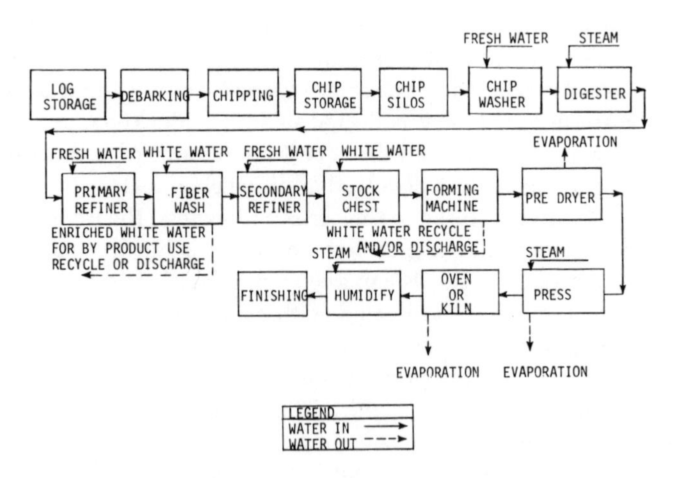

Figure 64: Flow diagram of a typical wet process hardboard mill S2S hardboard production line.

Wastewater Characteristics

Table 85 shows raw wastewater characteristics. Table 86 shows raw wastewater characteristics as a function of production. Most of the wood preserving plants employing waterborne salts are self-contained, or achieve zero discharge. Plants which treat with inorganic salts only are prohibited from discharging wastewater either to a navigable waterways or to a POTW. No pesticides or PCBs were detected in raw or treated wastewaters in the timber products industry at levels above the general detection limit of 1 ppb.

Control and Treatment Technology

In-Plant Control Technology

Water Reuse — Most wood preserving plants treating with salt-type preservatives (Subcategory A) reuse their process wastewater as makeup water for treating solution; several plants treating organic preservatives (Subcategories B and C) reuse treated wastewater for boiler makeup or cooling water. Segregation and recycling of cooling water at wood preserving plants that employ barometric condensers is essential. Reuse of steam coil condensate for boiler feed water also reduces the total volume of wastewater and conserves energy. Recent water reuse practices used by most insulation board (Subcategory D) and wet-process hardboard (Subcategory E) plants include the use of recycled process white water in place of freshwater at various points in the system. Process water can be reused for stock dilution, shower water, and pump seal water.

Table 85: Timber Products—Raw Wastewater Characteristics

Wastewater parameters[b]	Subcategory[a]						
	A[c]	B		C		D	E
Conventional							
BOD_5						275-7,080	200-4,495
Chemical oxygen demand	10-50	1,356-15,694		520-7,316			
Fluoride	4-20						
Nitrogen (NH_3-N)	80-200						
Oil and grease		11-1,902		12-1,357			
Phosphate	15-150						
Total suspended solids				81		200-4,400	41-2,500
Inorganics							
Antimony		L-47	L-13	L-47	L-13	0.67-3	0.5-8
Arsenic	13,000-50,000	3-142	L-1,000	3-142	L-1,000	1.6-3.3	1-1.3
Beryllium		L-19	L-2	L-19	L-2	0.5-0.83	5-0.67
Cadmium		L-10	L-5	L-10	L-5	0.5-1	0.5-5
Chromium (+3)	0-800						
Chromium (+6)	230-1,500						
Chromium (total)		1-98	L-13,900	1-98	L-13,900	1.3-11	1-470
Copper	50-110	31-1,600	60-3,910	31-1,600	60-3,910	200-450	33-530
Lead		1-91	1-16	1-91	1-16	1.3-21	2-55
Mercury		L-37	0.2-1.3	L-37	0.2-1.3	1-7.5	0.05-18
Nickel		3-210	20-100	3-210	20-100	8.8-240	3.3-270
Selenium		L-7	L-53	L-7	L-53	3.3-5	0.8-3.8
Silver		L-10	L-1	L-10	L-1	0.5-0.83	0.5-1.5
Thallium		1-98	L-13,900	1-98	L-13,900	1.3-11	1-470
Zinc		119-843	120-78,200	119-843	120-78,200	250-720	190-2,300
Organics							
Ancenaphthene		1.06-55		L-2.83			
Ancenaphthylene		L-1.21		L-2.06			
Benzene		0.003-2.8		L		1-0.27	L-0.9
Benzo (a) anthracene		L-7.7		L-0.034			

(continued)

Table 85: (continued)

Wastewater parameters[b]	Subcategory[a]				
	A[c]	B	C	D	E
Benzo (B) fluoranthene		L-1.68	L		
Benzo (K) fluoranthene		L-3.9	L		
Benzo (ghi) perylene		L-.315	L		
Benzo (A) pyrene		L-2.7	L		
Bis (2 ethylhexyl) phthalate		L-0.437	L-1.45		
2-Chlorophenol		L-0.042	L		
Chrysene	L-4.7	L-0.018			
Dibenzo (a,h) anthracene		L-0.43	L		
2,4-Dimethylphenol		L-6.6	L		
Ethylbenzene		0.037-2.1	L		L-0.02
Fluoranthene		0.633-35	L-0.282		
Fluorene		0.82-48	L-0.824		
Indero (1,2,3-CD) pyrene		L-5.5	L		
Methylene chloride		0.006-0.702	2.6		
Naphthalene		0.38-45	L-3.4		
Pentachlorophenol		L-306	L-27		
Phenanthrene		1.95-39	L-1.51		
Phenols	0.005-0.16	0.64-501.3	L-508.6	L-.8	L-8.9
Pyrene		0.36-22	L-0.194		
Toluene		0.027-3.2	L	L-0.06	0.01-0.07
1,1,2-Trichloroethane					L-0.09
Trichloromethane		0.02	0.09	L-0.02	L-0.02
2,4,6-Trichlorophenol		L-0.533	L		

[a] See Table 84 for identification of subcategories.

[b] Measurements are in mg/l, unless otherwise stated. "L" indicates concentrations below the detection limit.

[c] Plant treating with CCA-type and FCAP-type preservatives only.

[d] Wood preserving data for plants in subcategories B and C treating with organic preservatives.

[e] Wood preserving data for plants in subcategories B and C treating with both organic and inorganic preservatives.

Table 86: Timber Products—Raw Wastewater Characteristics—Production-Based Data[a]

Wastewater Parameter	Subcategory[b] B		C		D Mechanical refining	D Thermo-mechanical refining	E Smooth one side	E Smooth two sides
Conventional								
BOD_5[c]					1.27-20.9	17.0-43.2	1.89-68.5	43.2-116
Chemical oxygen demand[d]	10.2-283		3.44-208					
Oil and grease[d]	0.0627-30.4		0.718-38.6					
TSS[c]					0.46-45.2	6.25-42.8	0.56-22.5	11.7-20
TSS[d]			0.537[e]					
Inorganics[f]								
Antimony	0.00014[g]	0.00004[h]	0.00014[g]	0.00004[h]	0.000015		0.000052	
Arsenic	0.0353	0.00055	0.0353	0.00055	0.000029		0.000017	
Beryllium	0.00001	0.00001	0.00001	0.00001	0.0000067		0.000009	
Cadmium	0.00001	0.00002	0.00001	0.00002	0.0000065		0.000027	
Chromium	0.0002	0.00451	0.0002	0.00451	0.00016		0.0011	
Copper	0.00794	0.0032	0.00794	0.0032	0.0019		0.0053	
Lead	0.00034[c]	0.00014	0.00034[c]	0.00014	0.000063		0.00018	
Mercury	0.00001	0.00002	0.00001	0.00002	0.000042		0.000061	
Nickel	0.00153	0.0011	0.00153	0.0011	0.0005		0.00087	
Selenium	0.00003	0.00013	0.00003	0.00013	0.000038		0.000032	
Silver	0.00001	0.00001	0.00001	0.00001	0.0000056		0.000036	
Thallium	0.00003	0.00001	0.00003	0.00001	0.0000076		0.00001	
Zinc	0.00772	0.0457	0.00772	0.0457	0.0046		0.01	
Organics[i]								
Acenapthene	0.0332		0.0111					
Acenapthalene	0.0091		0.0081					
Anthracene	0.0959		0.0008					
Benzene	0.8200		0.0001					
Benzo (a) anthracene	0.0081		0.0002					
Benzo (A) pyrene	0.0060		0.0001					
Benzo (B) fluoranthene	0.0052		0.0001					
Benzo (ghi) perlene	0.0009		0.0001					
Benzo (k) fluoranthene	0.0073		0.0001					
Bis-2-ethylhexyl pthalate	0.0017		0.0093					
2-Chlorophenol	0.0004		0.0001					
Chrysene	0.007		0.0001					
Dibenzo (a,h) anthracene	0.0003		0.0001					
2-4-Dimethylphenol	0.0445		0.0001					

(continued)

Table 86: (continued)

Wastewater Parameter	Subcategory[b] B	C	D Mechanical refining	D Thermo-mechanical refining	E Smooth one side	E Smooth two sides
Ethylbenzene	0.0101	0.0201				
Fluoranthene	0.358	0.0012				
Fluorene	0.0292	0.0033				
Indero (1,2,3-CD) pyrene	0.0042	0.0001				
Methylene chloride	0.0049	0.0172				
Napthalene	0.111	0.0123				
Pentachlorophenol	0.552	0.0757				
Pesticides and PCB's[j]						
Phenol[c]	0.352	0.0066	0.004-0.0075		0.003-0.043	
Pyrene	0.0266	0.0008				
Toluene	0.0237	0.0201				
Trichloromethane	0.0001	0.0201				
2,4,6-Trichlorophenol	0.005	0.0001				

[a] Data is based on a composite of historical and field sampling (1975 - 1978) for wastewaters prior to end-of-pipe treatment processes.

[b] See Table 84 for description of subcategories.

[c] Pounds of pollutant per 1,000 pounds of product.

[d] Pounds of pollutant per 1,000 cubic feet of wood preserved.

[e] Single determination of one plant in 1978.

[f] For subcategories B and C, numbers shown are in pounds per 1,000 cubic feed of wood preserved; for subcategories D and E, numbers shown are in pounds of pollutant per 1,000 pounds of product.

[g] Numbers shown for inorganics in this column apply to plants treating only with organic preservatives.

[h] Numbers shown for inorganics in this column apply to plants treating with both organic and inorganic preservatives.

[i] Pounds of pollutant per 1,000 cubic feet of product.

[j] Concentrations of these pollutants were below detectable limits.

Process Changes — An average two-retort open steaming wood preserving plant (Subcategory C) can reduce its process wastewater flow by over 11,000 gallons per day, by converting its process from open steaming to modified or closed steaming. By closing the process white water system in an insulation board (Subcategory D) or hardboard plant (Subcategory E), it is possible to reduce the mass discharge of suspended solids in the raw waste load. Some of the measures which can be used to achieve close-up or maximum recycle of the process white water system are: (a) elimination of extraneous wastewater sources; (b) provision of sufficient white water equalization; (c) the installation of cyclones following the refiner; (d) clarification of white water; (e) extraction of concentrated wastewater; (f) corrosion control; and (g) control by press sticking. Specific recommendations for in-plant modifications on a plant-to -plant basis require a detailed and thorough working knowledge of each plant.

End-of-Pipe Treatment Technology: All existing nonpressure, wood preserving plants achieve zero discharge. For wood preserving inorganic salt plants, over 90 percent have achieved zero discharge, 8 percent discharge effluents to a POTW and about 2 percent discharge effluents directly to receiving waters. For wood preserving steaming plants, 67 percent achieve zero discharge, 32 percent discharge to a POTW, and about 1 percent have direct discharge.

For wood preserving Boulton plants, no plants have any direct discharges, 69 percent achieve zero discharge, and 31 percent discharge to a POTW. For insulation board plants, 13 percent achieve zero discharge, 19 percent have self-contained discharges, 37 percent discharge to a POTW, and 31 percent have direct discharge. For wet-process hardboard plants, only 13 percent are self-contained dischargers, 13 percent discharge to a POTW, and the remaining 74 percent are direct dischargers. End-of-pipe treatment technologies in the timber products processing industry are described below.

Preliminary and Primary Treatment — The treatment applied to the wastewater prior to biological treatment or its equivalent may consist of the following:

- Screening - Screens are common preliminary treatment facilities used by many insulation board and wet-process hardboard plants to remove bark, wood chips, and foreign materials from the wastewater prior to further treatment.
- Primary Settling - Most insulation board plants and many hardboard plants use gravity-type primary settling facilities to remove a major portion of the wood fibers from the wastewater.
- Oil-Water Separation - API-type oil-water separators are extensively used by wood preserving plants and are the preferred treatment technology for "primary oil separation." Secondary oil-water separation usually occurs in another API separator operated in series

with the first, or it may be conducted in any vessel or lagoon where the detention time is sufficient for further separation. Other methods of secondary chemical flocculation use polyelectrolytes, lime, ferric chloride, and clays to break up emulsions. Chemical flocculation is becoming unpopular because of secondary hazardous waste generation.

- Oil Absorption - A few wood preserving plants achieve almost complete removal of free oils by filtering the wastewater through an oil-absorbent medium. This practice generates hazardous wastes, which are expensive to dispose of, and this practice is not used very often.
- Slow-Sand Filtration - Many wood preserving plants which flocculate wastewater subsequently filter it through sand beds to remove the solids. It is highly efficient in removing both the solids resulting from the process as well as some of the residual oil, but is less frequently in use, due to the generation of hazardous waste by-products.

Secondary and Advanced Treatment —

- Biological Treatment - This technology is effective in reducing concentrations of COD, phenols, oil and grease, pentachlorophenol, and organic toxic pollutants from wood preserving wastewaters. Trickling filters, aerated lagoons, oxidation ponds, and activated sludge systems are commonly used. Wastewater generated by the insulation board and hardboard industries is also amenable to biological treatment.
- Spray or Soil Irrigation - Wastewaters generated by the timber products processing industry can be applied by sprinkling on medium to low permeability, vegetated soils. Treatment is by physical-chemical and microbial action, and by plant uptake as the wastewater travels through the soil matrix. High removals of color, COD, and BOD are achievable.
- Chemical Reduction and Precipitation - This is a method recommended for wood preserving wastewaters from processes other than those using waterborne preservatives. Sulfur dioxide or strong acid is used to reduce chromium from the hexavalent valence state to the trivalent state at pH 4 or below. The pH of the wastewater is then increased to 8.5 or 9.0 with lime or an equivalent, to precipitate not only the trivalent chromium, but also copper, zinc, fluorides, and arsenic. Care must be taken not to raise the pH above 9.5, since trivalent chromium is slightly soluble at higher pH values.
- Carbon Adsorption - Pretreatment of wood preserving wastes with lime, ferric chloride, or alum followed by pH adjustment and carbon adsorption are highly effective for the removal of chromium, copper, zinc, arsenic, phenols, COD, and TOC. Granular acti-

viated carbon adsorption can also be used for treating wastewaters from insulation board and hardboard plants. However, carbon adsorption is the most expensive technique for treating wood preserving wastewaters.

- Evaporation - Because of the relatively low volumes of wastewater generated by wood preserving plants, evaporation is a feasible and widely used technology for achieving zero discharge. Three types of evaporation processes are currently used. The first involves containing the wastewater in lined lagoons of sufficient size to accommodate several months of process wastewater and rainwater, and evaporating the wastewater in the atmosphere by spraying with nozzles.

  The second type is cooling tower evaporation, which is feasible for Boulton plants only. In this system, the condensed wood water is sent to an accumulator, and from there to an oil-water separator. The wastewater stream is then added to the cooling water, which recirculates through the surface condenser picking up heat, then through a forced-draft cooling tower where evaporation occurs.

  The third type of evaporation is thermal evaporation, using an external heat source. This method is energy-intensive and expensive, and is only feasible when used to supplement other treatment methods and at times of peak wastewater generation.
- Other Technologies - Other potentially applicable treatment technologies include tertiary metals removal systems by precipitation, membrane systems, adsorption on synthetic adsorbents, oxidation by chlorine, oxidation by hydrogen peroxide, and oxidation by ozone for wood preserving plants, and chemical coagulation and granular media filtration for insulation board and hardboard plants.

Table 87 shows the removal efficiencies of various treatment methods practiced by the timber products processing industry.

Type of Residue Generated

Residues generated in this industry are: screenings, oily skimmings, sludges, and brines. The wood preservation segment of the industry may generate one or more of these residues, depending on the manufacturing and wastewater treatment processes used; whereas the insulation board and hardboard segments generate mainly screenings and sludges.

Brines from water-based wood preservation processes are of such character that their discharge, even after treatment, is almost always prohibited. Recovery and reuse of such brines is common; evaporation is also practiced.

Oily skimmings that are generated by creosote or oil-based pentachlorophenol (PCP) preservation processes can generally be recycled and

Table 87: Timber Products—Removal Efficiencies of Treatment Technologies—Conventional and Toxic Pollutants

Wastewater parameter	Preliminary and primary treatment					Secondary and advanced treatment			In-place treatment[a,b]	
	Oil/water separation		Oil adsorption	Chemical floccu-lation	Clarifi-cation	Spray or soil irrigation	Carbon adsorption	Chemical flocculation/ carbon adsorption	Less than equivalent of BPT technology	Equivalent of BPT technology
	Primary	Secondary								
Conventional										
BOD_5					10-15	95				
Oil and grease	96.7	90	100	73-86					80	81
TDS				30-80		99	80		66	81
TSS					65-80					
Inorganics										
Antimony										63.6
Arsenic								84		44.8 (1.2)
Chromium								98		16.7
Copper								90		27.1
Lead										51.2 (3.2)
Mercury										(66.7)
Nickel										(82.5)
Zinc								76		41.1
Organics										
Acenaphthene										95.3
Acenaphthylene										97
Benzene										98.5
Benzo (a) anthracene										86
Benzo (A) pyrene										83.9
Benzo (B) fluoranthene										85.7
Benzo (ghi) perylene										78.9

(continued)

Table 87: (continued)

Wastewater parameter	Preliminary and primary treatment					Secondary and advanced treatment			In-place treatment[a,b]	
	Oil/water separation									
	Primary	Secondary	Oil adsorption	Chemical floccu-lation	Clarifi-cation	Spray or soil irrigation	Carbon adsorption	Chemical flocculation/ carbon adsorption	Less than equivalent of BPT technology	Equivalent of BPT technology
Benzo (K) fluoranthene										89.7
Bis (2-ethylhexyl) phthalate										66.7
2-Chlorophenol										75
Chrysene										83.1
Dibenzo (a,h) anthracene										83.3
2,4-Dimethylphenol										97.8
Ethylbenzene										99.9
Fluoranthene										85.4
Fluorene										95.9
Indero (1,2,3-CD) pyrene										89.6
Methylene chloride										12.2
Naphthalene										99.8
Pentachlorophenol									70	97.6
Phenanthrene										94
Phenol						99	7		43	99.9
Pyrene										93
Toluene										96.2
Trichloromethane										0
2,4,6-Trichlorophenol										98

[a] In-place treatment data are mainly obtained from wood preserving plants. BPT technology includes primary oil-water separation, chemical coagulation and sedimentation or filtration, and biological treatment.

[b] In some cases, two sets of toxic inorganics data are presented for the in-place treatment. Inorganic data without parentheses were obtained from wood preserving plants using organic preservatives only; while the data in the parentheses were obtained from wood preserving plants using both organic and inorganic preservatives.

reused in the manufacturing process, and therefore will not present a significant disposal problem.

Sludges from wood preservation industry come from the creosote or oil/PCP processes, and typically are "contaminated" with the preservative material or its components. In some cases, heavy metals such as zinc or chromium may also be present. Where chemicals are added to precipitate heavy metals or to coagulate suspended solids, the chemicals added will end up in the sludge. When a biological wastewater treatment process is used, the bulk of the sludge will be biomass. When powdered activated carbon is used in conjunction with either chemical precipitation or biological treatment, the carbon will also end up in the sludge. Pollutant characteristics of the materials adsorbed on the activated carbon will be more significant than the mass, since toxic organics or heavy metals requiring activated carbon for their removal will usually be the most offensive pollutants in the wastewater.

Screenings and primary sludges from the insulation board and hardboard segments of the industry will contain wood and bark chips and fibers, and foreign materials. Toxic "contamination" of these residues is possible, depending on the history of the wood source-material.

Secondary sludges from the insulation board and hardboard segments of the industry will usually be composed of the biomass resulting from biological treatment. Heavy metal "contamination" of such sludges does occur, but "contamination" by toxic organics will depend on the degree to which toxic organics are removed by biodegradation, volatilization (stripping), and photodecomposition during biological wastewater treatment. Although removal levels are somewhat unpredictable, toxic organic "contamination" of the biological sludges from insulation board and hardboard manufacturing is usually low.

Residual Management Options

Landfilling is the predominant method of sludge and screenings disposal for all segments of this industry. The main concern is the possible presence of toxic heavy metal or organic "contamination" of the sludge, and the related landfilling precautions that are required when such materials are present. Such landfills must be protected from erosion, must be covered with impervious material to minimize rainwater penetration, must be insulated from contact with groundwater, and seepage from the landfill, must be prevented by natural or artificial layers of impervious material, or by an impervious liner. Leachate should be collected and treated by an acceptable wastewater treatment technique. All the above measures must be designed for long-term, essentially permanent protection. Residues considered to be hazardous should only be disposed of at a "specially designated" landfill.

Where circumstances are favorable—that is, where toxic content is low and suitable agricultural users are nearby—biological sludges from the insulation board and hardboard segments of this industry have been applied to land as a soil conditioner.

Incineration of residues from this industry is also possible. Special precautions are necessary, however, if toxic organic compounds are present. The temperature and residence time in the furnace must be high enough to insure that complete combustion occurs, and/or suitable scrubbing of the flue gases must be practiced, to insure that only innocuous products are discharged into the air. If toxic heavy metals with relatively low boiling points are present—for example, zinc—then the residue is not suitable for incineration. However, this is a relatively uncommon occurrence.

APPENDIX

BIBLIOGRAPHY

Adhesives and Sealants

- *Technical Support Document for Miscellaneous Chemicals (contractor's draft report).* Contract 68-01-2932, USEPA, Washington, D.C. 20460, February 1975.

Auto and Other Laundries

- *Technical Support Document for Auto and Other Laundries Industry (contractor's draft report).* Contract 68-03-2550, USEPA, Washington, D.C. 20460, August 1979.
- *Status Report on the Treatment and Recycle of Wastewaters from the Car Wash Industry (contractor's draft report).* Contract 68-01-5767, USEPA, Washington, D.C. 20460, July 1979.

Coal Mining

- *Development Document for Interim Final Effluent Limitations Guidelines and New Source Performance Standards for the Coal Mining Point Source Category.* USEPA, 440/1-75/057 Group II, USEPA, Washington, D.C. 20460, October 1975.
- *Technical Assistance in the Implementation of the BAT Review of the Coal Mining Industry Point Source Category (contractor's draft report).* Contracts 68-01-3273, 68-01-4762 and 68-02-2618, USEPA, Washington, D.C. 20460, March 1979.
- Frank, A. "Mining Waste, Mine Reclamation and Municipal Sludge." *Sludge Magazine.* July–August 1978.

- *Dewatering of Mine Drainage Sludge.* Water Pollution Control Research Series 14010-FJX-12/71, USEPA, Washington, D.C. 20460, December 1971.
- *Studies on Densification of Coal Mine Drainage Sludge.* Water Pollution Control Research Series 14010-EJT-09/71, USEPA, Washington, D.C. 20460, September 1971.
- *Dewatering of Mine Drainage Sludge, Phase II.* EPA-R2-73-169, USEPA, Washington, D.C. 20460, February 1973.
- *Erosion and Sediment Control-Surface Mining in The Eastern US - Planning.* EPA-625/3-76-006, USEPA Technology Transfer, Cincinnati, Ohio 45268, October 1976.

Coil Coating

- *Development Document for Effluent Limitations Guidelines and Standards for the Coil Coating Point Source Category (contractors draft report).* EPA-440/1-79/07-a, USEPA Effluent Guidelines Division, Washington, D.C. 20460, August 1979.

Electroplating

- *Development Document for Electroplating Pretreatment Standard Final.* EPA-440/1-79/003, USEPA Effuent Guidelines Division, Washington, D.C. 20460, August 1979.
- *Management of Metal Finishing Sludge.* EPA-530/SW-561, USEPA Office of Solid Waste, Washington, D.C. 20460, February 1977.

Explosives Manufacturing

- *Technical Review of the BAT Analysis of The Explosives Industry (contractor's draft report).* USEPA, Washington, D.C. 20460, April 1979.
- Forsten, I. "Pollution Abatement In a Munitions Plant." *Environmental Science and Technology.* Vol 7, No. 9, September 1973.

Foundries

- *Foundry Industry (contractor's draft report).* Contract 68-01-4379, USEPA, Washington, D.C. 20460, May 1979.
- Throop W. M. "Ultimate Disposal of Foundry Waste Sludge." *American Foundry Society Transaction 77-39,* 1977.

Gum and Wood Chemicals Manufacturing

- *Development Document for Effluent Limitations Guidelines and Standards for The Gum and Wood Chemcials Manufacturing Point*

Source Category. EPA-440/1-79/078-b, USEPA Effluent Guidelines Division, Washington, D.C. 20460, December 1979.

Inorganic Chemicals Manufacturing

- *Development Document for Inorganic Chemicals Manufacturing Point Source Category - BATEA, NSPS, and Pretreatment Standards* (contractor's draft report). Contract 68-01-4492, USEPA Effluent Guidelines Division, Washington, D.C. 20460, April 1979.
- *Supplement for Pretreatment to the Development Document for The Inorganic Chemicals Manufacturing Point Source Category.* EPA-440/1-77-087-a, USEPA, Washington, D.C. 20460, July 1977.
- *Development Document for Effluent Limitations Guidelines and NSPS for the Major Inorganic Products Point Source Category.* EPA-440/1-74-007-a, USEPA, Washington, D.C. 20460, March 1974.

Iron and Steel Manufacturing

- *Steel Making: Hot Forming and Cold Finishing Segment Plus Addenda* EPA-440/1-74-024-a, USEPA, Washington, D.C. 20460, June 1974.
- Dawson, R.A. "Sludge Management in the Iron and Steel Industry." *Sludge Magazine.* January–February 1980.

Leather Tanning and Finishing

- *Development Document for Effluent Limitations Guidelines and Standards Leather Tanning and Finishing Point Source Category.* EPA-440/1-79/016, USEPA Effluent Guidelines Division, Washington, D.C. 20460, July 1979.
- Dawson R., "Leather Tanning Industry — Sludge Problems Ahead." *Sludge Magazine.* September - October 1978.

Nonferrous Metals Manufacturing

- *Development Document for Effluent Limitations Guidelines and Standards for the Nonferrous Metals Manufacturing Point Source Category (draft copy).* EPA-440/1-79/019-a, USEPA Effluent Guidelines Division, Washington, D.C. 20460, September 1979.

Ore Mining and Dressing

- *Development Document for Effluent Limitations Guidelines and New Source Performance Standards for the Ore Mining and Dress-*

ing Point Source Category, Volumes I and II. USEPA Effluent Guidelines Division, Washington, D.C. 20460, July, 1978.

- *Development Document for BAT Effluent Limitations Guidelines and New Source Performance Standards for The Ore Mining and Dressing Industry, Volumes I and II (contractor's draft copy).* Contract 68-01-4845, USEPA, Washington, D.C. 20460, February 1979.

Organic Chemicals

- *Development Document for Major Organic Products Segment.* EPA-440/1-74-009-a, USEPA, Washington, D.C. 20460, April 1974.
- *Development Document for Significant Organic Products Segment* EPA-440/1-75/045, USEPA, Washington, D.C. 20460, December 1979.
- Dawson, R.A. "Sludge Management in The Chemical Industry." *Sludge Magazine.* March - April 1980.

Paint and Ink Formulating

- *Development Document for Effluent Limitations Guidelines and Standards for the Paint Formulating Point Source Category.* EPA-440/1-79/049b, USEPA Effluent Guidelines Division, Washington, D.C. 20460, December 1979.
- *Development Document for Effluent Limitations Guidelines and Standards for the Ink Formulating Point Source Category.* EPA-440/1-79/090b, USEPA Effluent Guidelines Division, Washington, D.C. 20460, December 1979.
- Dawson, R.A. "Sludge Management in the Paint and Coatings Industry." *Sludge Magazine.* September - October 1979.

Pesticides Manufacturing

- *The Pesticide Manufacturing Industry — Current Waste Treatment and Disposal Practices.* Water Pollution Control Research Series 12020 FYE 01/72, USEPA, Washington, D.C. 20460, January 1972.
- *Disposal of Dilute Pesticide Solutions.* EPA 530/SW-519, USEPA Office of Solid Waste, Cincinnati, Ohio 45268, June 1976.
- "Mounting Quantities of Pesticide Wastes Require Safe Burial Sites." *Solid Waste Management.* December 1976.
- Ghassemi, M., S.C. Quinlivan, and H.R. Day. "Landfills for Pesticides Waste Disposal." *Environmental Science and Technology.* Vol. 10, No. 13, December 1976.

Petroleum Manufacturing

- *Development Document for Effluent Limitations Guidelines and Standards for the Petroleum Refining Point Source Category.* EPA-440/1-79/014-b, USEPA Effluent Guidelines Division, Washington, D.C. 20460, December 1979.
- Frank, A. "Oil Refinery Sludge." *Sludge Magazine.* May - June 1978.

Pharmaceutical Manufacturing

- *Development Document for Interim Final Effluent Limitations Guidelines and Proposed New Source Performance Standards for The Pharmaceutical Manufacturing Point Source Category.* EPA 440/1-75/060, USEPA, Washington, D.C. 20460, December 1976.
- *Effluent Limitation Guidelines for the Pharmaceutical Manufacture Industry (contractor's draft report).* USEPA Effluent Guidelines Division, Washington, D.C. 20460, May 1979.
- *Supplement to the Draft Contractor's Engineering Report for The Development of Effluent Limitations Guidelines for the Pharmaceutical Industry (BATEA, NSPS, BCT, Pretreatment).* USEPA Effluent Guidelines Division, Washington, D.C. 20460, July 1979.
- Swan, R. "Pharmaceutical Industry Sludge." *Sludge Magazine.* July - August 1978.
- *Pharmaceutical Industry — Hazardous Waste Generation, Treatment and Disposal.* SW-508. USEPA Office of Solid Waste, Washington, D.C. 20460, 1976.

Photographic Processing

- *Technical Support Document for Miscellaneous Chemicals (contractor's draft report).* Contract 68-01-2932, USEPA, Washington, D.C. 20460, February 1975.
- Frank, A. "Photo Processing Sludge: New Comstock Lode?" *Sludge Magazine.* January - February 1979.

Plastics and Synthetic Materials

- *Development Document Synthetic Resins Segment.* EPA-440/1-74-010-a, USEPA, Washington, D.C. 20460, March 1974.
- *Development Document Synthetic Polymers Segment.* EPA 440/1-74/036, USEPA, Washington, D.C. 20460, September 1974.

Porcelain Enameling

- *Development Document for Effluent Limitations, Guidelines and Standards for the Porcelain Enameling Point Source Category (draft report)*. EPA-440/1-79/072-a, USEPA Effluent Guidelines Division, Washington, D.C. 20460, August 1979.

Pulp, Paper and Paperboard

- *Preliminary Data Base for Review of BATEA Effluent Limitations Guidelines, NSPS, and Pretreatment Standards for the Pulp, Paper, and Paperboard Point Source Category.* Contract 68-01-4624, USEPA Effluent Guidelines Division, Washington, D.C. 20460, June 1979.
- Krzeminski, J. Jr. "Managing A 'CLEAN' Industrial Waste." *Sludge Magazine.* March - April 1979.

Rubber

- *Review of Best Available Technology for The Rubber Processing Point Source Category (contractor's draft report).* Contract 68-01-4673, USEPA, Washington, D.C. 20460, July 1978.

Soap and Detergents Manufacturing

- *Technical Support Document for Soap and Detergent Manufacturing.* EPA-440/1-74-018-a, USEPA, Washington, D.C. 20460, April 1974.

Steam Electric

- *Development Document for Proposed Effluent Limitations Guidelines and New Source Performance Standards for the Steam Electric Power Generating Point Source Category.* EPA-440/1-73/029, USEPA, Washington, D.C. 20460, March 1974.
- *Development Document for Effluent Limitations Guidelines and New Source Performance Standards for the Steam Electric Power Generating Point Source Category.* EPA-440/1-74/029-a, USEPA, Washington, D.C. 20460, October 1974.
- *Supplement for Pretreatment to The Development Document for The Steam Electric Power Generating Point Source Category.* EPA-440/1-77/084, USEPA, Washington, D.C. 20460, April 1977.
- *Draft Technical Report for Revision of Steam Electric Effluent Limitations Guidelines.* USEPA, Washington, D.C. 20460, September 1978.
- O'Donnel, F. "Scrubber Sludge" *Sludge Magazine.* March - April 1978.
- *Coal Ash Disposal Manual.* EPRI FP-1257. Electrical Power Research Institute, 3412 Hillview Avenue, Palo Alto, California 94304, December 1979.

- *Disposal of Polychlorinated Biphenyls (PCBs) and PCB - Contaminated Materials, Volume I.* EPRI FP-1207. Electrical Power Research Institute, 3412 Hillview Avenue, Palo Alto, California 94304, October 1979.
- *FGD Sludge Disposal Manual.* EPRI FP-977. Electrical Power Research Institute, 3412 Hillview Avenue, Palo Alto, California 94304, January 1979.
- *Sludge Dewatering for FGD Products.* EPRI FP 937. Electrical Power Research Institute, 3412 Hillview Avenue, Palo Alto, California 94304, May 1979.

Textile Mills

- *Development Document for Effluent Limitations Guidelines and Standards for The Textile Mills Point Source Category.* EPA-440/1-79/022b, USEPA Effluent Guidelines Division, Washington, D.C. 20460, October 1979.
- Frank, A. "Textile Mill Sludge Management." *Sludge Magazine.* May - June 1979.

Timber Products Processing

- *Development Document for Effluent Limitations Guidelines and Standards for The Timber Products Processing Point Source Category.* EPA-440/1-79/023b, USEPA Effluent Guidelines Division, Washington, D.C. 20460, October 1979.

Volume 2

Treatment Technology

-1-

INTRODUCTION

The *Handbook of Industrial Residues - Volume 1 and Volume 2* summarizes the available information on residual waste management options, data on categorical industries affected by Federal pretreatment standards, and pretreatment and sludge management technologies. The Handbook has been divided into two (2) volumes:

- Volume 1 - *Industries and Management Options*
- Volume 2 - *Treatment Technology*

Volume 1 contains information on management options. Several detailed scenarios are presented which depict options available to small, medium, and large POTWs. In addition, *Volume 1* provides detailed information to familiarize the POTW manager with specific categorical dischargers. The volume summarizes the current information on industries to be addressed by Federal categorical pretreatment standards. Included are industry characterizations, process descriptions, wastewater characterizations, control and treatment technologies, listing of residuals, and residual management options.

Volume 2 (this volume) provides information on possible pretreatment and sludge residual management technologies applicable to the categorical industries. Volume 2 is organized into two main units. Chapter 2 covers candidate technologies for the treatment and removal of pollutants from wastewater streams. Chapter 3 covers technologies and methods for dealing with residual sludges. Each pretreatment and sludge handling technology is evaluated in terms of advantages and disadvantages and applicability. This information should be useful in helping the POTW and industrial manager understand the capability and limitations of various residuals management options.

-2-

PRETREATMENT TECHNOLOGIES

INTRODUCTION

This chapter presents summaries of candidate technologies for pretreating industrial wastewaters before discharge to a POTW. These range from simple clarification in a settling pond to complex processes (reverse osmosis, ultrafiltration, carbon columns) requiring sophisticated control equipment and skilled operators. Table 1 can be used to quickly identify those technologies found applicable to the categorical industries discussed in Volume 1.

Pretreatment technologies can be divided into three broad categories: biological, physical, and chemical. Many pretreatment processes normally combine two or all three categories to provide the most economical treatment. In the following discussion, each technology is provided with a description, advantages and disadvantages, pollutants treated, and residuals generated. Additional information can be obtained from the references listed in Appendix A.

BIOLOGICAL PRETREATMENT TECHNOLOGIES

For many industrial wastewaters one of the most important concerns has to do with those constituents which exert an oxygen demand and have an impact on receiving waters. Although most industries are discussed in terms of the toxics they discharge, they are also significant sources of biochemical oxygen demand (BOD) and chemical oxygen demand (COD). In many instances, the most appropriate industrial pretreatment technology for removing this oxygen-demanding pollutant is biological treatment.

Table 1: Candidate Treatment and Control Technology for the 34 Categorical Industries

CANDIDATE TREATMENT AND CONTROL TECHNOLOGY / INDUSTRY TYPE	ADHESIVES & SEALANTS	ALUMINUM FORMING	AUTO & OTHER LAUNDRIES	BATTERY MANUFACTURING	COAL MINING	COIL COATING	COPPER FORMING	ELECTRICAL PRODUCTS	ELECTROPLATING	EXPLOSIVES MANUFACTURING	FOUNDRIES	GUM & WOOD CHEMICALS MANUFACTURING	INORGANIC CHEMICALS MANUFACTURING	IRON & STEEL MANUFACTURING	LEATHER TANNING & FINISHING	MECHANICAL PRODUCTS	NONFERROUS METALS MANUFACTURING
ACTIVATED CARBON ADSORPTION	●	●			●	●		●	●			●	●	●	●	●	●
AIR STRIPPING																	
ALKALINE PYROLYSIS									●								
BIOLOGICAL TREATMENT[a]	●					●				●		●		●	●		
CHEMICAL OXIDATION																	●
CHEMICAL REDUCTION						●			●								
CLARIFICATION PROCESSES			●		●	●	●				●			●		●	●
COAGULATION - PRECIPITATION	●	●		●	●	●		●	●		●	●	●	●	●	●	●
EQUALIZATION				●													
EVAPORATION PROCESSES									●								
FLOTATION SEPARATION PROCESSES	●		●			●			●							●	
FILTRATION PROCESSES			●	●	●	●		●					●		●		●
FOAM SEPARATION																	
INCINERATION																	
ION EXCHANGE					●	●			●								●
MEMBRANE PROCESSES	●	●	●			●			●		●					●	●
NEUTRALIZATION					●	●	●							●			●
OIL-WATER SEPARATION		●									●					●	
OZONATION									●								
POLYMERIC RESIN ADSORPTION																	
SCREENING																	
SOLVENT EXTRACTION																	
STARCH XANTHATE					●				●				●				
STEAM STRIPPING																	●
SULFIDE PRECIPITATION									●								●
WASTEWATER COOLING																	
WET AIR OXIDATION										●				●			

CANDIDATE TREATMENT AND CONTROL TECHNOLOGY / INDUSTRY TYPE	ORE MINING & DRESSING	ORGANIC CHEMICALS	PAINT & INK FORMULATING	PESTICIDES MANUFACTURING	PETROLEUM REFINING	PHARMACEUTICAL MANUFACTURING	PHOTOGRAPHIC PROCESSING	PLASTIC & SYNTHETIC MATERIALS	PLASTIC PROCESSING	PORCELAIN ENAMELLING	PRINTING & PUBLISHING	PULP, PAPER & PAPERBOARD	RUBBER	SOAP & DETERGENT MANUFACTURING	STEAM ELECTRIC	TEXTILE MILLS	TIMBER PRODUCTS PROCESSING
ACTIVATED CARBON ADSORPTION		●		●	●		●	●	●			●		●			
AIR STRIPPING		●			●												
ALKALINE PYROLYSIS						●											
BIOLOGICAL TREATMENT[a]		●	●	●	●	●		●	●			●	●	●		●	●
CHEMICAL OXIDATION				●													
CHEMICAL REDUCTION										●							
CLARIFICATION PROCESSES	●		●					●		●		●					●
COAGULATION - PRECIPITATION	●	●	●	●	●		●		●	●			●	●	●	●	
EQUALIZATION			●					●		●						●	
EVAPORATION PROCESSES				●													●
FLOTATION SEPARATION PROCESSES									●	●			●	●			
FILTRATION PROCESSES	●				●		●			●		●		●		●	●
FOAM SEPARATION												●					
INCINERATION		●		●		●		●									
ION EXCHANGE	●						●			●				●			
MEMBRANE PROCESSES										●		●		●	●		
NEUTRALIZATION				●			●			●						●	
OIL-WATER SEPARATION								●									●
OZONATION	●																
POLYMERIC RESIN ADSORPTION												●				●	
SCREENING																	●
SOLVENT EXTRACTION		●			●												
STARCH XANTHATE																	
STEAM STRIPPING		●		●			●	●									
SULFIDE PRECIPITATION																	
WASTEWATER COOLING															●		
WET AIR OXIDATION		●										●					

[a]IN THIS SECTION, BIOLOGICAL TREATMENT MEANS ANY OF THE FOLLOWING: AEROBIC SUSPENDED GROWTH PROCESSES; AEROBIC ATTACHED GROWTH PROCESSES; LAGOONS; ANAEROBIC PROCESSES; AQUACULTURE PROCESSES; LAND APPLICATION PROCESSES.

For discussion purposes, the biological treatment technologies have been divided into several general categories: aerobic suspended growth processes, aerobic attached growth processes, lagoons, anaerobic processes, aquaculture processes, and land application processes. Where necessary, each process is further subdivided to aid in the explanation.

Aerobic Suspended Growth Processes

Description: An aerobic suspended growth process is one in which microorganisms are kept in suspension in a liquid medium consisting of entrapped and suspended colloidal and dissolved organic and inorganic materials. This biological process uses the metabolic reactions of the microorganisms to attain an acceptable effluent quality by removing those substances exerting an oxygen demand. Depending on the type of material in the raw wastewater stream, this process may be preceded by one or several other treatment technologies (i.e., clarification, oil and grease removal, roughing filters, etc.).

In the suspended growth process, wastewater enters a reactor basin [i.e., concrete, steel, earthen tank(s)] where microorganisms are brought into contact with the organic components of the wastewater by some type of mixing device. This mixing device not only maintains all material in suspension but also promotes transfer of oxygen to the wastewater, thus providing oxygen necessary for sustaining the biological activities in the reactor basin. The organic matter in the wastewater serves as a carbon and energy source for microbial growth and is converted into microbial cell tissue and oxidized end products, mainly carbon dioxide. Contents of the reactor basin are referred to as mixed liquor and consist for the most part of microorganisms and inert and nonbiodegradable suspended matter.

When mixed liquor is discharged from the reactor basin, a means to separate the solids from the treated wastewater is normally provided. Usually a gravity settling basin is used. Concentrated microbial solids are recycled back to the reactor basin to maintain a concentrated microbial population for degradation of the wastewater. Because microorganisms are continually synthesized in this process, a means must be provided for wasting some of the generated microbial solids. Wasting of solids generally is done from the settling basin, although wasting from the reactor basin is an alternative.

The first suspended growth process, now called the conventional activated sludge process and shown in Figure 1 was developed to achieve carbonaceous BOD removal. However, since its inception, many modifications to the basic process have taken place. Some of the newer designs can accomplish phosphorus removal, nitrification, and toxic pollutant control. The multitude of variations in the activated sludge process are too numerous to be discussed in detail here. However, for information purposes some of these are: step aeration, contact stabilization, complete mix, extended aeration, aerated lagoon, and deep tank aeration.

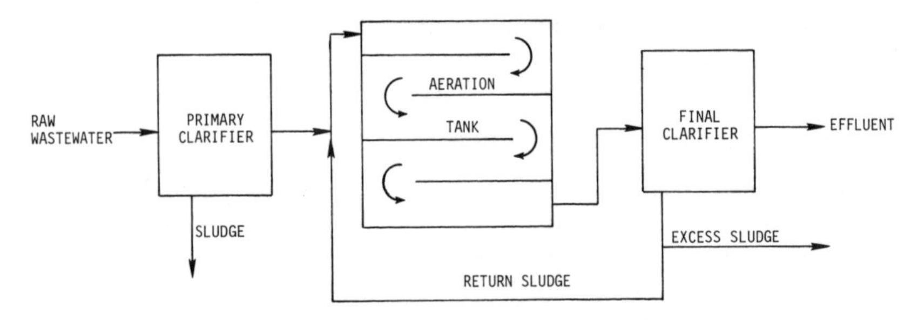

Figure 1: Schematic of first aerobic suspended growth process—conventional activated sludge.

Advantages and Disadvantages: The aerobic suspended growth processes have many variations with each variation having its own advantages and disadvantages. For the sake of brevity, this discussion of advantages and disadvantages is only from a general standpoint as compared with other technologies.

Aerobic suspended growth systems can be adapted to treat a wide range of industrial wastewaters. In fact, aerobic suspended growth systems have been continually refined to provide better treatment for more and more classes of pollutants. In addition, since these are biological processes relying on microorganisms for treatment, the procurement, handling, and storage of chemicals is very small or nonexistent. Its final advantage lies in the ease with which the process can be expanded to accommodate increased flows. Generally, this requires only the installation of additional reactor and settling basins.

Aerobic suspended growth systems do have some drawbacks. Suspended growth systems perform best under uniform hydraulic and pollutant loadings. For some industries, it is extremely difficult to maintain these conditions because of their manufacturing operations. A common event is a "shock" loading of high strength waste entering the treatment process with the end result being poor pollutant treatment. Also, certain toxic pollutants can kill the microorganisms in the reactor basin, causing a loss of treatment from this part of the overall system. Operational skills and controls required to effectively operate an aerobic suspended growth system are higher than most other biological processes. Finally, the energy cost involved in providing mixing and oxygen to the process to sustain microbial growth is increasing rapidly.

Pollutants Treated: Excluding those waste streams with toxic pollutants, aerobic suspended growth processes can be used to treat any wastewater containing biodegradable matter. (Toxic-laden waste streams would require additional preliminary treatment prior to the aerobic suspended growth process.) Its principal purpose is carbonaceous BOD removal with efficiencies over 90 percent possible, depending upon the characteristics of the wastewaters. The aerobic suspended growth processes are continually being modified. As a result, variations of the basic technology are now available which can remove ammonia, phosphorus, and even some toxic compounds. With experimental work still ongoing, even greater pollutant treatment, in terms of quality and quantity, can be expected from these processes.

Residuals Generated: Even though a portion is normally recycled back to the aerated basin, a considerable mass of waste sludge is produced and must be removed from the system on a regular basis – typically daily. The material is essentially excess microorganisms, though depending on the type of wastewater being treated and the modification of the process used, considerable amounts of inert material can also be present. The residuals are considered quite active, have a high oxygen demand, will generate odors if allowed to go septic, and are difficult to dewater. The residuals do require further processing by one or more of the processes described in Chapter 3.

Aerobic Attached Growth Processes

Description: An aerobic attached growth process is one in which the biological growth (microorganisms) are attached to some type of medium (i.e. rock, plastic sheets, plastic rings, etc.) and the wastewater either trickles downward over the surface or the medium is rotated through the wastewater. The process is similar to the aerobic suspended growth process, in that both depend upon biochemical oxidation of organic matter in the wastewater to convert to carbon dioxide in water, with a portion oxidized for energy used to sustain and promote growth of microorganisms. Depending on the type of material in the raw wastewater stream, this process may be preceded by one or several other treatment technologies (i.e. gravity sedimentation, oil and grease removal, etc.).

Three general types of aerobic attached growth systems will be described: conventional trickling filter, roughing filters, and rotating biological reactors.

Conventional Trickling Filter — Conventional trickling filters are typically round, though square or rectangular systems have been used in the past. They range from three to eight feet in depth and from 10 to 200 feet in diameter. The medium in the filter has generally been rock but since the early 1970's, plastic rings or corrugated plastic sheets have become popular.

Wastewater is applied to the filter, usually by a rotary distributor system and trickles downward through the media (Figure 2 shows a typical layout). Organic removal occurs by adsorption and assimilation of the soluble and suspended waste materials by microorganisms attached to the media. Oxygen for the process is supplied from passage of air through the interstices between the filter media which increases the dissolved oxygen in the wastewater. After the initial start-up period, microbial buildup creates an anaerobic interface with some of the filter media, thus causing growth of facultative and anaerobic organisms that, together with the aerobic organisms at upper microbial surfaces, form the basic mechanism for organic removal.

The quantity of biological slime produced is controlled by available food. Growth will increase as the organic load increases until a maximum effective thickness is reached. This maximum growth is controlled by physical factors including hydraulic dosage rate, type of media, type of organic matter, amount of essential nutrients present, temperature, and the nature of the particular biological growth. During filter operations, biological slime will be sloughed off, either periodically or continuously and collected and removed in the clarifier following the process.

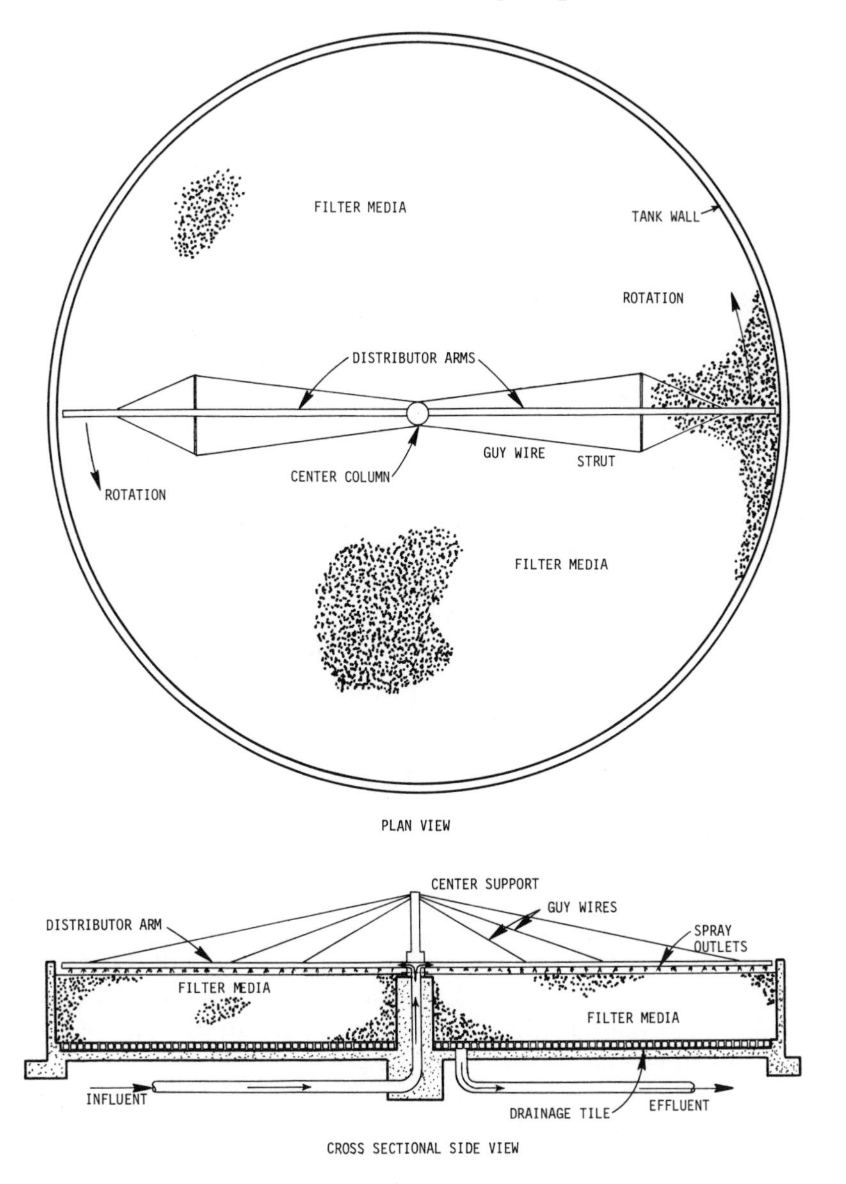

Figure 2: Typical conventional trickling filter.

Recirculation of trickling filter effluent has become a standard operational tool that has been used effectively in many instances to improve filter efficiency. Increasing the hydraulic flow rate reduces the likelihood of a dry surface during low flow periods, thereby maintaining maximum treatment capability. In addition, a higher flow rate maintains the shear force to slough excess growth, thereby reducing clogging problems. Finally, organic matter that may have missed exposure to the slime on its first passage through the filter may be treated on the second.

Roughing Filter — In some cases, it is necessary to provide an aerobic suspended growth system with some type of protection from "shock" loadings of organic pollutants or to provide an economical means of reducing the organic pollutant load 40 to 60 percent to make it more amenable to other pretreatment technologies. In these instances, a high rate trickling filter or roughing filter can be utilized.

Roughing filters are very similar to conventional trickling filters in appearances. Tanks are normally round with units being 12 to 18 feet high and up to 50 to 60 feet in diameter. At the greater heights, plastic is the preferred medium. Roughing filters are operated at very high hydraulic and organic loadings, sometimes called super-rate loadings, and usually do not provide for recirculation of filter effluent unless there is a chance of very low influent flows. Flow distribution is either by a rotary distributor or a spray nozzle system. Generally, clarification is not provided after the roughing filter, the flow going directly to the next treatment process.

Even though roughing filters can provide a high weight per unit volume of organic removal, their effluent will still contain a substantial amount of BOD and other pollutants. As a result, roughing filters should be considered as a first or intermediate step of a multistage biological treatment system.

Rotating Biological Reactors — In this process the biological growth is attached to a plastic disc which is rotated in a tank full of wastewater. The rotating biological reactor consists of large-diameter plastic media mounted on a horizontal shaft and placed in a tank. The contactor is slowly rotated while approximately 40 percent of the surface area is submerged in the wastewater. Organisms naturally present in the wastewater begin to adhere to the rotating surfaces and multiply until the entire surface area is covered with a thick layer of biomass. The biological population present on the plastic media is responsible for the treatment achieved.

In rotation, the reactor carries a film of wastewater into the air. Wastewater trickles down surfaces of the contactor and absorbs oxygen from the air. Organisms in the biomass remove both dissolved oxygen and organic materials from this film of wastewater. Further removal of organic materials and consumption of dissolved oxygen occur as the surfaces continue rotation through the wastewater in the tank. Unused dissolved oxygen in the wastewater film is mixed with the contents of the mixed liquor to maintain a mixed liquor dissolved oxygen concentration.

Shearing forces exerted on the biomass as it passes through the wastewater cause excess biomass to be stripped from the media into the mixed liquor. This prevents clogging of the media surfaces and maintains a constant microorganism population on the media. Mixing action of the rotating media keeps the stripped solids in suspension until the flow of treated wastewater carries them out of the process to the clarifier. The rotating media thus serves the following functions: provides surface area for development of a large, fixed biological culture; provides contact of biological growth with the wastewater; aerates wastewater; provides a positive means of continuously stripping excess biomass; and agitates the mixed liquor to keep sloughed solids in suspension and to provide thorough mixing.

Each shaft of media operates as an attached-film biological reactor, in which the rate of biological growth and rate of stripping excess biomass pass through subsequent stages of media. As wastewater passes from stage to stage, it undergoes a progressively increasing degree of treatment by specific biological cultures, including nitrifying bacteria, various types of protozoans, rotifers, and other predators. Generally, wastewater flows through the process just once, with no need for recycling of effluent. Also, because the attached biomass is continously growing, there is no need for sludge recycling.

Figure 3 shows a typical rotating biological reactor installation.

Advantages and Disadvantages:

Conventional Trickling Filters — There are several advantages for a trickling filter over other biological processes. First, trickling filters can more easily reinstitute microorganism growth after an accidental kill. Second, since oxygen is supplied naturally, the need for air or oxygen generating equipment is eliminated. This, along with a much more simple operation, lowers the requirements for highly skilled operating personnel. Both can result in substantial cost and energy savings.

Disadvantages are as follows. Trickling filters are subject to freezing in cold weather and clogging due to solids. Modifications in the treatment process are limited, and trickling filters generally do not provide as high a degree of treatment as suspended growth systems. Finally, in poorly operated systems, flies and other pests often infest trickling filters and create a nuisance in the vicinity.

Roughing Filters — Because roughing filters are employed as a pretreatment or conditioning step prior to further biological treatment, their advantages lie in improving the performance of subsequent treatment processes. Their best known use is to help minimize "shock" organic loads to downstream biological processes. In the case where a high temperature waste stream is present, they cool the flow, so as not to adversely impact the growth of microorganisms in other treatment processes.

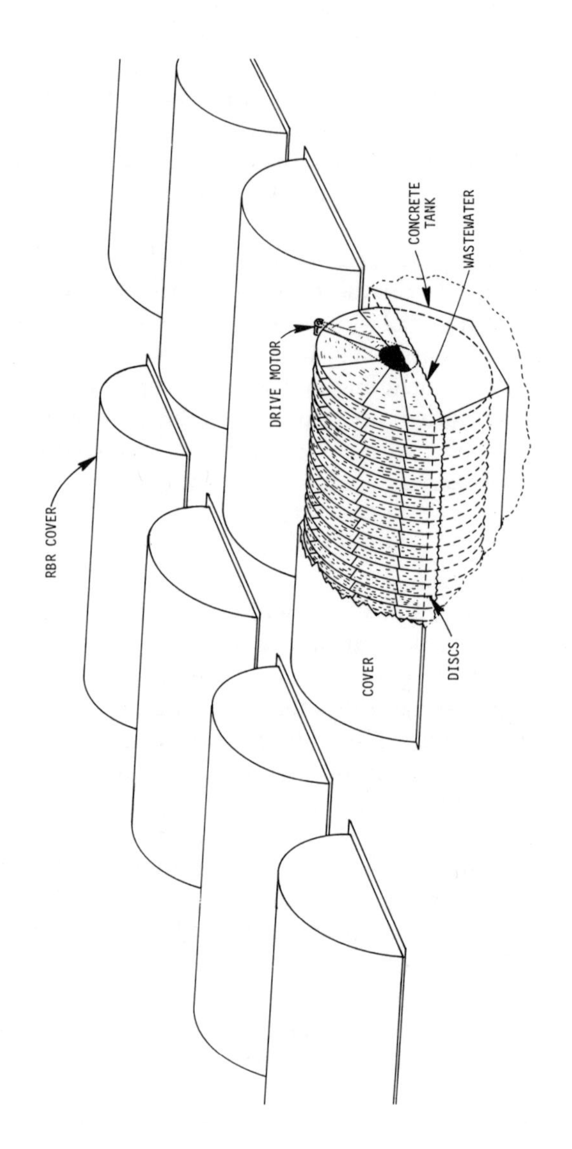

Figure 3: Typical rotating biological contactor.

The principal disadvantages of roughing filters are the same as trickling filters. They are subject to freezing in cold weather and plagued by flies and other pests if not operated properly. Also, roughing filters may become clogged due to high solids loadings.

Rotating Biological Reactors — Rotating biological reactors have the same advantages as a conventional trickling filter. In addition, their modular construction makes them easy to install for upgrading an existing treatment facility. Finally, the tank-in-series design allows the biological system a better chance of withstanding a "shock" or "toxic" load condition.

As with all attached growth systems, this treatment process has difficulty operating in cold climates. Enclosing the units for temperature protection can lead to other problems such as condensation. Another disadvantage of rotating biological reactors is that high organic loading can result in septicity in the units. Finally, this technology is susceptible to clogging, if dense media is used and/or high solids loadings are applied.

Pollutants Treated: Excluding those waste streams with toxic pollutants, aerobic attached growth processes can be used to treat any wastewater containing biodegradable matter. (Toxic-laden waste streams may require additional preliminary treatment prior to the aerobic attached growth process.) In general, the aerobic attached growth processes are not quite as good as aerobic suspended growth processes in removing biochemical oxygen demand, total suspended solids, and toxic pollutants.

Residuals Generated: For comparable pollutant removals, aerobic attached growth systems produce less sludge mass per unit of pollutant removed than aerobic suspended growth systems. For conventional trickling filters and rotating biological reactors, waste sludge must be removed from the system on a regular basis — typically daily. Roughing filters do not generate residuals per se since all wastewater flows to other downstream biological processes, where the solids are collected and disposed of.

Material generated is essentially excess microorganisms, though depending on the type of wastewater being treated, considerable amounts of inert material can also be present. Material from conventional trickling filters tends to be stringy and more dense than from aerobic suspended solids systems. Material from rotating biological reactors looks similar to that from aerobic suspended solids system. In general, residues generated are easier to manage but do require further processing by one or more of the processes described in Chapter 3.

Lagoons (Stabilization Ponds)

A lagoon is a relatively shallow body of water contained in an earthen basin of controlled shape, which is designed for the purpose of treating wastewater. Lagoons can be subdivided into five general types: aerated, aerobic, tertiary or polishing, anaerobic, and facultative. Aerated lagoons are classified as an aerobic suspended solids system and was discussed earlier in this chapter. The other four are discussed in the following.

Description:

Aerobic Lagoons — Aerobic lagoons are large, shallow, earthen basins that are used for wastewater treatment by using natural processes involving the use of both algae and bacteria. The objective is microbial conversion of organic wastes into algae. Aerobic conditions prevail throughout.

In aerobic, photosynthetic lagoons, oxygen is supplied by algal photosynthesis and by natural surface aeration. Oxygen produced by algae through the process of photosynthesis is used by bacteria for biochemical oxidation and degradation of organic waste. Carbon dioxide, ammonia, phosphate, and other nutrients released in the biochemical oxidation reactions are, in turn, used by the algae, forming a cyclic-symbiotic relationship.

High-rate aerobic lagoons usually have a liquid depth of 0.5 to 2 feet and receive organic loadings of 60 to 200 pounds BOD_5 per acre per day. Average detention time ranges from two to six days. Designing a shallow lagoon to allow maximum light penetration and atmospheric diffusion will maximize algae production and oxygen availability. To achieve best results with aerobic lagoons, their contents must be well mixed periodically using pumps or surface aerators. Algae concentration ranges from 100 to 250 mg/l.

Low-rate aerobic lagoons are generally 3 to 5 feet in depth and have a long detention time ranging from 10 to 40 days. Algae concentration ranges from 40 to 200 mg/l. Other aspects of low-rate aerobic lagoons are similar to that of high-rate aerobic lagoons.

Figure 4 shows a typical aerobic lagoon.

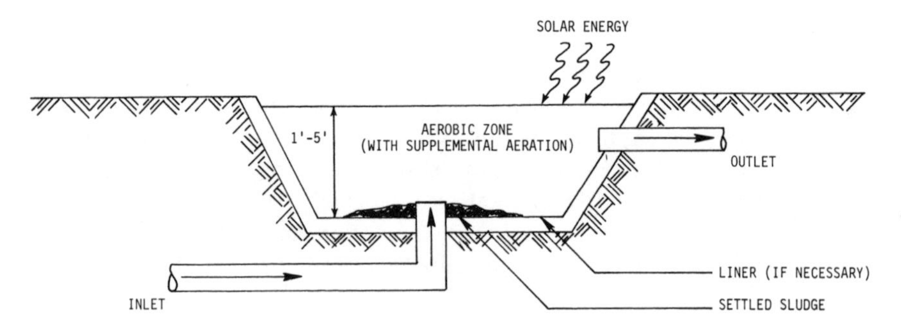

Figure 4: Typical aerobic lagoon.

Tertiary or Polishing Lagoons — The tertiary, maturation, or polishing lagoon generally is used for polishing (upgrading) other biological effluents. BOD_5, suspended solids, coliform count, nitrogen, and phosphorus are reduced. Algae and surface aeration provide oxygen for waste stabilization. Operationally, residual biological solids are eventually oxidized and nitrification is generally achieved. Organic loadings of biologically treated effluents range from 10 to 15 pounds BOD_5 per acre per day. A detention time of 18 to 20 days is generally required to provide for complete endogenous respiration of residual solids. The depth of polishing lagoons varies from one to five feet. Oxygen is supplied photosynthetically and from supplemental aeration. Stratification may or may not occur depending on the degree of vertical mixing. The flow regime is classified as "intermittently mixed." Algal concentration ranges from 15 to 100 mg/l. About 60 to 80 percent BOD_5 conversion can be achieved. Effluent suspended solids concentration ranges from 10 to 30 mg/l.

Polishing lagoons look very similar to aerobic lagoons and for all general purposes can be represented by Figure 4.

Anaerobic Lagoons — Anaerobic lagoons are earthen ponds built with a small surface area and a deep liquid depth of 8 to 20 feet. Usually these lagoons are anaerobic throughout their depth, except for an extremely shallow surface zone. Once greases form an impervious surface layer, complete anaerobic conditions develop.

Stabilization of high-strength organic wastes is brought about by a combination of precipitation and anaerobic conversion of organic wastes to carbon dioxide, methane, bacterial cells, organic acids, and other end products.

In a typical anaerobic lagoon, raw wastewater enters near the bottom of the lagoon (often at the center) and mixes with the active microbial mass in the sludge blanket, which is usually about six feet deep. The discharge is located near one of the sides of the lagoon, submerged below the liquid surface. Excess undigested grease floats to the top, forming a heat-retaining and relatively air-tight cover. Wastewater flow equalization and heating are generally not practiced. Excess sludge is washed out with the effluent. Recirculation of waste sludge is not required.

Organic loadings are from 0.36 to 10.5 pounds volatile solids per 1,000 cubic feet per day; however, loadings as high as 132 to 320 have been used successfully. Algae concentration is less than 5 mg/l. Detention time ranges from 1 to 50 days depending on climatic conditions.

Figure 5 shows a typical anaerobic lagoon.

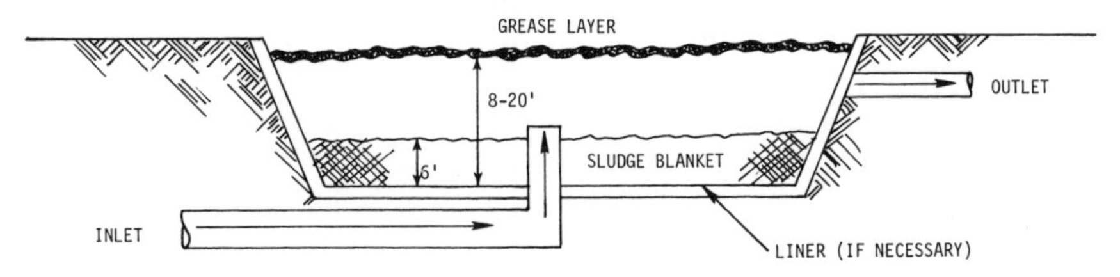

Figure 5: Typical anaerobic lagoon.

Facultative Lagoons — Conventional facultative lagoons are intermediate depth (three to eight feet) ponds in which wastewater is stratified into three zones: aerobic zone, intermediate or facultative zone, and anaerobic zone. Stratification is a result of solids settling and temperature-water density variations. Oxygen in the surface aerobic zone is provided by algal photosynthesis and atmospheric diffusion. In general, the aerobic surface layer serves to reduce odors while providing treatment of soluble organic byproducts of the anaerobic processes operating at the bottom. Photosynthetic activity at the lagoon surface produces oxygen diurnally, increasing dissolved oxygen during daylight hours, while surface oxygen is depleted at night. Sludge at the bottom of faculatative lagoons will undergo anaerobic digestion producing carbon dioxide, methane, cells, and other end products. Organic loadings generally range from 0.05 to 0.1 pounds BOD_5 per 1,000 cubic feet per day. BOD_5 removal efficiencies range from 70 to 95 percent. Algal concentration in the aerobic zone is about 20 to 80 mg/l. Detention time ranges from 7 to 30 days.

Aerated facultative lagoons are similar to conventional facultative lagoons in operational mode, detention time, depth, and BOD_5 conversion efficiencies, except that the aerobic zone is maintained by both the supplemental aeration system and the photosynthetic oxygenation, although the latter is reduced due to increased lagoon turbidity. Supplemental aeration is employed in facultative lagoons to relieve organic overload, to reduce time of treatment, or to reduce odor complaints. Organic loadings of 0.08 to 0.3 pounds BOD_5 per 1000 cubic feet per day are common. Aerated facultative lagoons are usually not stratified because of increased vertical mixing attributable to the supplemental aeration system. Heavy solids settle to the bottom and undergo anaerobic decomposition; the balance of influent organics are stabilized aerobically. Algal concentration ranges from 5 to 20 mg/l.

Facultative lagoons are often contained within earthen dikes and operated in series or parallel. When three or more lagoons are linked, effluent from either the second or third lagoon may be recirculated to the first for improvement of overall performance. Figure 6 shows a typical facultative lagoon.

Advantages and Disadvantages: Advantages in using any lagoon process discussed in this section are: low capital cost (excluding land if not readily available); simplicity of operation (little operator expertise is required); low operation and maintenance cost; and when designed and operated properly, high reliability.

Disadvantages in using any lagoon process discussed in this section are: high land requirements, possible noxious vegetative growths, possible odor emission, and the potential for seepage of wastewater into groundwater unless the lagoon is adequately lined. In addition, in most locales of the United States there are seasonal changes in both available light and temperature. Typically in the winter, biological activity decreases due to a reduction in temperature. In facultative lagoons, ice formations can cause operational problems. For anaerobic lagoons, efficient operation necessitates that liquid temperatures stay above 75°F. One last disadvantage is that effluents normally have suspended solids between 100 to 300 mg/l.

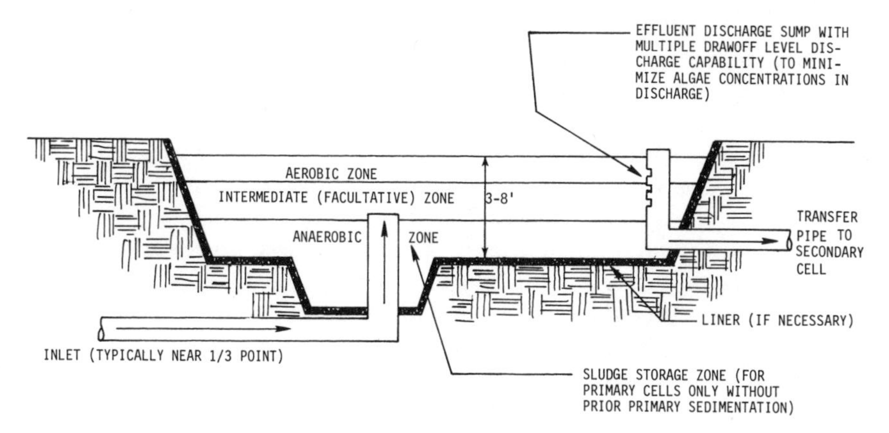

Figure 6: Typical facultative lagoon.

Pollutants Treated:

Aerobic Lagoons — Aerobic lagoons are used for treatment of weak industrial wastewater containing negligible amounts of toxic and/or nonbiodegradable substances.

Polishing Lagoons — Polishing lagoons receive the effluents from conventional biological processes, such as activated sludge, trickling filter, etc. for treatment.

Anaerobic Lagoons — Anaerobic lagoons are effective as roughing units prior to aerobic treatment of high strength organic wastewater that also contains a high concentration of solids. BOD_5 conversion efficiencies up to 70 percent are obtainable routinely. Under optimum operating conditions, BOD_5 removal efficiencies up to 85 percent are possible.

Facultative Lagoons — Facultative lagoons are effective for treatment of relatively weak industrial wastewaters which have been screened or clarified.

Residuals Generated: Aerobic, polishing, and facultative lagoons produce very small quantities of sludge. Generally, settled solids for these three processes may require cleanout and removal once very five to ten years or longer. In the anaerobic lagoon, sludge is usually washed out in the effluent. Since anaerobic lagoons are generally used for preliminary treatment before other biological pretreatment processes, this is not normally a problem.

Anaerobic Suspended Growth Processes

Description: An anaerobic suspended growth process is one in which the biological growth (microorganisms) are kept in suspension in a liquid medium consisting of entrapped and suspended colloidal and dissolved organic and inorganic materials. The basic biological difference between this process and aerobic suspended growth processes is that anaerobic processes function without the use of oxygen. Organic matter in the wastewater serves as a carbon and energy source for microbial growth and is converted into microbial cell tissue and oxidized end products, mainly methane and carbon dioxide.

Two general types of anaerobic suspended growth processes will be described: conventional and anaerobic contact.

Conventional — In the conventional process wastewater enters a closed reactor basin — normally a circular concrete tank — where microorganisms are brought into contact with the incoming wastewater by some type of mixing device. Heating of raw wastewater is not normally practiced but is a design option to be considered.

The reactor is typically of constant liquid volume though variable volume reactors have been utilized. Generally effluent leaves the tank through an overflow pipe and may require further treatment before being discharged to the POTW.

Digester gas is piped from the digester and is either flared or utilized within the system for firing a boiler or electrical power generation.

Figure 7 shows a typical conventional anaerobic suspended growth system.

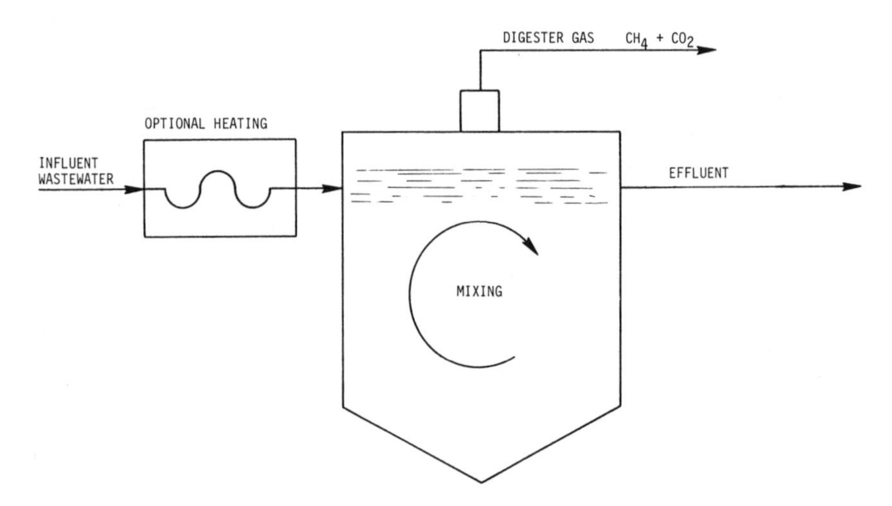

Figure 7: Typical conventional anaerobic suspended growth system.

Anaerobic Contact — The basic difference between this suspended growth process and the conventional process discussed above is that a positive solids-liquid separation device (vacuum degasifier, clarification centrifuge) is utilized on the reactor effluent. A portion of the solids removed are recycled back to the reactor. This allows for lower hydraulic detention times but long solids residence times, an important factor in anaerobic processes. When required, excess biomass is wasted to maintain a desired solids level within the reactor. Figure 8 shows a typical anaerobic contact process flow schematic.

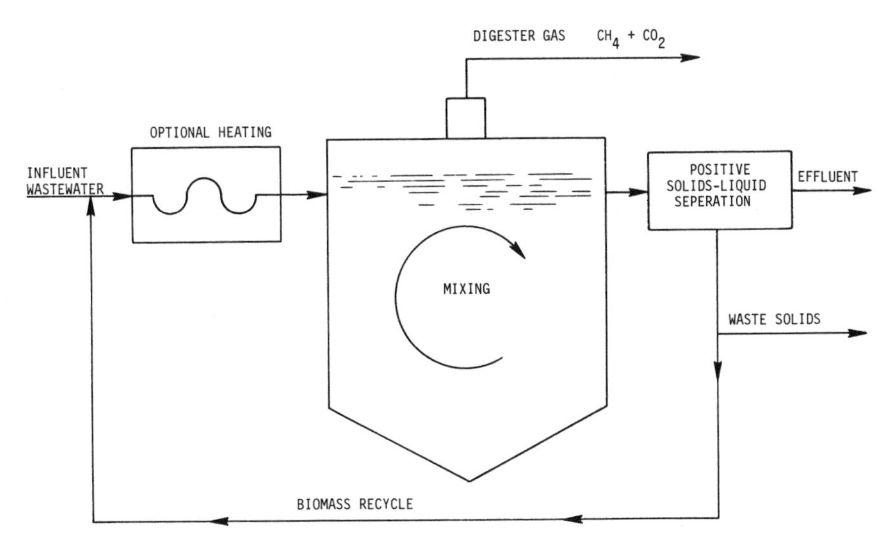

Figure 8: Typical anaerobic contact suspended growth system.

Advantages and Disadvantages:

Conventional — There are several advantages for an anaerobic system over aerobic processes. First, a high degree of wastewater stabilization may be accomplished with a relatively low production of biological solids and a usable byproduct, methane gas, is produced. Secondly, anaerobic processes do not require use of aeration equipment (main energy users in any aerobic operated process).

Finally, anaerobic processes utilize less land and have operating cost per pound of organic material removed, significantly less than aerobic processes.

Disadvantages are as follows. Anaerobic process requires long solids residence times generally 15 to 20 days minimum. The system is somewhat sensitive to operating conditions and there is potential aesthetic problem due to possible odor generation.

Anaerobic Contact — Besides having all the advantages of a conventional anaerobic suspended growth system, the anaerobic contact process allows for very short hydraulic detention times (four to six days) while still able to maintain long solids residence times. This reduced hydraulic volume capacity means smaller tanks and less land requirements. In addition, the larger biomass concentration within the reactor allows the system to operate over a large variation in organic feed rates without undue stress.

A disadvantage is that a positive solids-liquid separation device is required which can be depended upon to work consistantly. Typically this would be a vacuum degassifier followed by clarification or centrifugation. Both are added expense and operational cost over conventional systems. In addition, a residual is generated which must be managed.

Pollutants Treated: Excluding those waste streams with toxic pollutants, anaerobic suspended growth processes can be used to treat any organic wastewater with high solids content and over several thousand mg/l BOD_5. (Toxic-laden organic waste streams may require additional preliminary treatment prior to the anaerobic suspended growth process.) Categorical industries that have utilized this process for wastewater treatment are: pharmaceutical, petroleum refining, and pulp and paper.

Residuals Generated: Anaerobic processes generate much less biomass per unit of BOD_5 removed than aerobic processes. In addition, the long solids residence times required by anaerobic processes produce solids which are quite stable, have a low oxygen demand, and normally only produce a musty odor when allowed to sit in a open area for any length of time. Normally residuals would require further processing by one or more of the methods described in Chapter 3, although disposal to the POTW should be considered.

Anaerobic Attached Growth Process

Description: An anaerobic attached growth process is one in which the biological growth (microorganisms) are attached to some type of medium (i.e., rock, plastic rings, etc.) inside a sealed tank and wastewater flows upward through the medium. Organic matter in the wastewater serves as a carbon and energy source for microbial growth and is converted into microbial cell tissue and oxidized end products, mainly methane and carbon dioxide.

Figure 9 shows a schematic of an anaerobic attached growth process commonly called an anaerobic filter.

In this system soluble organic wastewater is pumped into the bottom of the reactor. As flow moves upward organic material in the wastewater comes in contact with the microorganisms attached to the supporting medium. Effluent leaves the tank at the top through an overflow pipe and normally does not require further treatment before being discharged to the POTW. Digester gas is piped from the digester and is either flared or utilized within the system as a prime energy source.

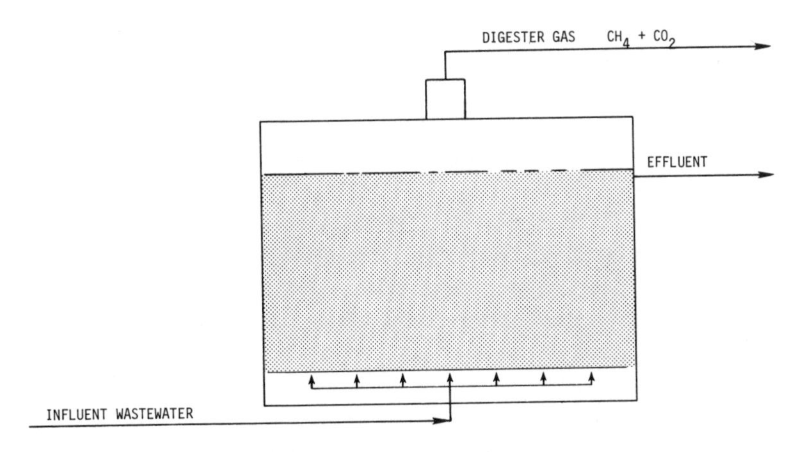

Figure 9: Typical anaerobic attached growth system (anaerobic filter).

Advantages and Disadvantages: In addition to having the same advantages as other anaerobic systems, attached growth systems have the following additional benefits. A internal mixing system is not required. In addition, long solids residence times and short hydraulic retention times are possible without solid-liquid separation and biomass recycle. These systems are extremely resistant to large organic input variations. Finally, this system produces the lowest amount of solids per unit of BOD_5 removed.

The principal disadvantage is that these type systems can only be utilized for soluble organic waste with very low suspended solids. Otherwise the system will "clog-up".

Pollutants Treated: Excluding those waste streams with toxic pollutants, anaerobic attached growth process can be used to treat soluble organic wastewater over 1000 to 1500 mg/l BOD_5 and very low suspended solids – under 100 mg/l. (Toxic laden organic waste streams may require additional preliminary treatment prior to the attached anaerobic growth process.)

Residuals Generated: Anaerobic attached growth systems generate the smallest amount of biomass per unit of BOD_5 removed for any anaerobic or aerobic process — due to the very long solids residence times. These solids are washed out with the system effluent and are not normally processed any further except as part of any downstream treatment.

Aquaculture Processes

Description: Aquaculture is a process involving production of aquatic organisms (both flora and fauna) under controlled conditions. Wastewater treatment is accomplished by sprinkling or flood irrigating wastewater into a wetland area or by passing the wastewater through a system of shallow ponds, channels, basins, or other constructed areas where aquatic organisms have been seeded or naturally occur and are actively growing. The aquatic organisms remove nutrients, BOD_5, suspended solids, metals, etc. Vegetation produced as a result of the system's operation may be composted for use as a source of fertilizer/soil conditioner, dried or otherwise processed for use as animal feed supplements, or digested to produce methane.

The water hyacinth, *(Eichornia crassipes)*, a fast-growing floating aquatic plant, appears to be the most promising organism for aquaculture treatment. Other promising organisms which can be used in aquaculture include floating duckweed, seaweed, rooted bulrushes and reeds, suspended algae, and midge larvae. A poly-culture system employing both aquatic plants and animals can also be used.

Figure 10 shows an aquaculture basin in which water hyacinth cover is planted for wastewater treatment. Figure 11 shows top and side view of an aquaculture wetland system in which duckweed are used for treatment.

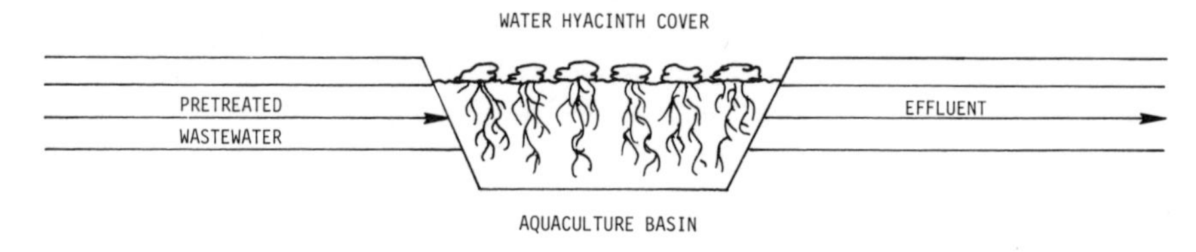

Figure 10: Aquaculture system using water hyacinth.

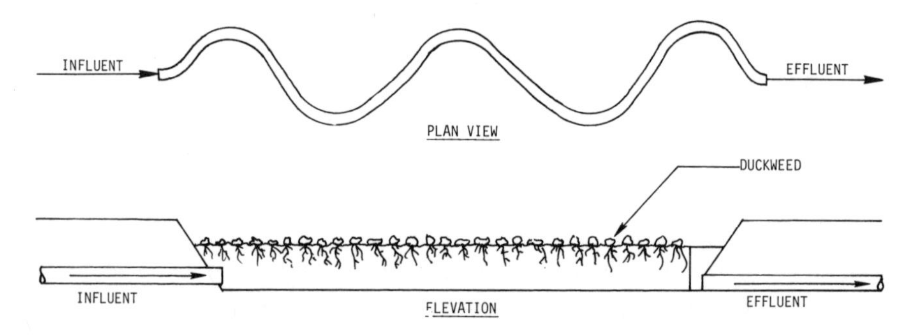

Figure 11: Aquaculture system using duckweed.

Advantages and Disadvantages: Aquaculture systems reduce pollutant levels of wastewater while enhancing available wildlife habitat, and/or providing resource recovery. It has high potential as a low cost, low energy consuming pretreatment system for small industrial flows.

Land requirement is a major limitation. In addition, temperature (climate) is a rather major limitation since effective wastewater treatment is linked to the active growth phase of the emergent vegetation. Protection of emergent vegetation, such as duckweed, from waterfowl, fish, and wind is needed. Aquaculture systems may cause an increase in evapotranspiration, and may be impractical for large facilities due to large land requirements (2 to 25 acres per million gallons per day of flow). Technology is still in the developmental stage. Many states prohibit wastewater discharge into wetlands.

Pollutants Treated: The process provides neutralization and some reduction of nutrients, heavy metals, organics, BOD_5, suspended solids, fecal coliforms, and pathogenic bacteria. It is useful for polishing treated effluents.

Residuals Generated: Residuals generated are dependent upon type of system and whether or not harvesting is employed. Duckweed, for example, yields 50 to 60 pounds per acre per day (dry weight) during peak growing period to about half of this figure during colder months.

Hyacinth harvesting may be continuous or intermittent. Average hyacinth production (including 95 percent water) is on the order of 1,000 to 10,000 pounds per acre per day.

Land Treatment Processes

Description: Land treatment is the application of wastewater, usually following some form of preapplication treatment, on land by one of several conventional methods involving irrigation or groundwater recharge. The purpose of land treatment is to recycle nutrients and water contained in wastewaters, often for productive use in agriculture or silviculture systems, in an ecologically sound manner, and to simultaneously provide a high degree of wastewater renovation. Treatment is provided by biological, chemical, and physical processes as the effluent moves through the natural filter provided by the plants, soil, and related ecosystem. The degree of treatment obtained can be influenced by the mode of land treatment utilized, site selection, loading rate, and other design variables.

The three principal land treatment methods are: slow-rate, rapid infiltration, and overland flow.

Slow-rate — The slow-rate process is controlled discharge of wastewater by spraying or surface spreading onto land to support plant growth. The slow-rate process is commonly referred to as irrigation with a portion of the flow percolating to ground water. Wastewater is primarily lost through crop consumption and by evapotranspiration. The slow-rate hydraulic pathway is illustrated in Figure 12. Objectives of slow-rate treatment include: treatment of applied wastewater; conservation of water by use of renovated water; revenue production by sale of irrigated crops; and the preservation and enlargement of greenbelts and open space.

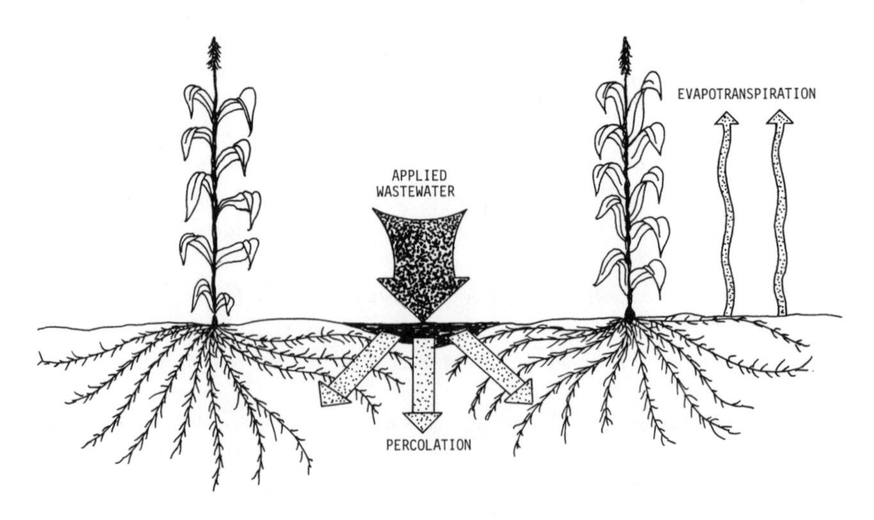

Figure 12: Slow rate hydraulic pathway.

The process is termed "slow-rate" because the rate is restricted by plant uptake, infiltration capacity of the soil, evapotranspiration, and climate conditions. Slow-rate systems require well drained soils of a uniform depth of five feet or more to allow for adequate root development. Mild climates are preferable, but more severe climates are acceptable, if adequate wastewater storage is provided for wet or freezing conditions. The slow-rate concept severely restricts the amount of wastewater which can be applied to the land and in turn it creates the need for a larger land area per unit volume of wastewater than rapid infiltration or overland flow systems.

There are a number of methods for applying wastewater to the land through irrigation systems. Each site has physical characteristics that will influence the choice of the method of application. The most commonly used methods are spraying, surface distribution by ridge and furrow, and flooding techniques. The method used depends on many factors, including soil characteristics, crops, labor requirements, maintenance, topography, and costs. Each method has distinct advantages for specific project conditions.

Spraying has become the predominant form of application of industrial wastewaters for several reasons. Most importantly, sprinklers distribute the organic and solids load of the wastewater uniformly over the site. This is important with high strength wastewater because it prevents portions of the site from becoming overloaded. Sprinkle systems are more readily adaptable to the widely varying wastewater flows which often occur in industrial processing. Additionally, there is less land preparation required with sprinkler application.

The ridge and furrow distribution method is accomplished by gravity flow. Wastewater is pumped to the head end of the furrows and is allowed to run until the tail end of the furrow is adequately irrigated. This results in a nonuniform application of water. This method is best suited to small systems where close control of large volumes of wastewater is not required. This method has been applied successfully to systems where the wastewater was toxic to vegetation.

The use of flooding with industrial wastewaters has limited applicability when a cover crop is used. Suspended solids contained in most industrial wastewaters are deposited at the head end of the strip or basin, overloading that area and choking out the vegetation.

Slow-rate land treatment systems convert most of the wastewater and nutrients into plant growth. Grasses are usually chosen for their vegetation because of their high nitrogen uptake capacities. However, when crop yield economic returns are emphasized, crops of higher values than grasses are usually selected. In areas where water does not limit plant growth, the nitrogen and phosphorus in wastewater can be recycled in crops. These nutrients can increase the yields of crops and decrease the expense of fertilizers.

Although slow-rate land treatment systems have gained popularity for agricultural uses, they have also been used in forested areas and silviculture operations. Wastewater is also used for the reforestation and reclamation of drastically disturbed areas such as those resulting from strip mining operations.

Rapid Infiltration — Systems utilizing the soil structure as the primary mode of treatment are considered rapid infiltration systems. In this type system, most of the applied wastewater percolates through the soil, and treated effluent eventually reaches groundwater, or is removed by wells or tile drainage. Wastewater is applied to rapidly permeable soils, such as sands and loamy sands, by spreading in basins or by spraying and sprinkling, and is treated as it travels through the soil matrix. Vegetation is not usually used, but there are some exceptions where crops are grown and harvested.

The principal objective of rapid infiltration is wastewater treatment. Other objectives for the treated water can include: groundwater recharge; recovery of renovated water by wells or under drains with subsequent reuse or discharge; recharge of surface streams by interception of groundwater; and temporary storage of renovated water in the aquifer. If groundwater quality is being degraded by salinity intrusion, groundwater recharge by rapid infiltration can help to reverse the hydraulic gradient and protect existing groundwaters.

Intermittent flooding in basins is the most common distribution method, although high-rate spraying (more than four inches per week) may also be used. Figure 13 shows the typical hydraulic pathway for rapid infiltraton. A much greater portion of the applied wastewater percolates to the groundwater than with slow-rate land treatment. There is little or no consumptive use by plants and less evaporation in proportion to a reduced surface area. In many cases, recovery of renovated water is an integral part of the system.

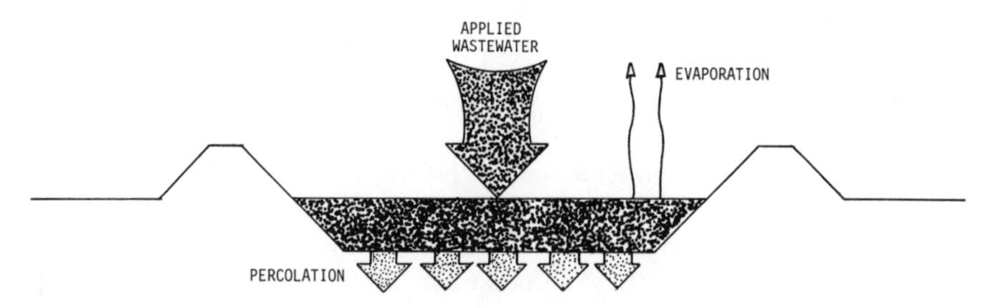

Figure 13: Rapid infiltration hydraulic pathway.

Return of renovated water to the surface by wells, underdrains, or groundwater interception may be necessary when discharge to a particular surface water body is dictated by water rights or when existing groundwater quality is not compatible with expected renovated water quality. Treated water can be withdrawn immediately.

Overland Flow — Overland flow is a land treatment process which can be used specifically on sites containing soils with limited permeability, or where the groundwater table is high so water cannot move into the soil profile. In the overland flow process, wastewater is filtered as it flows across the vegetative surface and soil to run off collection ditches. Wastewater is renovated by physical, chemical, and biological means as it flows in a thin film or sheet down over the sloping ground surface. Bacteria present on the grass and vegetative litter act to decompose the solids and organics. The system does not rely on percolation into the soil. Since overland flow is used on clay and silty-type soils with low infiltration capacity, the amount of vertical movement of wastewater into the soil is not an important factor. Figure 14 illustrates the overland flow hydraulic pathway.

Overland flow differs from spray irrigation primarily in that a substantial portion of the wastewater applied is designed to run off and must be collected for reuse or discharged. The objectives of overland flow are primarily wastewater treatment and, to a minor extent, crop production. The treatment objective is to improve effluent quality from screened primary treated or lagoon treated wastewater. Overland flow is also used for production of forage grasses and, in this manner, promotes the preservation of green belts and open space.

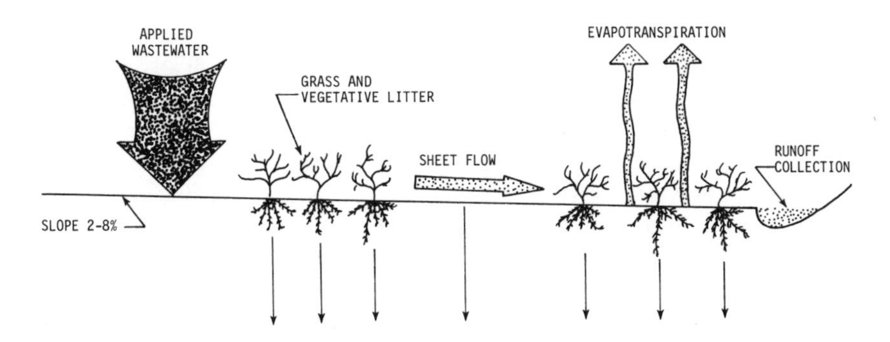

Figure 14: Overland flow hydraulic pathway.

Overland flow systems are extremely flexible and are capable of consistently producing the same quality effluent day after day, regardless of variations in wastewater quality. This uniform quality is attributed to the soil's tremendous buffering and stabilizing system.

Advantages and Disadvantages: Land application of industrial wastewaters offers many advantages. It is a cost effective, energy conserving wastewater treatment technology. Social, environmental, and economic benefits of land treatment may include the following: production of agriculture; silviculture; and aquaculture; creation of greenbelts; preservation of open space; energy savings; conservation of potable water resources; groundwater recharge; resource recovery and conservation; and profits from the sale of crops.

In addition, land application offers advantages over many other wastewater pretreatment technologies, since it does not require the addition of chemicals and does not produce sludge. Operation and maintenance cost of land treatment systems are relatively low.

The major disadvantage of land application systems is that they require large areas of land. For industries in metropolitan areas, the high price and limited availability of land are often prohibitive for land treatment. Land treatment has limited industrial application, since many wastewaters contain toxic constituents which may adversely affect the soil and groundwater. Since land treatment relies on and affects natural systems, its development in some locations may be subject to climatic, geologic, hydrologic, and topographic constraints.

Pollutants Treated:

Slow-rate — Land treatment systems are particularly successful in removing BOD, suspended solids, nitrogen, and phosphorus from wastewater. In general, slow-rate systems produce the best results of all the land treatment systems. When requirements for discharge to POTW are very stringent for nitrogen, phosphorus, and BOD, they can be met with slow-rate land treatment. Organics are reduced substantially in slow-rate systems by biological oxidation within the top few inches of soil. Suspended solids are removed by filtration. Volatile solids are biologically oxidized, and fixed or mineral solids become part of the soil matrix. Nitrogen is removed primarily by crop uptake which varies with the type of crop grown and the crop yield. To remove the nitrogen effectively, the portion of the crop that contains the nitrogen must be physically removed from the field. Denitrification can also be significant even if the soil is in an aerobic condition most of the time. Phosphorus is removed from solution by fixation processes in the soil, such as adsorption and chemical precipitation. A small but significant portion of the phosphorus applied is taken up and removed with the crop.

Rapid Infiltration — BOD, suspended solids, and fecal coliforms are almost completely removed in most cases. Nitrogen removals are generally poor unless specific operating procedures are established to maximize denitrification. Increased denitrification can be achieved by adjusting application cycles, supplying an additional carbon source, using vegetated basins, recycling a portion of the renovated water containing high nitrate concentrations, and reducing application rates. Phosphorus removals can range from 70 to 99 percent depending on the physical and chemical properties of the soil. As with slow-rate systems, the primary removal mechanism is adsorption with some chemical precipitation.

Overland Flow — In overland flow systems, biological oxidation, sedimentation, and grass filtration are the primary removal mechanisms for organics and suspended solids. Nitrogen removal is extremely successful in overland flow due to the nature of the soil surface. An aerobic-anaerobic layer exists at the surface of the soil and allows both nitrification and denitrification to occur. Plant uptake of nitrogen is another important removal mechanism, but permanent nitrogen removal by plant uptake is only possible if the cover crop is harvested and removed from the field. Ammonia volatilization can be an important removal mechanism if the pH of the water is above 7. Phosphorus and heavy metals are removed principally by adsorption, precipitation, ion exchange in the soil, and plant uptake. However, since there is only limited contact of wastewater with the soil matrix, phosphorus removal efficiencies are usually about sixty percent. Increased removals may be obtained by adding alum or ferric chloride prior to application. Dissolved solids are removed by plant uptake, ion exchange, and leaching.

Residuals Generated: Land treatment processes do not generate sludge. Residuals in the form of harvested plant material may result from slow-rate and overland flow systems where vegetation is an integral component of the treatment process.

Many systems report cutting but not harvesting the vegetative cover. This procedure may be of benefit for systems that handle nutrient deficient wastewaters — for example, the pulp and paper industry. Decay of cut vegetation will provide for the recycling of plant nutrients. Vegetative litter may also prove beneficial by providing additional media for microorganisms. There is danger in generating too much vegetative litter, which can exert a substantial BOD loading of its own on the system and thus limit its capacity.

PHYSICAL PRETREATMENT TECHNOLOGIES

Activated Carbon Adsorption

Description: Limitations of conventional treatment processes with regard to removal of some organic compounds are a significant factor in the treatment of industrial wastewaters. In most instances, because of an industry's wastewater characteristics, many pretreatment technologies prove to be inadequate in removing certain organic compounds. Adsorption using activated carbon, activated alumina, or peat has been demonstrated to be effective for removal of a wide range of organic materials and has begun to receive widespread attention as an industrial pretreatment technology. Its most important aspect is that adsorption can remove conventional pollutants, as well as toxic organics, and thus serves a dual purpose. Activated carbon is by far the most common material utilized and will be the focus of the following discussion. Removal mechanisms using activated alumina and peat are similar to activated carbon.

Activated carbon removes dissolved materials through the action of two distinctly different mechanisms. The first of these is adsorption, which actually removes the dissolved organics from the wastewater. Organic molecules in solution are drawn to the porous surface of the carbon granule by inter-molecular attraction forces, where the organics become substrates for biological activity. It is theorized that adsorption is the principal mechanism by which dissolved organics are removed and that biological activity functions as a regenerant of the adsorption sites by reopening the porous surfaces of the activated carbon. The exact mechanism is unknown, but an analysis of available data shows that the biological contribution to the removal capacity is significant.

Industrial wastewater treatment with activated carbon involves two major and separate process operations. First is the contacting system. Here, wastewater to be treated passes through a vessel filled with the activated carbon, where dissolved impurities are removed by adsorption. Generally, a series of vessels is used to provide optimum performance and allow for regeneration of spent carbon. Contacting systems may be comprised of either pressure or gravity-type vessels.

The regeneration system is the second process operation. After a period of time, adsorption capacity of the activated carbon in the contacting system is exhausted. Then carbon must be taken out of service and regenerated thermally by combustion of the organic adsorbate. Fresh activated carbon is added as necessary to replace that lost during hydraulic transport and regeneration. These losses include both attrition due to physical deterioration and burning of some of the carbon during the actual regeneration process.

Figure 15 shows a typical carbon adsorption unit. Figure 16 shows a typical carbon regeneration furnace.

Advantages and Disadvantages: Activated carbon's biggest advantage lies with its wastewater treatment performance. Not only is it applicable for treating a very wide range of pollutant types, activated carbon also has a high removal efficiency for all of them. Activated carbon systems require little space and can easily accommodate increases in wastewater flow by installation of additional units. It also requires no treatment chemicals, thus eliminating the problems and cost associated with them.

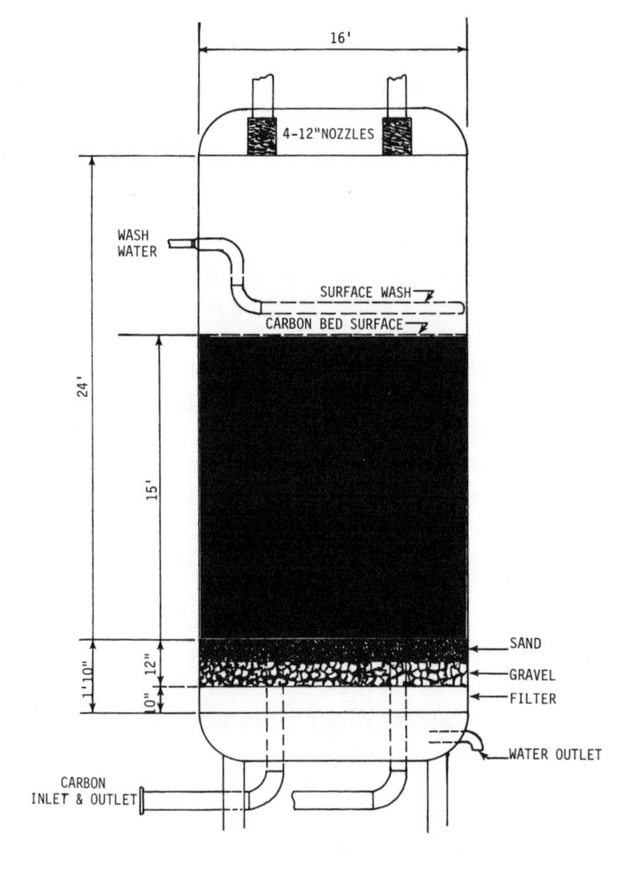

Figure 15: Typical carbon adsorption unit.

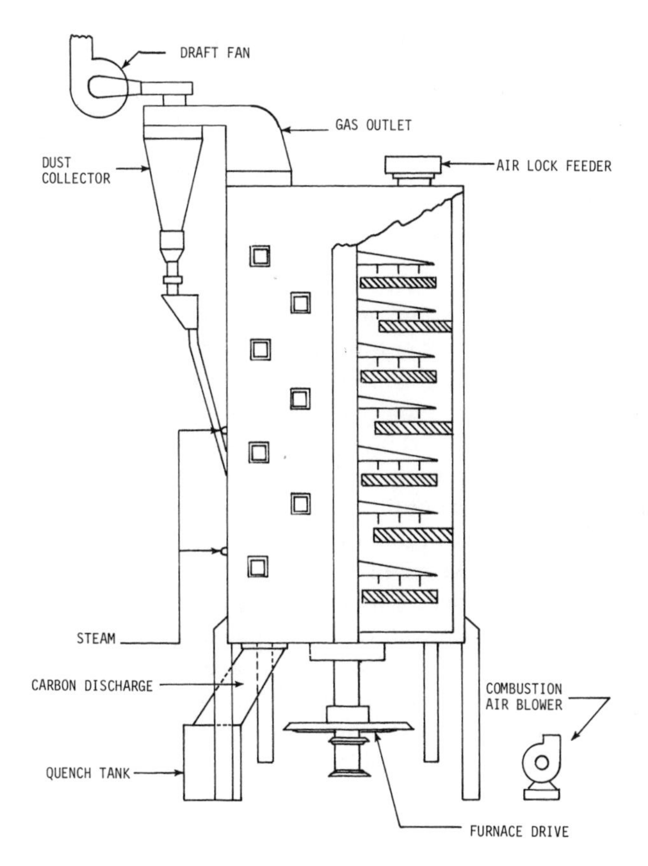

Figure 16: Typical carbon regeneration furnace.

Although it sometimes appears that activated carbon treatment may be the best technology available, it is not without drawbacks. This technology requires a high level of operative skills. Also, to eliminate fouling and to increase optimum performance, upstream treatment processes, often including primary and secondary treatment, are usually required. Finally, operating and maintenance costs can be relatively high due to the need for carbon reactivation and/or replacement.

Pollutants Treated: In terms of wastewater treatment, activated carbon adsorption is one of the most versatile pretreatment technologies. Activated carbon can be used to remove classical pollutants, such as BOD, COD, TSS, nitrogen, as well as toxic pollutants, like refractory organic compounds and metals. It is for treatment of the latter that activated carbon has received most attention, particularly for industrial applications. Although a few pollutants are not effectively removed by this technology, generally it will provide an effluent of exceptional quality.

Residuals Generated: Wastes generated by this technology warrant careful consideration not because of their volume, but rather their toxicity. Since refractory organics and metals are removed, spent carbon can contain considerable amounts of these compounds. Whether the carbon is regenerated or replaced, proper disposal is imperative to avoid environmental contamination.

Air Stripping

Description: The removal of ammonia and other gases (H_2S, CO_2, etc.) from industrial wastewaters can be accomplished through air stripping. The solubility of ammonia in wastewater is a function of temperature and pH — increasing either will lower the solubility. In the air stripping process the air wet bulb temperature limits effectiveness of heating wastewater. Therefore ammonia removal efficiency is enhanced by increasing the pH, usually by addition of lime. Ammonia containing wastewater and lime slurry are pumped to a rapid mix tank, flocculated, and then allowed to settle. Clarified, lime-treated, wastewater is pumped to the top of packed towers. In each tower, fans draw air up through the tower counter current to the falling wastewater. Tower packing is generally a series of bundles of plastic pipe or redwood slats spaced two to three inches on center. The sections are horizontal, and the direction of each row alternated. After stripping wastewater flows into a recarbonation basin, Figure 17 shows a typical ammonia stripping tower.

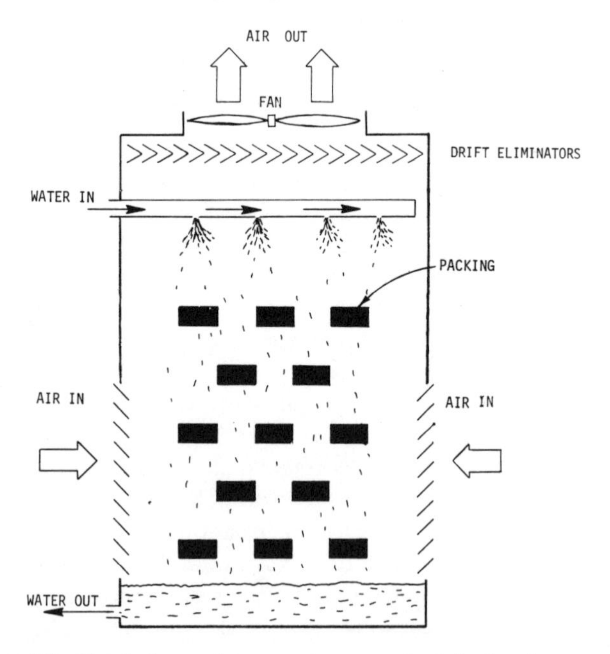

Figure 17: Typical countercurrent ammonia stripping tower.

Advantages and Disadvantages: The primary advantage is that where practicable, ammonia stripping is the cheapest, simplest, and easiest method to remove nitrogen from industrial wastewaters.

There are several disadvantages. The process is almost impossible to operate when ambient air temperatures drop below a wet bulb temperature of 32°F. Another problem is possible development of air emission problems such as fog, odor, or plumes. Noise from motors and fan drive equipment, fans, and water is another problem. Finally the problem of scaling — calcium carbonate scale — results in loss of efficiency from reduced air circulation and droplet formation and may possibly plug the tower.

Pollutants Treated: Air stripping of ammonia from treated wastewater for dilute solutions (under 100 mg/l) is practical. Air stripping of high concentrations — over 2000 mg/l — or other volatiles may cause significant air emission problems.

Residuals Generated: Air stripping of ammonia produces significant quantities of calcium phosphate, magnesium carbonate, and calcium carbonate sludge which must be disposed of. Though significant in volume, the sludge is not considered to pose any environmental hazard.

Alkaline Pyrolysis

Description: Removal of cyanide from industrial wastewaters can be accomplished without use of strong oxidizing chemicals, such as chlorine or ozone. An alkaline pyrolysis system is an alternative. In this technology the principal treatment action is based upon the application of heat and pressure. First, a caustic solution is added to the cyanide-bearing waste stream to raise its pH. Next, the wastewater is transferred to a continuous reactor, where it is subjected to high temperatures and pressures. Although specific temperatures and pressures required are dependent upon the particular characteristics of the waste stream, temperatures over 300°F and pressures around 100 psig are common. Under these conditions, the complete breakdown of cyanide generally occurs within a couple of hours. Figure 18 shows a schematic of an alkaline pyrolysis unit.

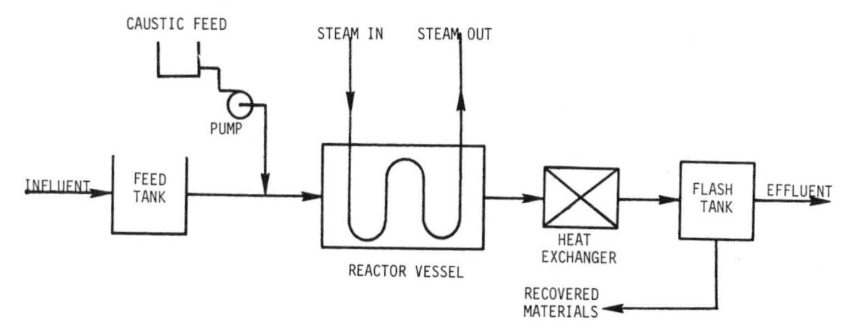

Figure 18: Schematic of a typical alkaline pyrolysis system.

Advantages and Disadvantages: Absence of chemicals in this process eliminates procurement, storage, and handling problems. The process is well suited to automatic control. However, since the process employs heat and pressure (and related equipment), it has a relatively higher cost. Also, the system tends to be more appropriate for smaller wastewater flows.

Pollutants Treated: To date, this technology's principal usage has been for the removal of cyanide. However, in theory it could be used to treat a number of pollutants that are not readily handled by other physical, chemical, or biological processes. Existing information shows that alkaline pyrolysis can achieve excellent effluent levels with respect to cyanide.

Residuals Generated: Alkaline pyrolysis breaks down cyanide compounds into their elemental form; i.e., carbon, nitrogen, and other elements. These are either vented off as gases or remain in the waste stream, where they are removed by a downstream treatment process. Therefore, residuals are not directly generated by this technology.

Clarification (Sedimentation)

Description: When an industrial wastewater, containing a suspension of solid particles that have a higher specific gravity than the transporting liquid is in a relatively quiescent state, particles will settle out because of gravity. This process of separating the settleable solids from the liquid is called clarification or sedimentation. In some pretreatment systems employing two or more stages of treatment and clarification, the terms primary, secondary, intermediate, and final clarification are used. Primary clarification is the term normally used for the first clarification process in the system. It is used to remove the readily settleable solids prior to subsequent treatment processes, particularly biological treatment. This treatment step results in significantly lower pollutant loadings to downstream processes and is appropriate for industrial wastewaters containing a high suspended solids content. Secondary clarification is normally the term used for a clarifier following the second stage in the pretreatment process. Historically the second process had been a biological process. If the secondary clarifier is followed by a third process and another clarifier, the secondary clarifier is sometimes called an intermediate clarifier. The last clarifier in the system is called a final clarifier. In many cases, the secondary and final clarifier are the same units.

Actual physical sizing (depth, surface area, inlet structure, etc.) is highly dependent upon the quantity and composition of the wastewater flow to be treated. Because these criteria will vary substantially among industries and even among plants, a detailed discussion is not possible here. Instead, only a general overview of the clarification process is presented.

Clarification units can be either circular or rectangular and are normally designed to operate on a continuous flow-through basis. Circular units are generally called clarifiers, while rectangular units are commonly referred to as sedimentation tanks. However, the nomenclature is interchangeable. Figure 19 shows a typical circular unit and Figure 20 shows a typical rectangular unit.

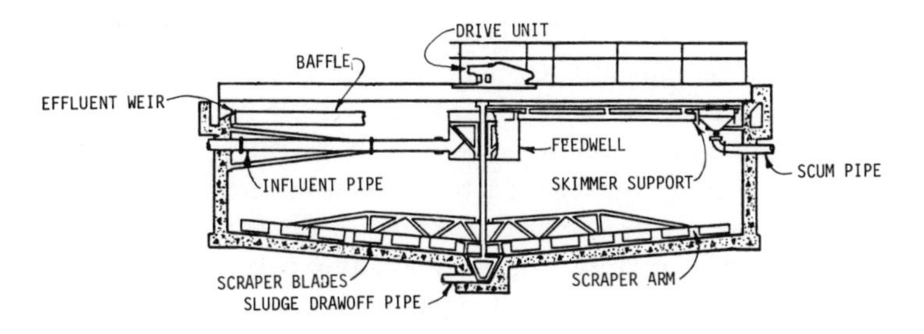

Figure 19: Typical circular clarifier.

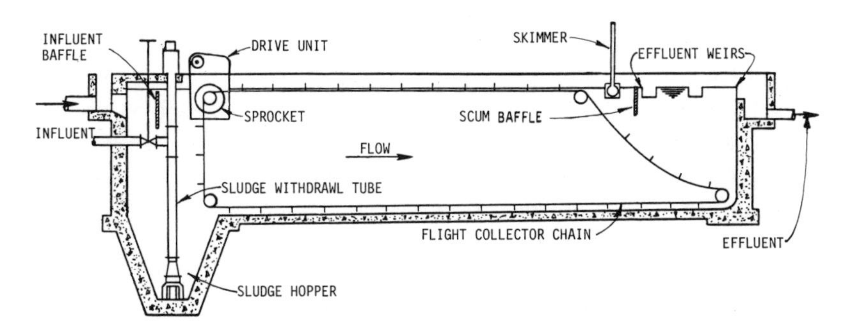

Figure 20: Typical rectangular clarifier.

The clarification or sedimentation process is divided into four distinct zones: inlet, clarification, sludge, and outlet. Wastewater with its suspended solids is admitted to the treatment unit through the inlet zone. Separation of these solids from the wastewater occurs in the relatively quiescent clarification zone. From here, clarified wastewater is removed from the unit via the outlet zone. Separated solids are allowed to accumulate and compact in the sludge zone and then removed by mechanical means from the bottom of the tank. Sludge is then pumped to the residual management system. Also any scum which is accumulated on the surface is skimmed off and also sent to the residual management system.

Design of clarification units involves many complex theories. Generally, clarifiers are designed on the basis of the type of suspended solids to be removed from the waste stream. There are three types. Class I solids are discrete particles that will not readily flocculate. They are typical of raw influents. Class II particles are characterized by a relatively low solids concentration of flocculated material. These types of particles are usually found in wastewaters that have been subjected to chemical addition. Either of the above two particle classes may be found in industrial wastewaters. Class I particles are typical of physically-manufactured operations, while Class II is more common to chemically-manufactured operations. The last type are Class III particles. These are normally the solids generated from a biological treatment process. Because of their poor settling properties, special care must be given in the design for their removal.

Although the fundamental principles of clarification are relatively simple, there are many intricate factors which must be considered when designing a unit for the real world. Because of the tremendous variations in industrial waste characteristics, detailed analyses on a case-by-case basis should be conducted to insure that clarifiers will perform efficiently and reliably.

Advantages and Disadvantages: Some advantages of the clarification process are as follows. Clarification is effective in removing a substantial portion of the suspended solids and BOD in an industrial wastewater. Not only does this aid in the performance of other downstream treatment processes, but it also protects them from negative effects, such as clogging. This technology is relatively receptive to modifications, such as chemical addition, which allows for alterations to compensate for changes in wastewater characteristics. Finally, clarification is a relatively simple operation, which requires neither sophisticated equipment nor highly skilled operators.

Although a very beneficial component of a wastewater treatment system, clarification does have some disadvantages. Construction cost of clarifiers can represent a large expenditure. Design and operation are usually based on specific pollutant loadings and wastewater flows. As a result, it is difficult for these units to accommodate large variations or increases in wastewater flow or pollutant loadings.

Pollutants Treated: Any industrial wastewater which contains solids that will settle under quiescent conditions is a candidate for this process.

Residuals Generated: The type of residual generated depends on the pretreatment process before clarification. Sludge mass may be dense, gelatinous, stringy, slimy, oily, etc. In addition, it will contain the majority of suspended matter that was in the wastewater but in a more concentrated form.

Equalization

Description: Equalization is not a treatment technology, per se, but rather a system to enhance the performance of downstream treatment processes. The varying nature of industrial wastewater flows in terms of volume and strength is widely recognized. Nearly all manufacturing facilities, whether they employ batch or continuous-type operations, experience fluctuating flows at different periods in time. Since improved efficiency, reliability, and control of downstream physical, biological, and chemical processes are possible when they are at or near uniform conditions, flow equalization is used.

The primary objective of equalization is simply to dampen the impact of varying wastewater flows, and thus achieve a constant or nearly constant mass flow through the downstream treatment processes. However, a secondary objective is to lower the concentration of wastewater pollutants by blending the wastewaters in the equalization basin. This results in a more uniform and constant loading of wastewater pollutants to subsequent treatment processes.

Although some other mechanisms are available, the most positive and effective means to maximize the benefits possible with equalization is through the use of specially designed equalization basins. These basins should normally be located near the head of the pretreatment system. Also, adequate aeration and mixing must be maintained to keep equalization basins aerobic and prevent solids deposition.

Equalization basins may be designed as either in-line or side-line units. For the in-line method, all of the wastewater flow passes through the equalization basin. This results in significant concentration and mass flow damping, although pumping requirements are higher. Figure 21 shows a typical in-line system design. In the side-line method, only that amount of wastewater flow above the daily average is diverted through the equalization basin. This scheme minimizes pumping requirements, but at the expense of less effective damping of pollutant concentrations. Figure 22 shows a typical side-line system design.

Advantages and Disadvantages: Flow equalization has a positive impact on all downstream treatment processes. The most beneficial impact is reduction of peak overflow rates, resulting in improved performance and effluent quality. Likewise, flow equalization permits use of smaller sized first stage clarifiers than in the case of nonequalized flow. For systems employing biological treatment, equalization can protect biological processes from impact or failure due to shock loads of treatment-inhibiting substances. Also, it can minimize the size of biological reactors and aeration equipment requirements.

Flow equalization is not without drawbacks. Installation of an equalization basin will result in additional construction, operation, and maintenance costs. Similarly, associated equipment such as pumps, mixers, etc., are required. Finally, improperly monitored equalization basins can become septic, causing odor problems and impacting downstream treatment processes.

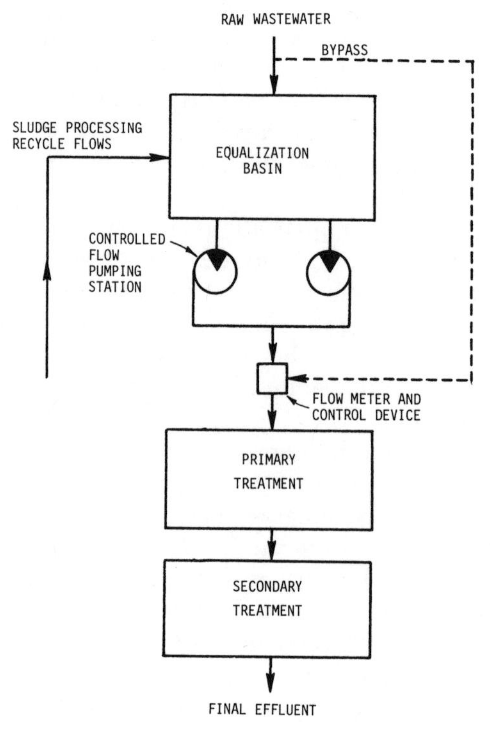

Figure 21: Pretreatment system using in-line equalization.

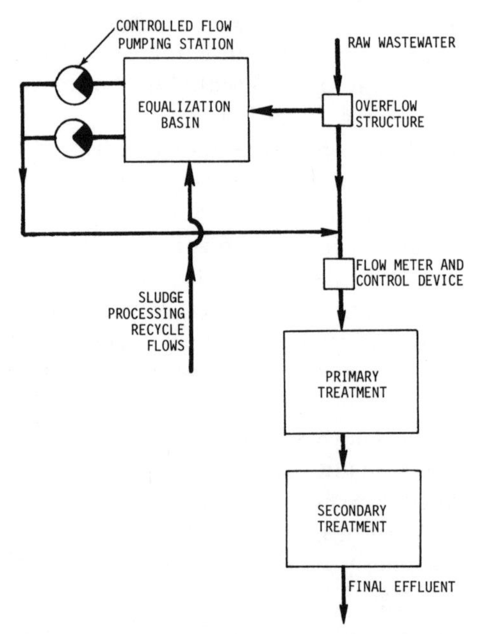

Figure 22: Pretreatment system using side-line equalization.

Pollutants Treated: Equalization is not a treatment process, per se, but rather an initial processing step to stabilize and enhance the operation and performance of downstream treatment processes. Therefore, equalization does not directly treat any wastewater pollutants.

Residuals Generated: Since equalization does not remove any wastewater pollutants, no residuals are generated.

Evaporation Processes

Description: Evaporation is a concentration process. Water is evaporated from a solution, increasing the concentration of solute in the remaining solution. If the resulting water vapor is condensed back to liquid water, the evaporation-condensation process is called distillation. However, to be consistent with industry terminology, evaporation is used in this report to describe both processes.

Both atmospheric and vacuum evaporation have been employed in a number of industries. Atmospheric evaporation could be accomplished simply by boiling the liquid. However, to lower the evaporation temperature, heated liquid is sprayed on an evaporation surface, air is blown over this surface and then released to the atmosphere. Thus evaporation occurs by humidification of the air stream, similar to a drying process.

For vacuum evaporation, the evaporation pressure is lowered to cause the liquid to boil at reduced temperature. All of the water vapor is condensed and, to maintain the vacuum condition, noncondensible gases (air in particular) are removed by a vacuum pump. Vacuum evaporation may be either single or double effect. In double effect evaporation, two evaporators are used, and the water vapor from the first evaporator (which may be heated by steam) is used to supply heat to the second evaporator, condensing as it does. Roughly equal quantities of wastewater are evaporated in each evaporator; thus, the double effect system evaporates twice the water that a single effect system evaporates, at nearly the same cost in energy but with added capital cost and complexity. The double effect technique is thermodynamically possible because the second evaporator is maintained at lower pressure (higher vacuum) and, therefore, lower evaporation temperature. Figure 23 shows schematics of the various types of evaporation systems that can be utilized.

Advantages and Disadvantages: Advantages of using evaporation for treating industrial wastewaters are as follows. Where valuable raw ingredients or products are present in a waste stream, it provides an effective means for their recovery. Also, water produced by this technology is of high purity and, therefore, can be recycled back to the production operations. Finally, this treatment process will provide for the removal of pollutants which cannot be treated by any other means.

The major disadvantage of evaporation is that it is a large energy consumer and, thus, relatively expensive to operate. For some applications, pretreatment may be required to remove solids, which tend to cause fouling in the condenser or evaporator. Finally, the buildup of scale on the evaporator plates reduces the heat transfer efficiency and may present a maintenance problem and increased operating costs.

Pollutants Treated: Since evaporation involves only removal of water from a solution, there is no fundamental limitation on its use for industrial wastewater treatment, i.e., it can be used to remove any type of pollutant.

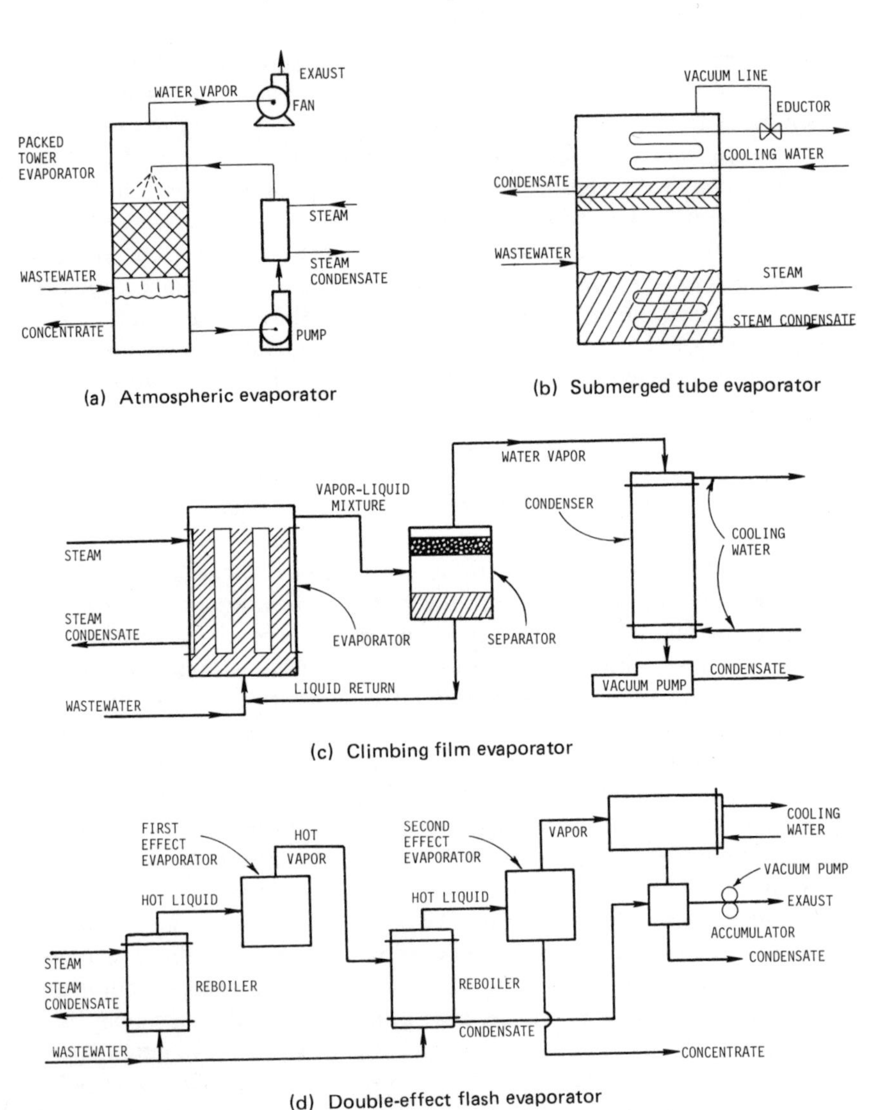

Figure 23: Schematics of various types of evaporation systems.

However, practical considerations generally restrict its use to recovery/recycling applications or where other technologies are not effective. This technology probably provides the highest quality effluent.

Residuals Generated: Because of its cost, evaporation is usually applied to selected, small waste streams. Even though these may have high waste concentrations, the volume of residuals generated is generally small. As indicated above, evaporation is principally used for raw ingredient or product recovery. Thus, the residual materials are often recycled back to production. If not, there will be a highly concentrated, small volume waste.

Flotation Separation Processes

Description: As opposed to clarification, which separates suspended particles from liquids by gravitational forces, flotation separation separates these particles by reducing their density by the introduction of air into the system. Fine bubbles either adhere to, or are absorbed by, the solids, which are then lifted to the surface.

There are several methods of achieving flotation. In one flotation method, dissolved air flotation, small gas bubbles (50 to 100 mu) are generated as a result of the precipitation of a gas from a solution supersaturated with that gas. Supersaturation occurs when air is dispersed through the sludge in a closed, high pressure tank. When the sludge is removed from the tank and exposed to atmospheric pressure, the previously dissolved air leaves solution in the form of fine bubbles.

In a second method, dispersed air flotation, relatively large gas bubbles (500 to 1,000 mu) are generated when gas is introduced through a revolving impeller or through porous media.

In vacuum flotation, supersaturation occurs when sludge is subjected initially at atmospheric pressure, to a vacuum of approximately nine inches of mercury in a closed tank. Although all three methods have been used in industrial pretreatment systems, the dissolved air flotation process is the dominant method used in the United States. In this process, float material is removed from the surface of the tank and clear liquid is removed from the middle of the tank.

Flotation separator tanks can be either rectangular or circular in shape and constructed of either concrete or steel. Tanks are provided with equipment to provide uniform flow distribution at entrance, pressurize gas, and skim off float material. In designing a dissolved air flotation separator system, the following variables are typically considered: full, partial, or recycled pressurization; feed characteristics; surface area; float characteristics; hydraulic loading; chemical usage; type of pressurization equipment; and operating pressure. Figure 24 shows a typical dissolved air flotation separator.

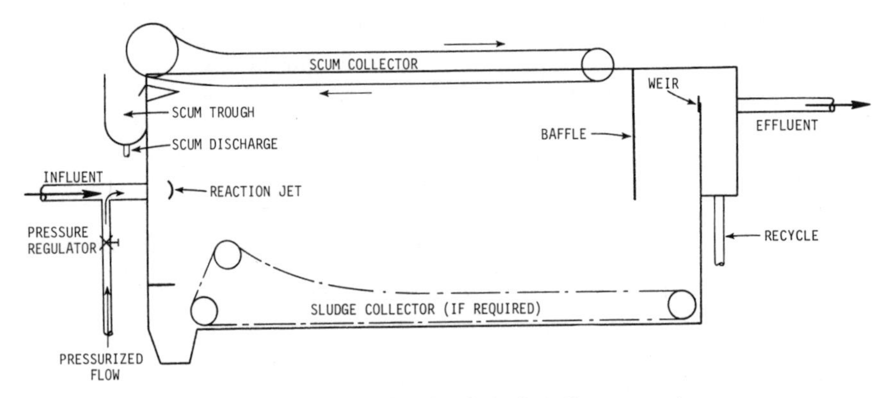

Figure 24: Typical dissolved air flotation separator.

Advantages and Disadvantages: Flotation separation has the following advantages. Oil and grease, finely divided suspended solids, grit, etc., are effectively removed in one unit and the resulting float may be usable as an auxiliary fuel source. Construction costs can be minimized because the higher overflow rates and shorter detention times mean that smaller tanks are required. Also, because of shorter detention periods and presence of dissolved oxygen in the wastewater, odor problems are kept at a minimum.

Some disadvantages associated with this type of treatment process are as follows. Additional equipment required results in higher operation and maintenance costs than for other solids removal processes. Also, extra power and chemicals are usually needed. Flotation separator units generally do not effectively treat the "heavier" suspended solids, as do settling units. Flotation separation is somewhat more technical than conventional settling. Thus, more highly skilled personnel are needed to operate the process.

Pollutants Treated: Flotation separation is an appropriate technology for treating suspended solids and oil and grease in industrial wastewaters. Generally, the process will achieve 40 to 65 percent suspended solids removal and 60 to 80 percent removal of oil and grease. The addition of chemicals, such as alum, ferric chloride, and polymers prior to actual flotation can significantly improve the performance of this technology. In the above situations, removal of suspended solids will also result in a reduction in BOD levels. The potential exists for dissolved air flotation to strip volatile organics into the atmosphere. However, this has never been documented.

Residuals Generated: The float or sludge layer, which is skimmed off the top of the tank, will contain the same particulate pollutants that were in the wastewater but in a much more concentrated form. Because of the presence of oil and grease and possible chemicals, the subsequent processing of the sludge can be difficult.

Filtration Processes

Description: Long applied in the treatment of water supplies, filtration in the 1970's started to become widely used in wastewater treatment. Thus, for those industries using this technology to purify their process waters, its transfer to waste treatment should be easily accommodated. Filtration can be utilized as an intermediate step to prepare wastewater for further treatment or as a final polishing step, depending upon the nature of the waste stream to be treated.

Filtration in the most general sense is the process of straining, and generally only suspended insoluble material can be removed. In this Handbook, the word filtration will refer to the removal of solids in the general range of one micron to the larger suspended solids. The section on Screening covers the straining of larger materials and the section on Membrane Processes, covers the straining of liquids to remove solids smaller than one micron.

Filtration processes can be either pressure filtration (i.e., liquid is pumped through the filter media) or gravity filtration (i.e., liquid flows through the filter media by gravity). Filter media is either granular material (sand, coal, etc.) or woven wire or nylon cloth. Granular filters are either single, dual, or multi-media and can be either downflow, upflow, cross flow, or radial flow.

Filtration involves the capture of suspended solids by a selected medium. Eventually, the pressure-drop across the bed becomes excessive or the ability of the medium to remove the particles becomes impaired. When this occurs, cleaning is necessary to restore operating pressures and maintain effluent quality. Generally, multiple filtration units are employed so that treatment can be maintained while one unit is undergoing cleaning. Microscreens are supplied with continuous cleaning spray nozzles so that the unit does not have to be taken out of service. It is common practice to return the backwash water from cleaning back to the head of the pretreatment system.

Figure 25 shows several types of granular filter configurations. Figure 26 shows a typical micro screen installation.

Advantages and Disadvantages: Filtration technologies offer several advantages. Generally filtration units have low capital and operating costs. Normally no treatment chemicals are required, which eliminates procurement, storage, and handling problems and costs. Also, most units require very little space, and increases in wastewater flow can easily be accommodated by installing additional filters. Finally, filtration units are one of the best performers in terms of solids removal.

With respect to disadvantages, filters require some degree of operator skill, due to control and backwashing requirements. If proper operation of the units is not maintained, fouling of the filters can be a problem. Also, this technology is generally limited to the treatment of the pollutants named below.

Other pollutants usually pass through the filters and in some instances may actually deteriorate the filter media.

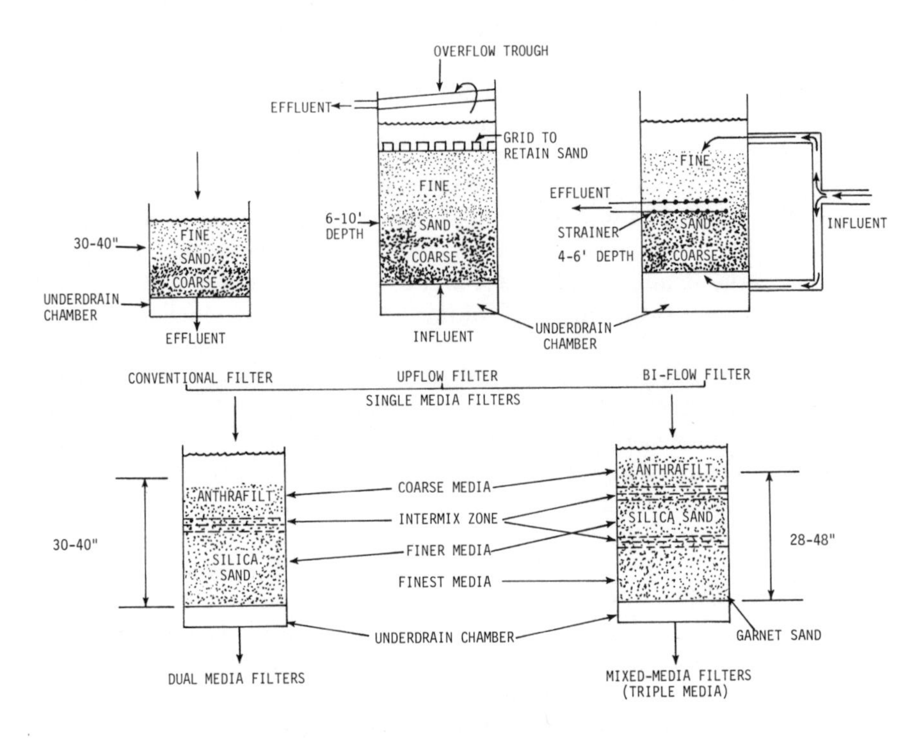

Figure 25: Several granular filter configurations.

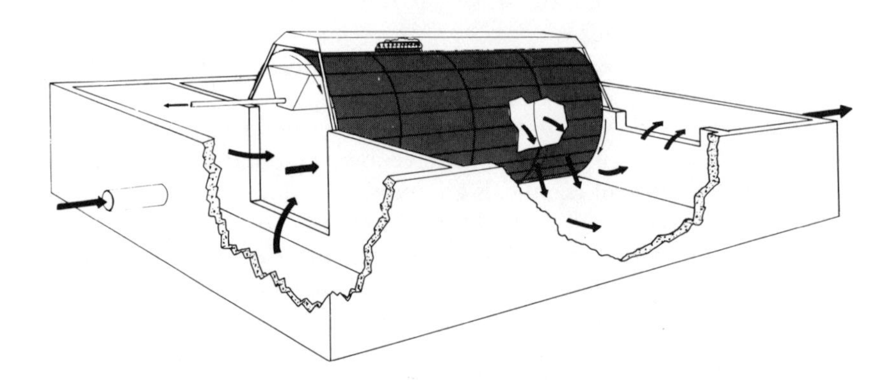

Figure 26: Typical microscreen unit.

Pollutants Treated: Filtration technologies are generally employed in the removal of residual biological floc in settled effluents from biological treatment, algae from lagoons, and removal of residual chemical biological floc after alum, iron, or lime precipitation.

Generally filters cannot function properly with solids such as greases, gels, and oils. These materials tend to coat, blind, or clog a filter, making it inoperative.

Residuals Generated: Since filters are used as a polishing step, large volumes of solids are not generated by this technology. That which is generated is contained in the backwash water used to clean the filter units. This water is usually returned to the head of the pretreatment system. Thus, sludges from this technology are virtually nonexistent.

Foam Separation Processes

Description: In foam separation processes, small bubbles are generated either using a gas, normally air forced through a sparger, or by mechanical mixing, which rise through the liquid-adsorbing surface-active solutes and collecting suspended solids. On emerging from the liquid, these bubbles form a foam which is forced out of the foamer and subsequently collapsed to yield a waste concentrate.

Foam separation processes can be grouped into four general categories: foam fractionation, ion flotation, microflotation, and precipitate flotation. In many industrial applications, a surface active agent is already present in the wastewater which can be utilized as a collector of particulates, drawing and attaching particulates to generated air bubbles and as a frother to produce a stable foam in which the particulate is concentrated and removed. In other applications, a surface active agent must be added to the wastewater, with the required charge of the surfactant ion determined by the nature and charge of the particulate to be foam-separated.

Figure 27 shows a schematic of a foam fractionation process.

Advantages and Disadvantages: Foam separation processes offer two significant advantages over other particulate removal processes: low level utilization requirements and low capital cost.

Disadvantages include the following. Process variables are difficult to analyze and therefore the process has inherent process control problems; the addition of a surface active agent can be expensive; and thirdly, very little demonstration work has yet been done with this process.

Pollutants Treated: Foam separation processes are able to remove soluble surface-active materials and suspended particulate matter. Generally, this process is a possible candidate for any industrial process with high concentrations of particulate matter. Applications include foam fractionation of black liquor from sulfate pulping, flotation of oily iron dust, and foaming of acid mine water combined with domestic wastewater.

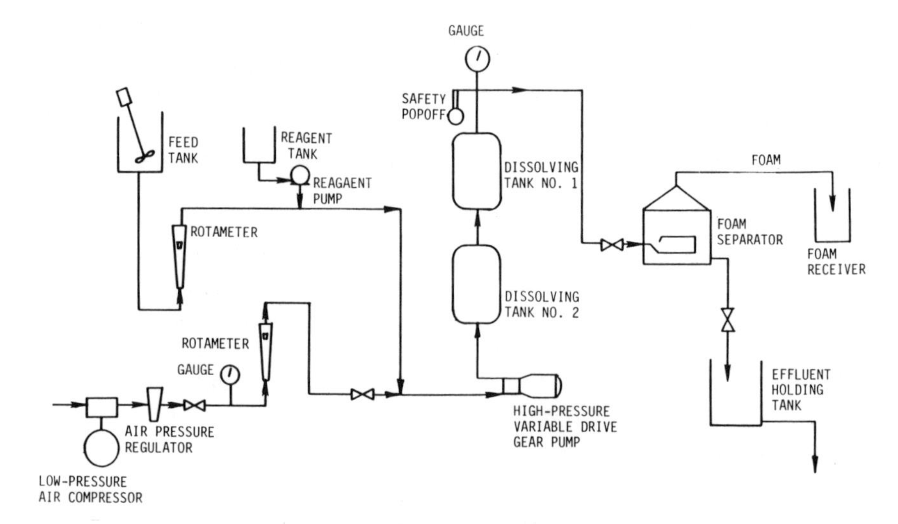

Figure 27: Schematic of a foam fractionation process.

Residuals Generated: Foam separation processes generate a concentrated stream of soluble surface-active materials and suspended particulate material removed from the wastewater. The wastewater will contain between 2 to 7 percent solids, is gelatinous and sticky, and can be difficult to manage and dispose of.

Incineration

Description: Incineration is a two-step process involving first drying and then combustion. Drying and combustion may be accomplished in separate units or successively in the same unit, depending upon temperature constraints and control parameters. In all incinerators, drying and combustion processes follow the same phases: raising the temperature of the feed stream to 212°F, evaporating the water in the feed stream, increasing the temperature of the water vapor and air, and increasing the temperature of the dried organic material to the ignition point. Although presented in simplified form, incineration is a complex process involving thermal and chemical reactions which occur at varying times, temperatures, and locations in the incinerator.

Advantages and Disadvantages: Incineration has one major advantage — when done properly, incineration of hazardous and toxic organic liquid wastes leaves nothing but carbon dioxide, water, and possibly some sterile ash.

Major disadvantages are that it consumes enormous, though partially recoverable, amounts of energy, since most incineration temperatures will have to be maintained at 1500°C or more, requires extensive monitoring of off gas, high capital cost, and high maintenance cost.

Pollutants Treated: Incineration can be utilized for the destruction of any organic wastewater.

Residuals Generated: Incineration of aqueous organic waste streams generally do not produce any or very little residuals. Those residuals that are generated will contain only trace amounts of hazardous organic material and may also contain nonorganic hazardous materials depending on the origin of the waste stream.

Ion Exchange

Description: Ion exchange is a technology in which ions, held by electrostatic forces to charged functional groups on the surface of ion exchange resin, are exchanged for ions of similar charge in a solution in which the resin is immersed. Ion exchange is classified as an adsorption process because the exchange occurs on the surface of the solid, and the exchanging ion must undergo a phase transfer from solution phase to surface phase. This technology has been used extensively for wastewater treatment of a variety of industrial wastes to allow for removal and/or recovery of specific waste materials or byproducts, particularly certain metal ions.

In general, a synthetic ion-exchange resin consists of a network of hydrocarbon radicals to which are attached soluble ionic functional groups. The hydrocarbon molecules are linked in a three-dimensional matrix to provide strength to the resin.

Behavior of the resin is determined by ionic groups attached to the resin. The total number of ionic groups per unit of resin determines exchange capacity, and the group type affects both equilibrium and selectivity. Cation exchangers, those resins carrying exchangeable cations, contain acid groups. The term "strongly acidic" is used in reference to a cation exchange resin containing ions from a strong acid, such as H_2SO_4, and "weakly acidic" designates cation exchange resins made from a weak acid, such as H_2CO_3. Anion resins containing certain ammonium compounds are referred to as "strongly basic", and those with weak base amines are referred to as "weakly basic."

The majority of cation exchangers used in wastewater treatment operations are strongly acidic, and they are able to exchange all cations from the solution. Both types of anion exchangers are employed. Strongly basic anion resins are capable of exchanging all anions, including weakly ionized material such as silicates and dissolved carbon dioxide, and weakly basic resins exchange only strongly ionized anions such as chlorides and sulfates. Characteristic selectivities of commercial resins are well known and useful for determining which resin is most suitable for a specific application. Further, it is possible to construct a resin with high selectivity for the polluting ions involved in a particular operation.

Ion exchange resins are regenerated in at least three different ways: by resin removal and replacement service, by conventional in-place regeneration, and by rapid cyclic operation and regeneration. Development of moving bed and fluidized bed approaches is also underway.

Figure 28 shows a typical ion exchange system.

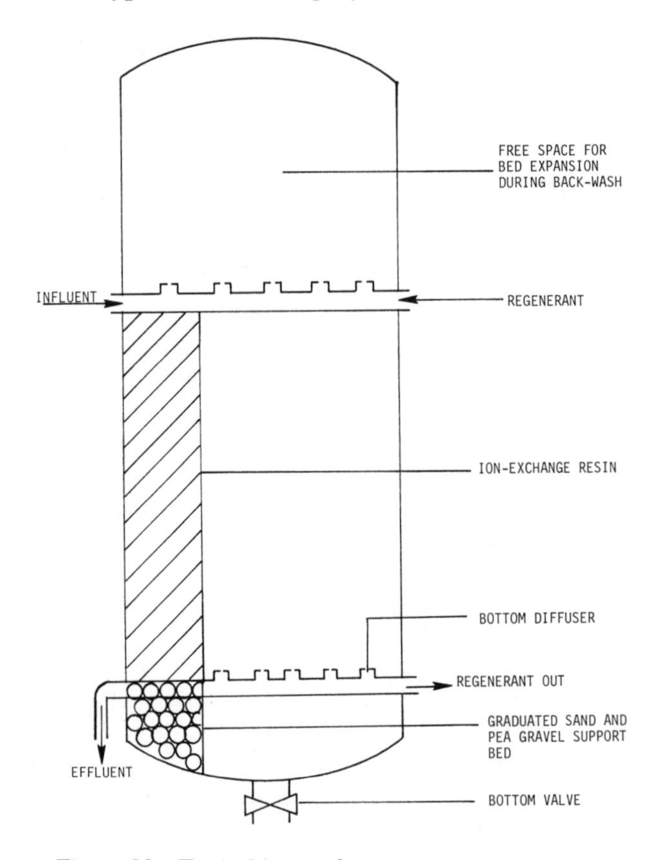

Figure 28: Typical ion exchange system.

Advantages and Disadvantages: Ion exchange systems are compact, relatively inexpensive and easily installed. Most ion exchange resins are chemical resistant and are stable at high temperatures. This makes them particularly attractive for treating very chemically strong wastewater streams. Also, selectivity characteristics of ion exchangers can be adjusted through the use of special resins to capture specific compounds. This is very important when it is desirable to recover select materials for recycling back to production.

There are several disadvantages. First, the presence of certain wastewater constituents may cause fouling of the resins and even clog the exchanger system. When highly selective resins are utilized, the exchange reaction may form tightly-bonded materials which can make regeneration difficult. Finally, in the packed column type of ion exchangers, solids in the wastewater may cause excessive pressure losses and fouling resin surfaces, thus significantly reducing removal efficiencies.

Pollutants Treated: By choosing the proper resin, ion exchange units can be fine-tuned to remove specific pollutants. Generally, ion exchangers have been employed to remove ammonia, organic and inorganic salts, and metals. As development continues on better and cheaper resins, ion exchange will receive more and more applications.

Residuals Generated: Wastes from this treatment process result from regenerating the spent ion exchange resins. Generally, this is a small waste stream that is further processed to remove excess water. Depending upon the particular materials removed, they may either be recycled or disposed of. If disposal is employed, careful consideration must be given to the metals that may be present in the waste.

Membrane Processes

Description: As was discussed in the Filtration section, membrane processes in this Handbook cover the straining of liquid to remove solids less than one micron. In general, four types of membrane processes are used in industrial pretreatment of wastewaters: electrodialysis, reverse osmosis, ultrafiltration, and microfiltration.

Electrodialysis — In electrodialysis, cation-exchange membranes are alternated with anion-exchange membranes in a parallel manner to form compartments 0.5 to 1.0 millimeter thick. The entire membrane assembly is held between two electrodes. Figure 29 shows a general process diagram.

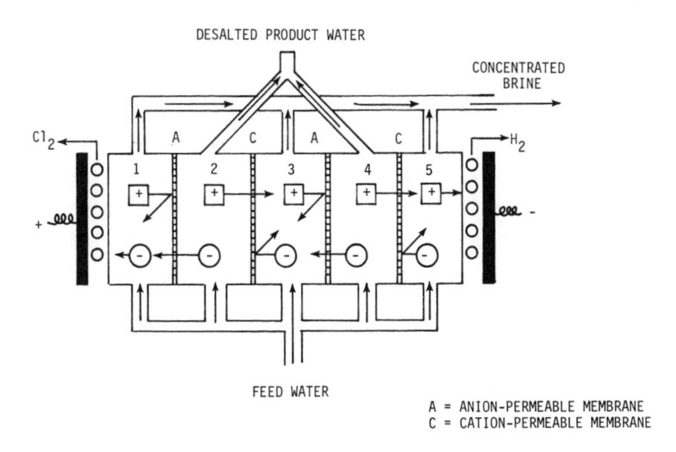

Figure 29: Electrodialysis general process diagram.

When an electrical potential is applied to the electrodes, all positive ions (cations) tend to travel towards the negative electrode (cathode), and all negative ions (anions) tend to move toward the positive electrode (anode). By this mechanism, ions are removed from solution on one set of compartments (even number) and transferred to the other (odd number) compartments.

Reverse Osmosis — In this process, the membrane is mounted in an apparatus so that a two-solution compartment is formed, as shown in Figure 30. Wastewater to be treated is pressurized and circulated through the high pressure solution compartment. Water permeates to the low pressure side and is removed. The concentrated brine is removed.

In reverse osmosis the membrane allows passage of water but impedes the passage of salts and small molecules. Of the three membrane filtration processes — reverse osmosis, ultrafiltration, and mircofiltration — reverse osmosis requires the highest pressure.

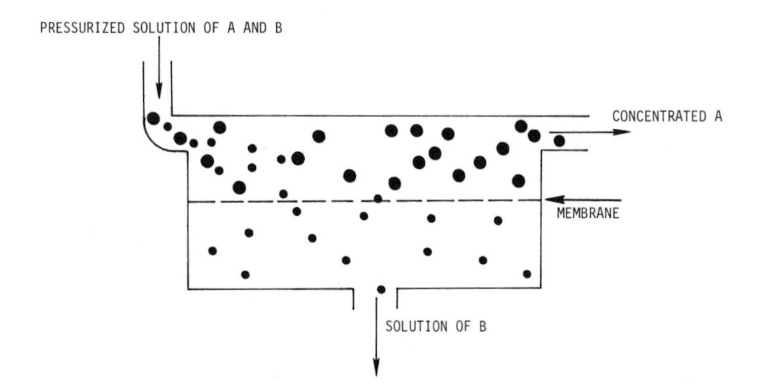

Figure 30: Reverse osmosis general process diagram.

Ultrafiltration — Mounting of the membrane is similar to reverse osmosis and Figure 30 would also be representative of this process. Membranes used in this process allow passage of a solvent molecule but impede the passage of molecules with a molecular weight of about 100 or higher.

Microfiltration — Mounting of the membrane is similar to reverse osmosis and Figure 30 would also be representative of this process. Membranes used in this process allow passage of solvent molecules but impede the passage of large celloids and small particles.

Advantages and Disadvantages: The ability to concentrate dilute wastewaters for recovery of chemicals is the main advantage of membrane processes, especially where these chemicals represent a significant cost to an industrial operation. Another advantage is its low operating cost. Since no latent heat of vaporization or fusion is required, membrane process units can operate with very low power requirements. The ability to be installed in a very small space is another advantage.

The membrane is the principal component that contributes to limitations or disadvantages. In cases where high temperature waste streams must be treated, deterioration of the membrane is common. Also, certain strong oxidizing compounds in the wastewater can cause a similar problem. Finally, some treatment upstream of the membrane process unit is usually required to prevent the membrane from becoming fouled.

Pollutants Treated: Membrane processes are generally employed to treat those wastewaters which contain a high content of dissolved inorganic materials, such as chlorides, sulfates, etc. Also, in those cases where metals are present in various complexes, membrane processes may provide effective treatment. Pollutant removals are highly dependent on the specific membrane process unit.

Residuals Generated: A very concentrated waste stream is produced by the membrane processes and can be a difficult residual to dispose of. If employed for product recovery, the residuals are returned to production.

Oil-Water Separation

Description: In practically all manufacturing industries, oil and grease can be found in a plant's wastewater. They generally result from equipment lubrication, accidental spills, etc. However, for some industries, such as petroleum and edible oil refining, oil compounds can represent a significant constituent of their flows. This is due to use of these oils as the raw material in production. Not only can the oil have a detrimental impact on the performance of a wastewater treatment system, but it also represents a valuable raw material being wasted. Thus, specific treatment processes have been developed to separate the major portion of the oil from the waste stream. These are called oil-water separators. Figure 31 shows a typical unit.

The basic principle by which oil separators work is the differential between specific gravities of water and the oils to be removed. Since oils have lower specific gravities, they will rise to the top of the unit, while the heavier water sinks to the bottom. An important consideration in separator design is oil globule size, since Stoke's law for terminal velocity of spheres in a liquid medium will determine the rate at which oil rises.

An oil-water separation unit consists of two basic sections: inlet section and oil-water separator channels. Both of these will be discussed briefly. Major components of the inlet section are the preseparator flume and forebay. The preseparator flume serves two functions: reduction of influent flow velocity and collection of flotable oil. As the waste stream enters the flume, velocity is reduced so as to avoid turbulent mixing of the flow. At the end of the flume a retention baffle and skimmer capture some floating oils and remove them from the surface. From here wastewater discharges to the forebay. This acts as a header which distributes the flow to the separator section.

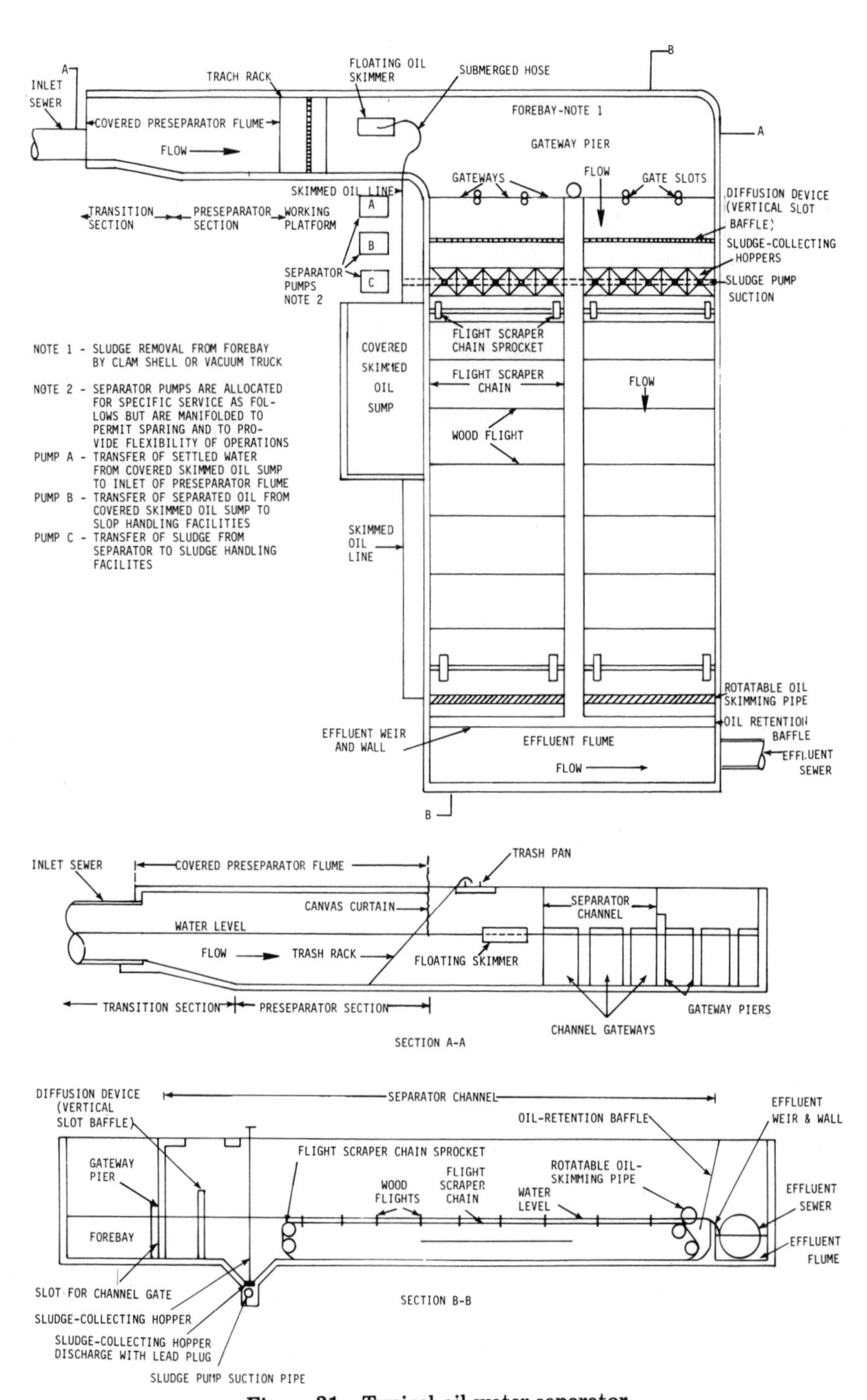

Figure 31: Typical oil-water separator.

As flow travels from the forebay, it passes through shutoff gateways, which control the amount of flow. Immediately downstream of the gateways are vertical diffusion devices, such as vertical slot baffles or reaction jets. These function to further reduce flow turbulence and to distribute flow equally over the channel cross-sectional area. In the separator channel, natural forces are allowed to work on the wastewater in a quiescent zone. Oil and other floatable material rises to the top, while water, solids, etc., remain on the bottom. At the end of the channels, retaining baffles and skimmers capture and recover the oily compounds as the separated water leaves the tanks. Sludge collectors also scrape the bottom of the tank to remove any deposited solids.

Advantages and Disadvantages: The major advantage of oil-water separators is their ability to treat wastewaters which are heavily laden with oil compounds. Because of their design, they represent a very simple treatment operation which minimizes personnel requirements. Since this technology relies on natural forces, rather than chemicals or aeration, its operating costs are minimized.

Oil-water separators' disadvantages are as follows. Its ability to remove only specific size oil globules can result in very small particles passing through the system. Also, if the wastewater has a high solids content and the oil and solids become mixed, the resultant globules may be too heavy to float to the surface. Finally, improperly maintained units are subject to odor problems.

Pollutants Treated: This technology is applicable for treatment of wastewaters containing large amounts of oil compounds, and it is very effective in removing a major portion of these constituents. Some settling of suspended solids does occur in an oil-water separator, although this is not its primary purpose.

Residuals Generated: A substantial volume of waste is generated by this technology. The waste is very oily, making management difficult. For most applications, the captured materials represent a valuable raw material, and thus are generally recycled back to production.

Polymeric Resin Adsorption

Description: Polymeric resin adsorption involves two basic steps: contacting the wastewater stream with resin and allowing the resin to adsorb solutes from solution; and, subsequently regenerating the resin by removing adsorbed chemicals.

Generally, a typical system will consist of two fixed beds of resin. One bed will be on stream for adsorption while the second is being regenerated. In cases where the adsorption time is very much longer than regeneration time — when solute concentrations are very low — one resin bed plus a hold-up storage tank could suffice.

The adsorption bed is normally fed downflow at flow rates between 0.25 to 2 gallons per minute per cubic foot of resin. Linear flow rates are in the range of 1 to 10 gallons per minute per square foot. Adsorption is stopped when the bed is fully loaded and/or the concentration in the effluent rises above a certain level.

Regeneration of the resin bed is performed in situ with basic, acidic, or salt solutions or regenerable nonaqueous solvents being most commonly used. As a rule, about three bed volumes of regenerant will be required for resin regeneration.

Figure 32 shows a resin adsorption system for removal and recovery of phenol from wastewater.

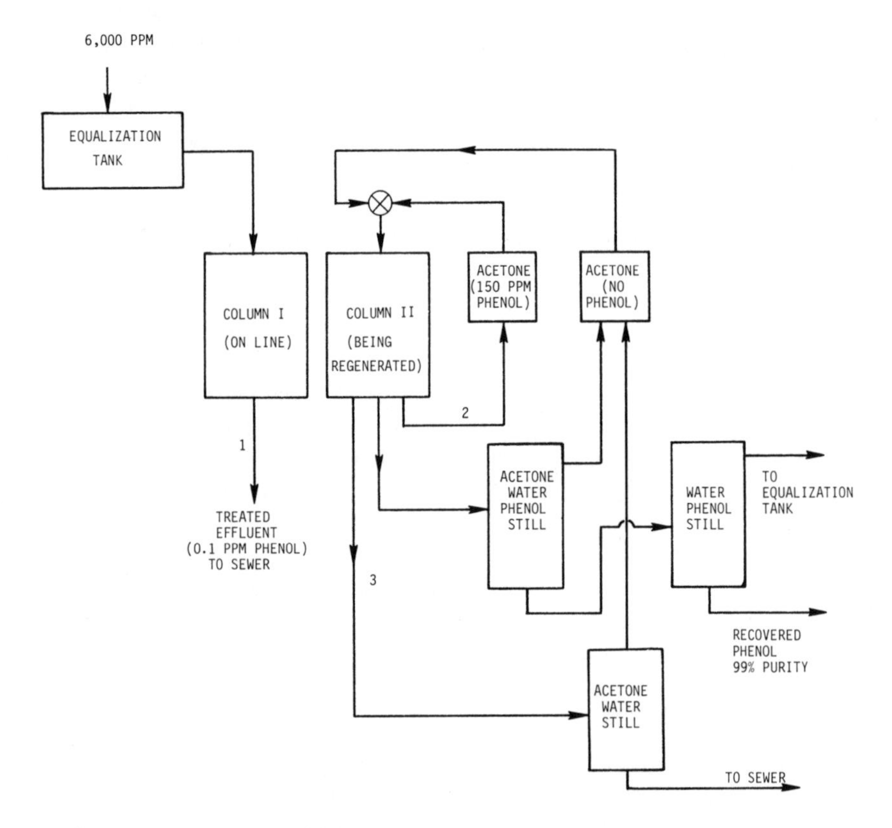

Figure 32: Flow diagram of a polymeric resin adsorption system for the removal and recovery of phenol from wastewater.

Advantages and Disadvantages: Technology is relatively new and as yet little information is available. Resin systems are very compact and require little space. High levels of total dissolved solids do not interfere with the action of resin adsorbents on organic solutes. Resin systems are easily and economically regenerated. Finally it is frequently possible to "tailor" a resin for specific application.

The principal disadvantage is the general lack of design information, startup problems, and realistic operating costs. In addition, resins systems are only applicable to single liquid phase feed streams with suspended solids concentrations under 50 mg/l.

Pollutants Treated: Existing applications of this technology have been successfully applied to removing color from a wide variety of wastewater streams. Other pollutants currently being treated are: phenol recovery and antibiotic recovery from fermentation broth, removal of organics from brine, and removal of fats from meat production wastewater.

Residuals Generated: Residuals generated in this technology are related to disposal of the used regenerant solution when it can not be recycled. In most cases residuals would be classified as hazardous and would need to be treated accordingly.

Screening

Description: In any wastewater treatment system, potential exists for very large pieces of material to be present in the wastewater flow. Domestic treatment plants are often plagued by such things as rocks, logs, rags, etc. in their influent. These types of items can be found in industrial wastewaters also, although they are generally less common. When discussing industrial pretreatment systems, large pieces of waste materials generally result from incompletely processed raw materials, equipment failures, leaks and spills, etc., with the end result being their presence in the plant's wastewater flow.

Regardless of their nature or source, these very large pieces can clog the treatment system, damage equipment, and/or adversely impact treatment plant performance. Also, in some industries, these pieces may represent a valuable raw ingredient. In any event, they must be removed from the waste stream.

The most common industrial pretreatment method is screening. Screening devices come in a variety of sizes and shapes, depending upon the specific application. However, they all have one common characteristic: they are constructed to allow the passage of liquid and small materials, but will capture large pieces of materials of a preselected size.

Screening devices can be hand cleaned or mechanically cleaned. Mechanically-cleaned units are predominantly used in industrial applications, where large volumes of materials are either returned to production in the case of recovered raw materials or stored for subsequent disposal. Figure 33 shows a drawing of a mechanically-cleaned bar screen.

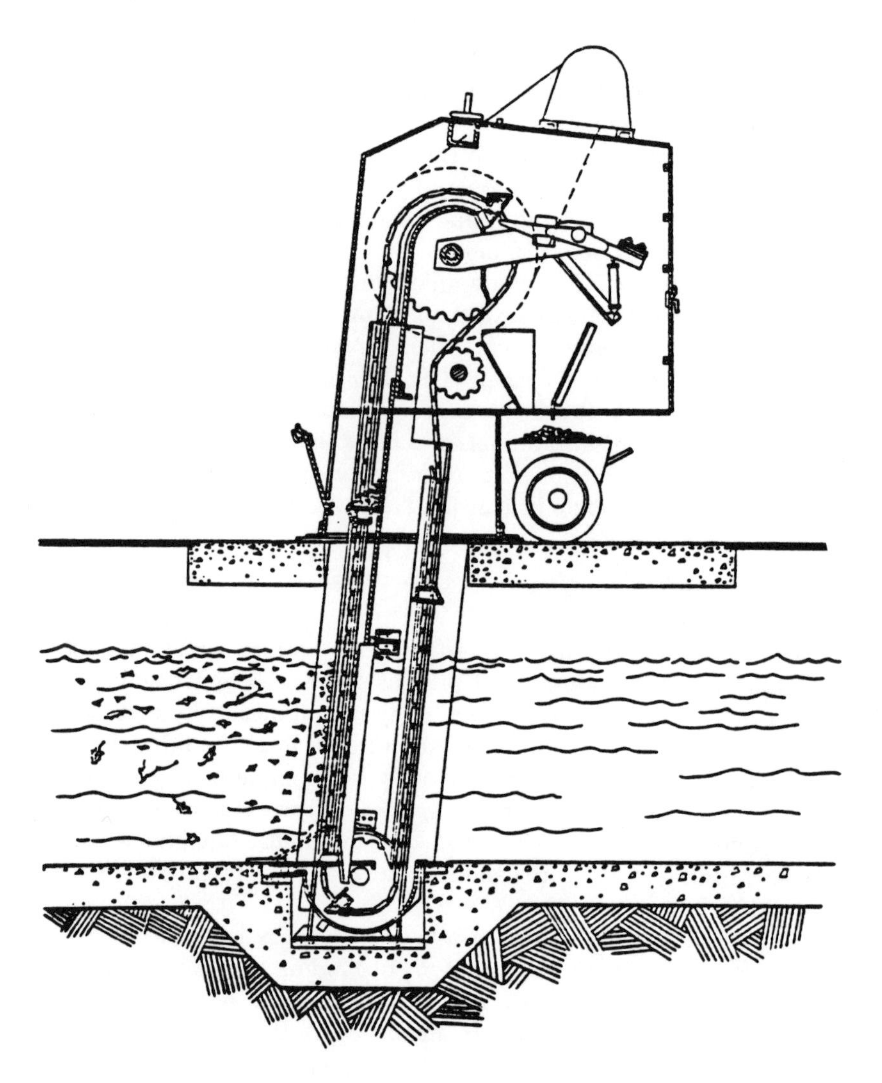

Figure 33: Typical mechanically cleaned bar screen.

Advantages and Disadvantages: Screening has the following advantages. By installing screens at the front end of a wastewater treatment system, removal of large undesirable pieces from a waste stream is centralized at a single location and the need for at-source control or prevention can be minimized. In those cases where captured materials represent portions of the raw ingredients for production, their recovery and recycling may be advantageous from an economic standpoint. Screening devices provide valuable protection to all downstream treatment equipment and generally improve performance of the overall treatment system.

Although not overly significant, screening does have some disadvantages. First, they do require operation and maintenance at the treatment plant. Hand-cleaned screens, if used, need frequent cleaning, and mechanically-cleaned ones would require inspections to insure they are operating properly. If this is not done, plugging of the system could occur. Finally, if captured wastes are not handled properly, odor and nuisance problems may arise.

Pollutants Treated: Large objects normally greater than 0.25 inches in diameter or stringy material are the primary pollutants removed.

Residuals Generated: The amount of residuals which could result from screening are highly dependent upon type of industry and the nature of its wastewater. For these industries where recovered materials represent unprocessed raw ingredients, their recycling will minimize the amount of residuals generated. If most of the materials are foreign matter, such as rocks, logs, rags, etc., the amount of wastes could be more substantial. In any event, most of the removed materials can be easily managed and disposed of.

Solvent Extraction

Description: Solvent extraction is the separation of constituents of a liquid solution by contact with another immiscible liquid. The process will typically include three basic steps: the actual extraction, solute removal from the extracting solvent, and solvent recovery from the treated stream.

In the extraction step the two liquid phases (feed and solvent) are brought into intimate contact to allow solute transfer either by forced mixing or by countercurrent flow. The extractor will also be able to provide separation of the two phases after mixing. One output stream from the extractor is the solute-laden solvent; some water may also be present. Solute removal may be via a second solvent extraction step, distillation, or some other process. Solvent recovery from the treated stream may be required if solvent losses would otherwise add significantly to the cost of the process, or cause a problem with a discharge of the treated stream. Solvent recovery may be accomplished by stripping, distillation, adsorption, or other suitable processes.

Figure 34 shows a schematic of a single solvent extraction unit operating on a wastewater stream.

Advantages and Disadvantages: Solvent extraction offers significant energy and other operating cost advantages in those situations where material recovery is possible to offset process cost.

Solvent extraction on concentrated waste streams will not generally produce an effluent which can be discharged to the POTW. Additional treatment is normally required. Finding a suitable solvent can be extremely difficult and expensive. Finally, the process is not very applicable to large quantities of very dilute wastewaters.

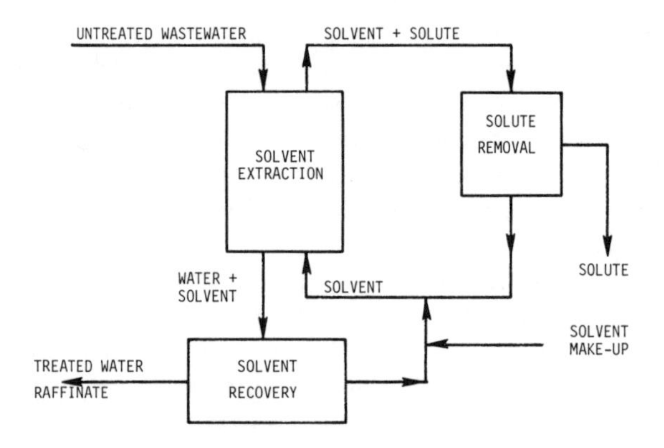

Figure 34: Schematic of a single solvent extraction process.

Pollutants Treated: Solvent extraction is a proven technology for recovery of organics from liquid solutions, with removal of phenol being the most frequent application.

Residuals Generated: Generally solvent extraction will be used for material recovery for resale or reuse. Solvent extraction systems seldom produce a discharge suitable for direct discharge. Normally the discharge receives biological treatment, thus biological sludges are generated. When mixed organic liquids are treated principally for the recovery of just one component, the other components will become residuals which require disposal.

Steam Stripping

Description: Distillation is the process of removing gases or vapors from a liquid or solid by evaporation and subsequently condensing the removed compounds. Steam stripping is a variation of distillation whereby steam is used as both the heating medium and driving force for removal of volatile materials from an industrial waste stream. Steam enters at the bottom of a tower and the wastewater being treated is fed at either the middle or near the top of the unit. As steam passes through the wastewater, volatile materials are vaporized and driven off with the steam, which exits at the top of the tower.

In packed columns, the unit is packed with materials that are inert, corrosion resistant, and able to withstand high temperatures. Also, these materials are specifically shaped in order to maximize the surface area for a given volume. In tray towers, the unit contains a series of trays which contain bubble caps or sieve perforations to allow for vapor-liquid contact. Generally, packed columns afford more flexibility than tray towers and are, therefore, usually used in treating industrial wastewaters that may change in composition and content.

The tower overheads (top steam) will contain the removed volatile materials and may contain some of the condensed steam, while treated wastewater will exit at the bottom of the unit. Depending upon composition of the waste stream and operational mode of the steam stripping unit, multiple compounds may be removed. In this case, further separation of the compounds may be desirable, especially if some of the removed materials are to be recycled. Additional separation techniques available include: selective condensation, decantation, extraction, etc.

Figure 35 shows a schematic of a steam stripper.

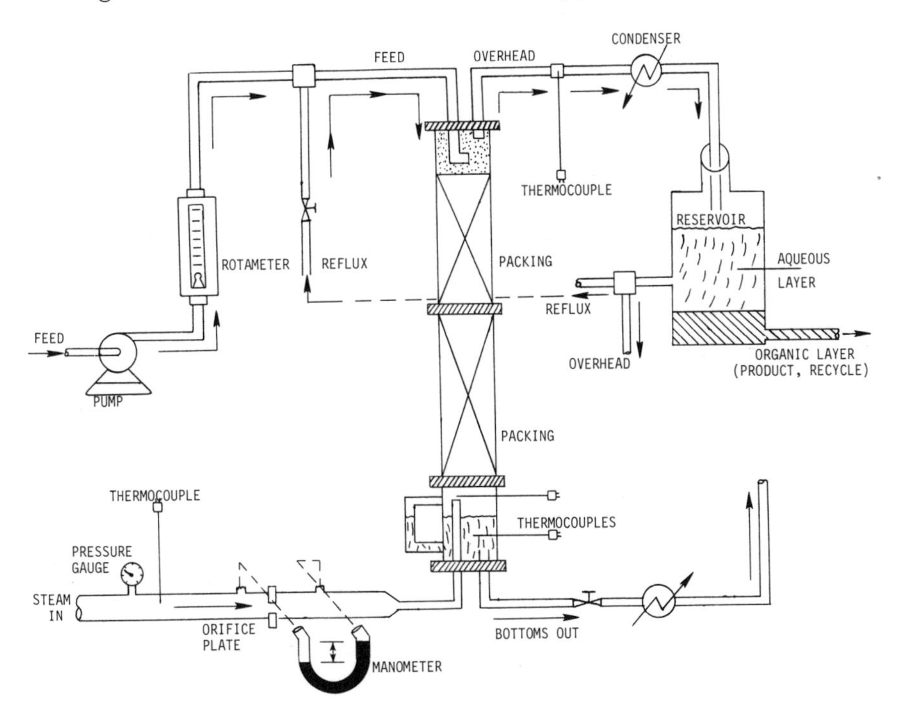

Figure 35: Schematic of a steam stripper.

Advantages and Disadvantages: Steam stripping is very effective in reducing a wide range of pollutants, especially volatile organics, to very low effluent levels. It is a flexible technology which can be adjusted to compensate for changes in wastewater flow and composition. Finally, since almost every manufacturing facility requires process steam, a large component of this technology — the steam generating equipment — will already be in place at the plant site. This will help minimize equipment costs.

The principal drawback to steam stripping is the relatively high operating cost associated with making steam. Also, an improperly maintained unit is subject to leaking, which can cause the emission of pollutants to the atmosphere. Upstream treatment may be necessary to prepare the wastewater for passage through the stripping unit.

Pollutants Treated: For industrial waste treatment, steam stripping is used for the removal of volatile organic compounds. Presently, developmental work on this technology is still continuing. However, in instances where it has been used, effluent concentrations of the magnitude of less than one mg/l have been achieved.

Residuals Generated: Pollutants treated by this technology will be volatized and carried off with the steam from the treatment operation. After condensing the steam, the waste compounds are concentrated by decantation, extraction, etc. Residual management can be difficult.

Wastewater Cooling Processes

Description: Wastewater cooling processes can be subdivided into the following categories: mechanical draft, natural draft, atmospheric wet, dry, and once-through systems.

Mechanical Draft Cooling Towers — There are two types of mechanical-draft cooling towers: forced draft (high pressure air) and induced draft (low pressure air). Both types allow air through the tower and into intimate contact with heated water. In the mechanical forced-draft cooling tower, the fan is mounted at the tower base and air is forced in at the bottom and discharged at a low velocity through the top. The mechanical induced-draft cooling tower can be further classified into counter-flow or cross-flow design, depending on the relative flow directions of water and air. Counter-flow towers have incoming air contact the coldest water first, and the most humid air contact the warmest water by moving the two fluids in opposite directions. Warm water falls from the top of the tower; and cold air enters the bottom of the tower and leaves at the top. Cross-flow towers have the incoming air contact the outer edge of the tower and move horizontally through the falling water. Figure 36 shows schematics of the various types of mechanical cooling towers.

Natural Draft Cooling Towers — Natural draft wet cooling towers are hyperbolic in shape, and involve the movement of air in the towers by natural draft due to a difference in density between inlet and exit air streams. The driving potential is proportional to the ambient temperature and the height of the towers. In natural draft counterflow towers, the air stream flows vertically upwards while water is sprayed downward. In natural draft crossflow towers, the packing is set up around the periphery of the base so that the air stream flows horizontally while water is sprayed downward. Figure 37 shows a representation of a natural draft wet cooling tower.

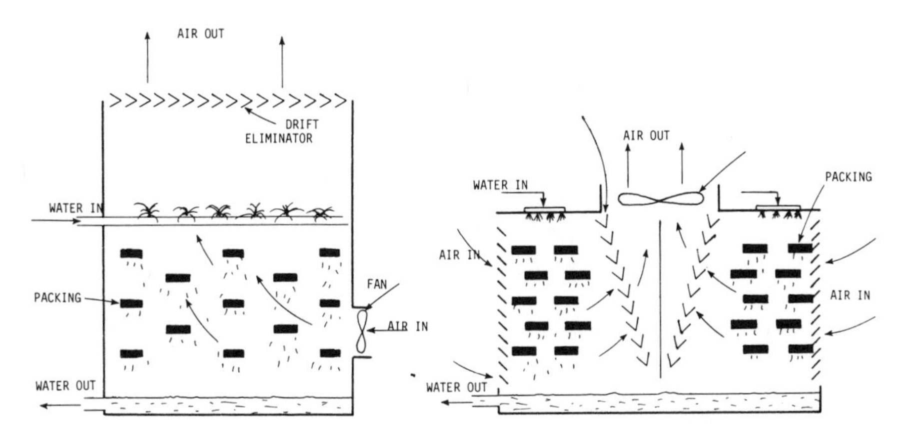

(a) Mechanical-forced draft cooling tower

(b) Mechanical-induced draft—crossflow cooling tower

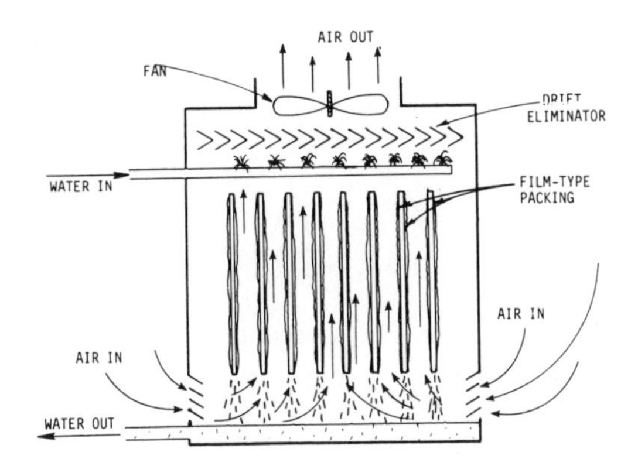

(c) Mechanical-induced draft counterflow cooling tower

Figure 36: Representation of various type mechanical cooling towers.

Atmospheric Wet Cooling Tower — Atmospheric wet cooling towers utilize naturally occuring winds by being built with a high, broad side exposed to the prevailing wind. Water is sprayed down into the air stream. Figure 38 is a representation of an atmospheric type wet cooling tower.

Dry Cooling Tower — The dry cooling tower is a relatively recent development. The tower consists of an air-cooled heat exchanger mounted inside a cooling tower chimney. The airflow past the heat exchanger may be driven by mechanical fans, or it may result from the natural draft of the heated air inside the tower. In the towers, there is no intimate contact between air and water, and no evaporation taking place. Figure 39 shows a representation of the basic idea of a dry cooling tower.

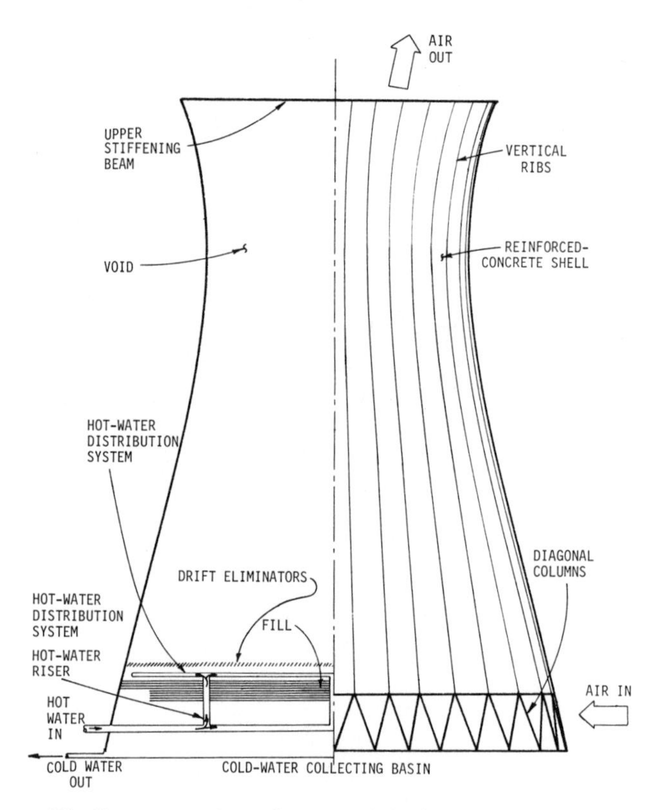

Figure 37: Representation of a natural draft wet type cooling tower.

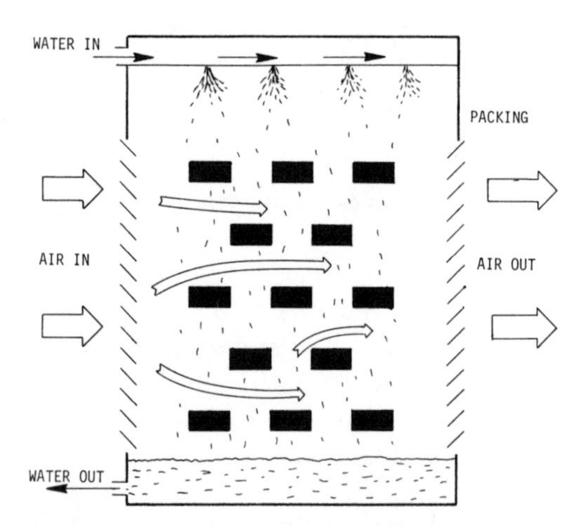

PACKED ATMOSPHERIC TOWER SPRAYS WATER DOWNWARD OVER FILLING OR PACKING, WHICH INCREASES COOLING EFFICIENCY BY FURTHER BREAKING UP FLOW OF WATER AND EXPOSING MORE WETTED SURFACE. OPEN LOUVERS LET OUTSIDE AIR PASS THROUGH TOWER OVER ITS FULL HEIGHT.

Figure 38: Representation of an atmospheric wet type cooling tower.

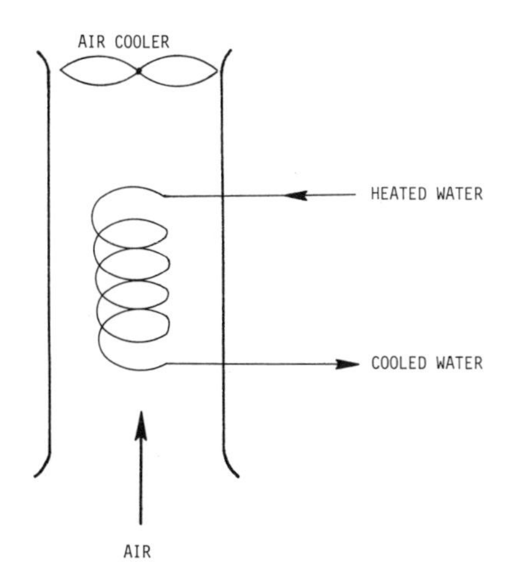

Figure 39: Basic idea of a dry cooling tower.

Once-through Cooling Systems — In once-through cooling systems, cool water is diverted from a river or estuary and pumped through the heat exchanger tubes, where it picks up heat from the warm stream on the outer surface of the tubes. The warmed water is then discharged to a cooling pond, spray pond, or to a river at a point where recirculation of the warm water will not occur.

A cooling pond is a body of water in which the warm water discharge is pumped so that it may be cooled and eventually recirculated. There is a net positive heat exchange across the water surface effected by conduction, evaporation, and radiation. There are three types of cooling ponds: (a) the completely mixed pond, in which pond temperature is almost uniform except in a small region near the plant discharge; (b) the flow-through pond, in which pond temperatures decrease away from the plant discharge; and (c) the internally circulating pond in which warm inlet water is discharged below the cooled water, causing a "top-to-bottom" mixing of water. Figure 40 shows a schematic of a cooling pond system.

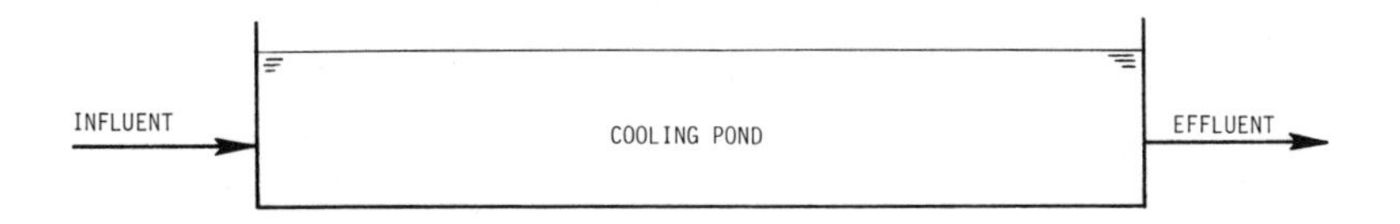

Figure 40: Schematic of a once-through cooling pond.

A spray pond uses a number of nozzles which spray water into contact with the surrounding air. Spray ponds lower the water temperature mainly by evaporative cooling. Figure 41 shows a schematic of a spray pond system.

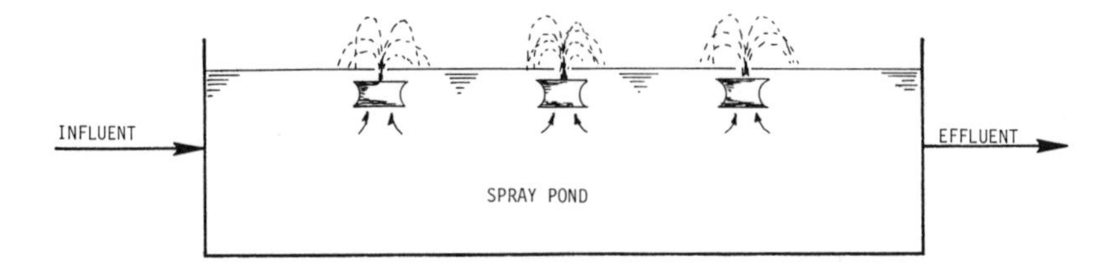

Figure 41: Schematic of a once-through spray pond system.

Advantages and Disadvantages

Mechanical Draft Cooling Towers — The initial cost of a mechanical draft wet cooling tower is less than that of a comparable capacity natural draftwet cooling tower. However, this is offset by higher operating and maintenance costs over the life of the tower.

A mechanical-induced draft-counterflow tower has several advantages: (a) the warmest water contacts the most humid air, while the coldest water contacts the driest air, thus utilizing the evaporative process most effectively; (b) warm air circulation is minimized by pushing the discharge straight up; (c) very large fans (60 feet in diameter) can be used; and (d) close approach and long cooling range is possible at reasonable cost. Its disadvantages are: (a) the typically small open area at the base causes high velocity inlet air and therefore increases fan horsepower; (b) baffles must be provided to achieve uniform air velocities through the packing; and (c) it requires greater pumping heat than a counterflow unit due to taller packing.

A mechanical-induced draft-crossflow tower has the following advantages when compared to a counterflow unit: (a) lower pumping head is required due to lower packing heights; (b) lower pressure drop through the packing is required; (c) higher water loadings are possible for a given height; and (d) a lower overall height of the tower is required. Its disadvantages are: (a) a substantial crossflow correction factor must be applied; (b) the packing is not as efficient and more air flow is required for an equivalent capacity tower; (c) winter operation is sometimes a problem since ice may form on the structure; and (d) the packing is in contact with the warmest water causing more rapid deterioration.

The advantages of mechanical-forced draft wet towers are: (a) vibration and fan noise are reduced, since the fan is mounted near the base; (b) problems of blade erosion and condensation in gearboxes are greatly reduced; (c) forced draft towers are more efficient than induced draft type. On the other hand, forced draft towers have some disadvantages: (a) practical fan size is limited to 12 feet in diameter, thus multiple fans must be used for high air flow rates; (b) baffles are necessary to insure uniform air distribution; (c) recirculation of the hot, humid discharge may be a problem; and (d) in cold weather, ice may form on fan blades, clogging air inlets and causing breakage.

Natural Draft Wet Cooling Tower — The volume of air that passes through the natural draft wet tower is dependent upon atmospheric conditions. The design point is always based on a "worst" set of atmospheric conditions. Somewhat oversized towers result, but satisfactory operation under adverse conditions is assured. For a given height and diameter, the natural draft-crossflow tower will typically have a smaller pressure drop and lower cooling capacity than that of a comparable counterflow tower.

Atmospheric Wet Cooling Towers — Atmospheric wet cooling towers are simple and relatively trouble free; packing and filtration equipment are the only important items requiring maintenance. Their disadvantages are: (a) the cooling range is low; (b) a large open surrounding area is required; (c) the system is highly sensitive to the velocity and direction of winds; (d) pumping losses are high; and (e) material costs are relatively high.

Dry Cooling Towers — The dry cooling tower technique appears to be a very promising alternative to once-through cooling. Due to the high cost of a dry cooling tower, it does not appear to be economical to replace the cooling equipment of an existing plant with a dry cooling tower, because either the plant would not operate at its design point, or the tower would have to be extremely large.

Once-through Cooling Systems — When water is available at essentially no cost, once-through cooling is the least expensive method, provided a suitable site can be found. Capital, operating, and maintenance costs are all low. However, various forms of aquatic life are forced through the condenser tubes, which can result in high organism mortality rates.

For cooling ponds, maintenance problems are minimal. However, large amounts of surface area are required since the cooling process is slow. Land cost is the major factor in cooling pond cost.

Spray ponds require little maintenance other than occasional cleaning of pipes and nozzles and routine maintenance on the pump. The danger of freezing and/or poor heat transfer due to climatic conditions may be present. Low heat transfer coefficients, dead zones where the circulation is almost zero, and high consumption are the undesirable alternatives.

Pollutants Treated: Cooling technologies are mainly used for treating hot discharge water from thermal power plants. In some cases, they may be used to cool hot industrial wastewater before biological treatment.

Residuals Generated: The use of wet cooling towers may have some undesirable side effects. Toxic chemicals used for water treatment may be discharged to nearby receiving waters. Evaporation of water into the air stream results in an increase in the dissolved solids concentration in the remaining water within the tower. Scale formation or corrosion in various components of the cooling system will eventually occur depending upon the pH of the water.

The dry cooling tower system does not give any significant secondary pollution. Cooling water flows in a closed loop through the plant and heat exchanger; therefore, the only external effect is to increase the temperature of the air passing across the heat exchange surface.

Wet Air Oxidation

Description: Wet air oxidation refers to a process in which air or oxygen is injected into a wastewater stream containing dissolved or suspended organic material at elevated temperatures and pressures. Under these conditions the organic material is oxidized. Figure 42 shows a schematic of a basic wet air oxidation system.

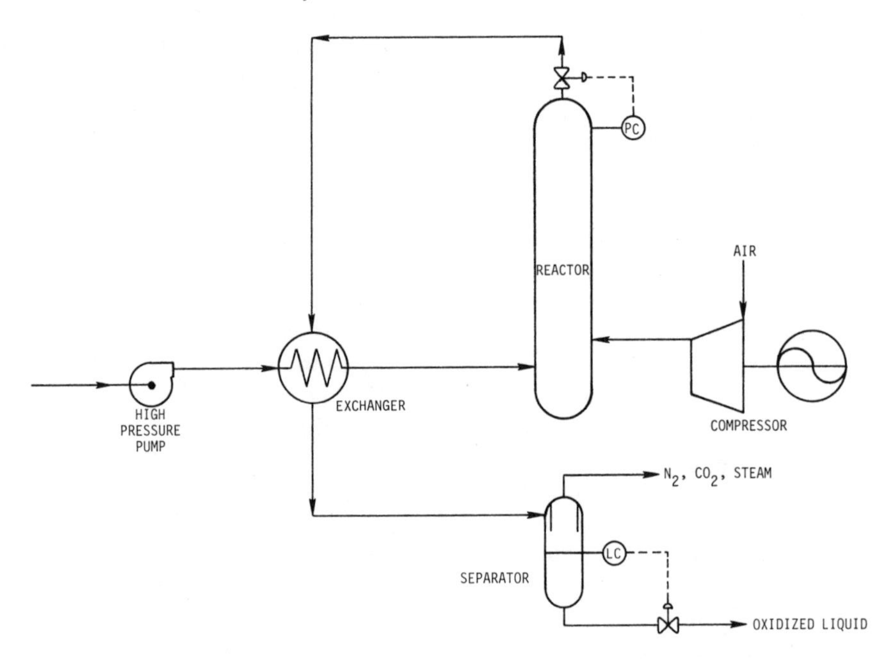

Figure 42: Schematic of a basic wet air oxidation process.

Wastewater is supplied to the oxidation unit by a high-pressure pump, and oxygen-containing gas is mixed with the pressurized wastewater stream. The mixture flows through a pressure vessel designed to provide adequate retention time for the reaction to take place. In the heat exchanger, the mixture is heated to a temperature at which the oxidation reaction will begin. The oxidation reaction is exothermic, and the heat liberated further raises the temperature of the reaction mixture to the design temperature.

Advantages and Disadvantages: The primary advantage of wet air oxidation is that it allows an economical means of detoxifying many types of hazardous organic wastes which are neither burnable nor biodegradable. Also land requirements are small.

Disadvantages to this process are high captial, operating, and maintenance costs associated with the specialized equipment. In addition, effluent from the process is not generally suitable for discharge to the POTW. Usually another treatment process — generally biological — is utilized to produce an effluent acceptable for discharge.

Pollutants Treated: Wet air oxidation is capable of oxidizing most organic compounds (the process cannot treat certain polysubstituted halogenated organic compounds such as kepone and PCB's). Existing installations presently treat explosive wastes, gas waste liquors from coke ovens, pulping liquor, acrylonitrile waste liquors, and caprolactam waste liquors. In addition, some inorganic compounds are also treatable. Most multivalent metallic cations are converted to their highest oxidation state and appear as dissolved compounds or as insoluble hydroxides or oxides.

Residuals Generated: The wet air oxidation process itself does not produce any residuals. Since the effluent from the process is normally not suitable for discharge, further processing is required, generally biological treatment. Biological sludges are generated and can be managed similiar to other biological sludges.

CHEMICAL PRETREATMENT TECHNOLOGIES

Chemical Oxidation

Description: Chemical oxidation is a chemical reaction in which one or more electrons are transferred from the chemical being oxidized to the oxidizing agent. Elemental chlorine, hypochlorite salts, etc., are strong oxidizing agents in aqueous solution and are the principal chemical used in this industrial pretreatment technology.

Chemical oxidation by chlorine or chlorination is primarily used in industrial treatment systems to treat cyanide-bearing wastewaters. This classic procedure can be approximated by the following two-step chemical reaction:

$$Cl_2 + NaCN + 2NaOH \rightarrow NaCNO + 2NaCl + H_2O$$
$$3Cl_2 + 6NaOH + 2NaCNO \rightarrow 2NaHCO_3 + N_2 + 6NaCl + 2H_2O$$

The reaction indicated by the first equation represents the oxidation of cyanides to cyanates. This step is most significant since it results in a marked reduction in volatility and a thousand-fold reduction in toxicity. The second equation represents the oxidation of cyanate, and is the final step in the total oxidation of cyanide to nitrogen and bicarbonates.

Continuous flow units are generally used by plants with large volumes of cyanide-bearing wastewaters, while batch units are more common at small facilities. Regardless of the type of unit, careful attention must be given to pH, chemical addition, retention time, etc. If not, potential exists for intermediate cyanide compound to revert back to cyanide and, thus, nullify the performance of the treatment process.

Figure 43 is a schematic of a chlorination process for treating cyanide.

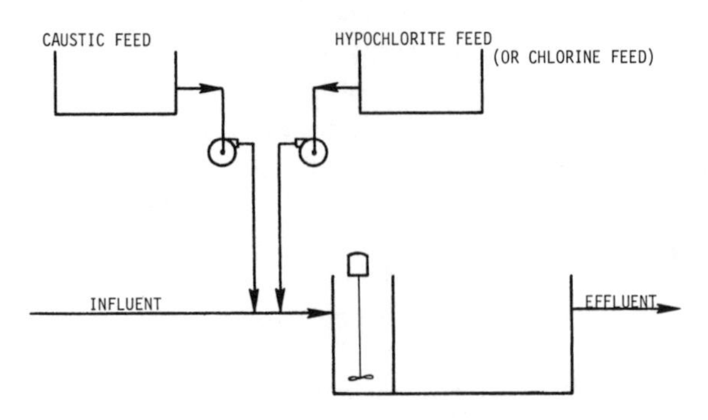

Figure 43: Schematic of a cyanide chlorination oxidizing system.

Advantages and Disadvantages: Advantages of chlorination for industrial wastewater treatment are as follows. First, it is a well proven process that can be adapted to treat many different types of wastewater flow. Second, it is a relatively simple treatment process that does not require sophisticated equipment. Third, chlorination is well suited for automatic operation which can minimize personnel requirements. Because of the above, oxidation has low capital and operating costs.

However the chlorination process is not without disadvantages. If not properly operated, dangerous intermediate reaction products could be formed. In some cases these products are toxic. Depending upon the individual characteristics of a wastewater, interferences are possible which can consume the oxidizing chemicals and result in ineffective treatment. Finally, since strong oxiding chemicals are required, potential handling and storage problems may exist.

Pollutants Treated: This technology can be used to treat any oxidizable pollutant. Industrial use of chlorination is principally directed toward the treatment of cyanide-bearing wastewaters. With some exception, chlorination is practically the only treatment process which will adequately oxidize cyanide to harmless end products, such as nitrogen and bicarbonates. This technology has received widespread use to date and is capable of achieving excellent effluent levels.

Residuals Generated: If properly operated, this technology will completely oxidize cyanide to bicarbonates, nitrogen, etc. In most of the operations, gases are vented to the atmosphere and the salts remain in the wastewater to be removed by other downstream treatment processes. Therefore, residuals generated by chlorination will not require direct management.

Chemical Reduction

Description: Certain metals must be reduced from their high valence state before they can be precipitated or settled. The most common method in use today is to perform this reduction by chemical means. Chemical reduction is a reaction in which one or more electrons are transferred to the chemical being reduced from the reducing agent.

The main application of chemical reduction in treatment of industrial wastewaters is in reduction of hexavalent chromium to trivalent chromium. Thus, it will be discussed here. The reduction enables the trivalent chromium to be separated from solution in conjunction with other metal salts by precipitation. Sulfur dioxide, sodium bisulfite, sodium metabisulfite, and ferrous sulfate form strong reducing agents in aqueous solution and are, therefore, useful in industrial waste treatment facilities for the reduction of hexavalent chromium to trivalent chromium. Gaseous sulfur dioxide is probably the most widely used agent in this process. The reactions involved may be illustrated as follows:

$$3SO_2 + 3H_2O \rightarrow 3H_2SO_3$$
$$3H_2SO_3 + 2H_2CrO_4 \rightarrow Cr_2(SO_4)_3 + 5H_2O$$

The above reactions are favored by very low pH. In fact the speed and completeness of the reduction is directly related to the pH level of the waste stream. Care must also be given to ensure that oxidizing compounds such as dissolved oxygen and ferric iron in the wastewater to be treated are minimized. This is because they will interfere with the chemical reactions by consuming the reducing agent.

Figure 44 shows a schematic of a chemical reduction system for treating chromium-bearing wastewaters.

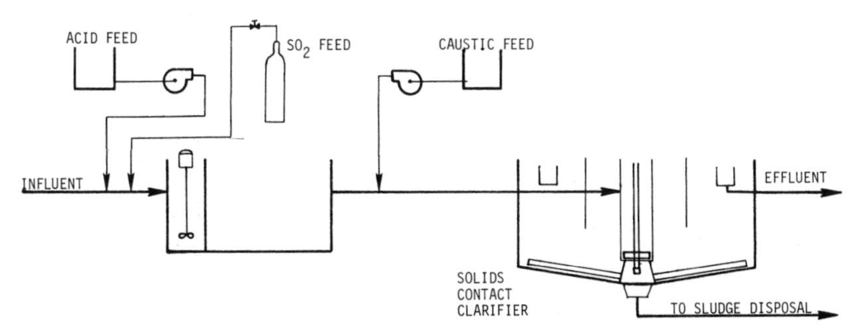

Figure 44: Schematic of chemical reduction system for treating chromium-bearing wastewaters.

Advantages and Disadvantages: The principal advantage of this process is its demonstrated effectiveness. In all of its applications within industry, chemical reduction has successfully treated high valence metals. In addition, the process is well suited to automatic control. Chemical reduction processes also operate at ambient conditions.

However, chemical reduction is not without some limitations. Careful pH control is required for effective reduction. In addition, when waste streams contain other oxidizing matter, the reducing agent may be consumed by these materials and interfere with the treatment of the metals. Finally, for those systems using sulfur dioxide, a potentially hazardous situation exists when it is stored and handled.

Pollutants Treated: This technology is appropriate wherever high valence compounds are present in an industrial wastewater flow. Its most prevalent use has been in the reduction of hexavalent chromium to trivalent chromium, where it has been demonstrated to be very effective. Chromium reduction is not a pollutant removal process in the strict sense, but rather a preparatory step prior to the ultimate removal of all metals. By reducing hexavalent chromium to its trivalent form, this treatment process enables trivalent chromium to be separated from solution, along with other metals, by alkaline precipitation in a clarification unit.

Residuals Generated: If one were to look at only the chemical reduction portion of this technology, it could be said that no residuals are generated. However, chromium sludges result from the clarification phase, a downstream chemical precipitation step. This results in a chromium containing sludge. Because of chromium's toxicity, special management procedures are required. Disposal to a controlled landfill suitable for hazardous waste is usually required.

Coagulation-Precipitation

Description: Often the nature of an industrial wastewater is such that conventional physical treatment methods will not provide an adequate level of treatment. Particularly, ordinary settling or flotation processes will not remove ultrafine colloidal particles and metal ions. In these instances, natural stabilizing forces, such as electrostatic repulsion and physical separation, predominate over the natural aggregating forces and mechanisms, van der Waal forces and Brownian motion, which tend to cause particle contact. Therefore, to adequately treat these particles in industrial wastewaters, coagulation-precipitation may be warranted.

The first and most important part of this technology is coagulation, which involves two discrete steps. Rapid mixing is employed to ensure that the coagulating chemicals are thoroughly dispersed throughout the wastewater flow for uniform reaction. Next, the wastewater undergoes flocculation which provides for particle contact, so that they can agglomerate to a size large enough for removal.

The mechanisms of coagulation which are involved in the above steps are the result of four destabilizing actions. Electrostatic charge reduction is accomplished by adsorption of oppositely charged ions. Adsorption of certain chemical groups may cause interparticle bridging. Gelatinous hydrolysis products will be formed as a result of particle solubility changes, due to pH changes. And finally, fine particles may become enmeshed in the products of coagulation.

The final part of this technology involves precipitation. This is in reality, the same as settling and thus can be performed in a unit similar to a clarifier.

A typical chemical mixer for coagulation is shown in Figure 45. Figure 46 shows examples of the flocculation basin that must follow the mixing system. The precipitation portion of this technology is essentially clarification. See Figures 19 and 20 for examples of this process component.

Advantages and Disadvantages: Coagulation-precipitation is a derivative of the clarification process. Its principal advantages are as follows. Coagulation-precipitation is a proven technology that can utilize existing treatment processes, i.e., clarification units, and at the same time provide better pollutant removals. Thus, treatment can be maximized, while costs are minimized. Also, this technology is well suited to automatic control and can operate at ambient temperatures. Changes in wastewater characteristics can be handled through adjustments in the addition of coagulating chemicals.

Some disadvantages of using coagulation-precipitation for the treatment of industrial wastewaters are as follows. Depending upon the specific components in a waste stream, chemical interferences are possible which could adversely affect the process's performance. Also, a substantial quantity of sludge is generated by this technology, resulting in increased sludge management and disposal problems.

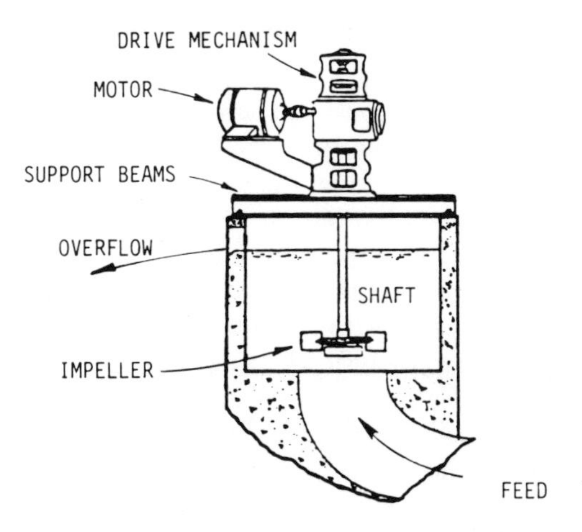

Figure 45: Chemical mixer for coagulation.

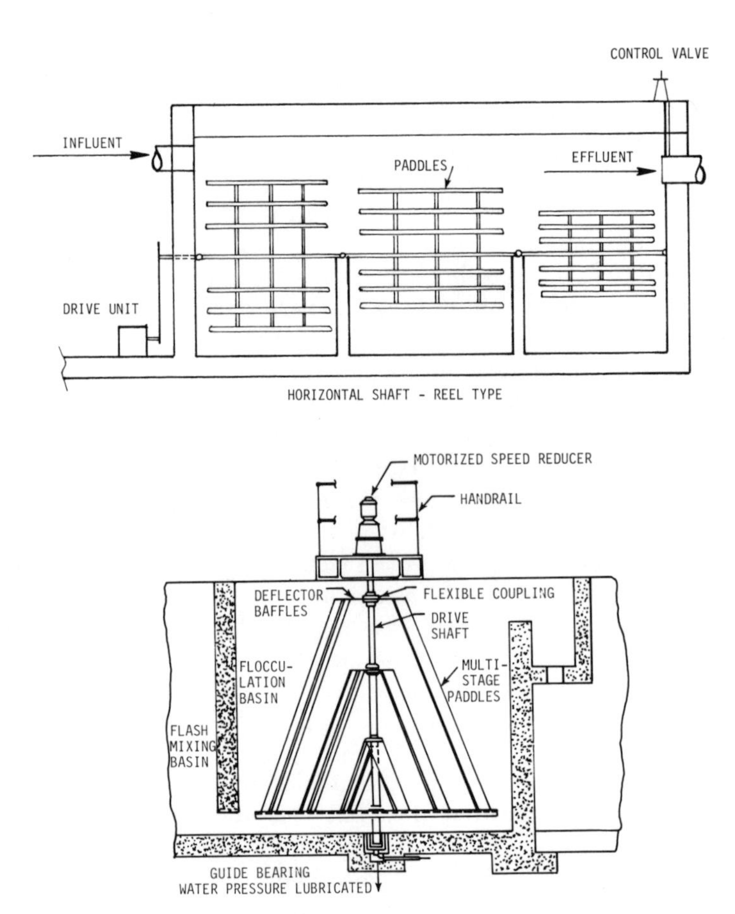

Figure 46: Mechanical flocculation basins.

Pollutants Treated: Coagulation-precipitation is capable of removing from industrial wastewater the classical pollutants such as BOD, COD, and TSS. In addition, depending upon the specifics of the wastewater being treated, coagulation-precipitation can remove additional pollutants such as phosphorous, nitrogen compounds, metals, etc. This technology is attractive to industry because excellent classical and toxic pollutants removal can be combined in one treatment process.

Residuals Generated: Significant quantities of residuals are created by this technology, because coagulating-precipitating chemicals are removed along with the captured pollutants. The degree of difficulty in residuals management depends on the type of chemical(s) utilized and can range from easy to difficult.

Neutralization

Description: In virtually every type of manufacturing industry, chemicals play a major role. Whether they result from raw materials or various processing agents used in the production operations, some residual compounds will ultimately end up in a process wastewater. Thus, it can generally be expected that most industrial waste streams will deviate from the neutral state, i.e., be acidic or basic in nature.

Highly acidic or basic wastewaters are undesirable for two reasons. First, they can adversely impact aquatic life in receiving waters and, second, they might significantly affect performance of downstream treatment processes at the plant site or at a POTW. Therefore, in order to rectify these potential problems, one of the most fundamental treatment technologies, neutralization, is employed at industrial facilities.

The basic principles of neutralization have been documented and are generally well known in the field of water pollution control. Thus, they do not warrant a detailed discussion here. It is sufficient to say that in the simplest sense, neutralization involves adding an acid or a base to a wastewater to offset or neutralize the effects of its counterpart in the wastewater flow, namely adding acids to alkaline wastewaters and bases to acidic wastewaters.

The most important considerations in neutralization treatment are: a thorough understanding of the wastewater constituents so that the proper neutralizing chemicals are used; and, proper monitoring to ensure that the correct quantities of these chemicals are used and that the effluent is in fact neutralized. For acid waste streams, lime, soda ash, and caustic soda are the most common base chemicals used in neutralization. In the case of alkaline-waste streams, sulfuric, hydrochloric and nitric acid are generally used for neutralization. Some plants or industries have operations that produce separate acid and alkaline waste streams. If properly controlled, these waste streams can be mixed to produce a neutralized effluent without the need for adding neutralizing chemicals.

Figure 47 shows one type of neutralization system.

Advantages and Disadvantages: Eliminating adverse impacts on water quality and wastewater treatment system performance is not the only benefit of neutralization. Acidic or alkaline wastewaters can be very corrosive. Thus, by neutralizing its wastewaters, a plant can protect and extend the life of its treatment units and associated piping. This can result in improved operations and substantial savings.

There is one major disadvantage to neutralization. The chemicals used in the treatment process are corrosive themselves and can be dangerous. As a result, it is imperative that the proper facilities exist for their handling and storage, not only to protect plant personnel, but also the surrounding plant operations.

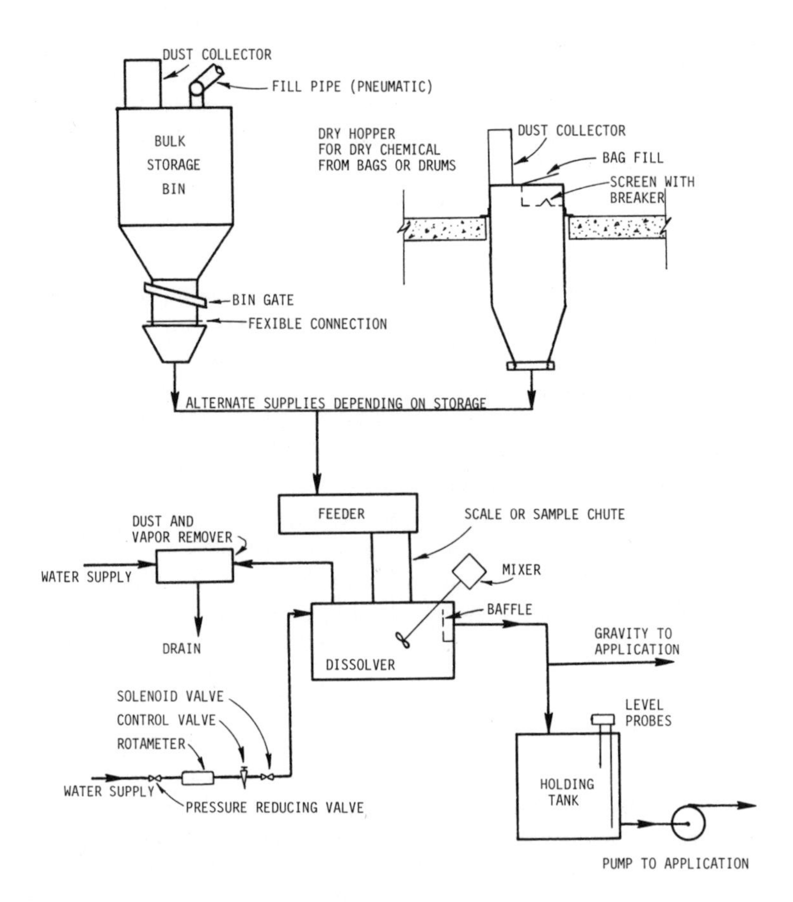

Figure 47: Dry feed system for neutralization.

Pollutants Treated: Neutralization treats the pH of a wastewater flow. Although most people do not think of pH as a pollutant, it is in fact designated by the USEPA as such. Since many subsequent treatment processes are pH dependent, neutralization can be considered as a preparatory step in the treatment of all pollutants.

Residuals Generated: No residuals are generated by this process.

Ozonation

Description: Ozone (allotropic form of oxygen) is a good oxidizing agent and can be used to treat process wastewaters which contain cyanide. In fact, it oxidizes many cyanide complexes that are not broken down by chlorine, for instance, iron and nickel complexes. Ozonation is primarily used to oxidize cyanide to cyanate and to oxidize phenols and chromophores to a variety of nontoxic products. The ozonation process is illustrated by the following ionic equation:

$$CN^- + O_3 \rightarrow CNO^- + O_2$$

The reaction indicated by the above equation represents oxidation of cyanides to cyanates. Oxidation of cyanate to the final end products, nitrogen and bicarbonate, is a much slower and more difficult process, unless catalysts, such as copper or manganese, are present. Therefore, since ozonation will not readily affect further oxidation of cyanate, it is often coupled with other downstream treatment processes to ensure the complete removal of cyanide.

This technology is similar to oxidation by chlorine. Figure 43 would be representative of an ozonation system except the chemical being added would be ozone instead of chlorine.

Advantages and Disadvantages: The ozonation process is well suited to automatic control and will operate effectively at ambient conditions. Also, the reaction product (oxygen) is beneficial to the treated wastewater. Since the ozone is generated on-site, procurement, storage, and handling problems are eliminated.

However, the ozonation process does have its drawbacks. First, it has relatively higher initial and operating costs than chlorination. And like chlorination, interference is possible, if other oxidizable matter is present in the waste stream. Finally, in most cases, the cyanide is not effectively oxidized beyond the cyanate level.

Pollutants Treated: Ozonation can be used to remove any pollutant which is amenable to treatment by ozone. However, for industrial applications, ozonation is principally used in the treatment of cyanide-bearing wastewaters. With the proper catalysts, such as manganese or copper, this technology is capable of achieving excellent removal efficiencies. Without these catalysts, however, subsequent treatment may be required to ensure the complete treatment of cyanide.

Residuals Generated: Ozonation operates on the same basic principles as chemical oxidation. Thus, the gases and salts produced by this technology during treatment are vented to atmosphere and remain in the wastewater, respectively. As a result, residuals generated by ozonation do not require direct management.

Starch Xanthate

Description: Precipitation of heavy metals from industrial wastewaters using starch xanthate polymers is a relatively new application.

Advantages and Disadvantages: This process for heavy metal removal offers several advantages over other metal removal processes. Control of influent pH and suspended solids before treatment is not required. The sludge produced settles faster and is not gelatinous. Finally, only a small excess of reagent is necessary to drop the metal residual level to below most accepted discharge limits.

There are two major disadvantages: operating cost and operating experience. Operating costs are high because the polymer is quite expensive. Full-scale operating experience is very limited as of this time (summer, 1980).

Pollutants Treated: This process has shown that it will remove most heavy metals to very low residuals. Some metals such as Fe^{+2}, Mn^{+2}, and Zn^{+2} are removed at large initial concentrations but the residuals normally are not below acceptable effluent discharge limitations.

Residuals Generated: Of all the metal removal precipitation methods, this process generates the least amount of sludge. Sludge is not gelatinous and settles faster than other precipitates. It is expected that management of this residual may be easier than other metal-precipitated sludge, but there are no operational data to date.

Sulfide Precipitation

Description: Sulfide precipitation has been singled out from the precipitation technology because of its capabilities for removing heavy metals from wastewaters. In this process, a sulfide source — H_2S, Na_2S, NaHS, or FeS — is added to the wastewater containing the metal. It is important to note that the sulfide source added must be more soluble than the metal sulfide to be precipitated. As the added sulfide dissolves, the dissociated sulfide ion reacts readily with the heavy metal that has a lower sulfide solubility. When equilibrium is attained, the metal of lower solubility is precipitated, and the one of higher solubility remains dissolved.

Advantages and Disadvantages: Precipitation of metals as sulfides offers several distinct advantages over hydroxide precipitation treatment.

1. Solubilities for most heavy metals are many orders of magnitude less for the sulfide form than for the hydroxide form.
2. Metal sulfides tend to retain low solubility at pH values of 7.0 and less where most hydroxides tend to become appreciably soluble.
3. Many of the metal hydroxides tend to redissolve upon increasing the pH value above a certain critical value for each metal.

Historically, sulfide precipitation has not been utilized by industry because of the problems experienced in generating noxious hydrogen sulfide gas. This problem has been eliminated by several new processes using different sources of sulfide. Another problem has been the forming of colloidal particles which do not settle very well. Finally, operating costs tend to be higher than hydroxide-removal processes.

Pollutants Treated: Sulfide precipitation can essentially be used to remove any heavy metal.

Residuals Generated: Significant quantities of hazardous residuals are created by this technology, because of the chemicals being added and the pollutants being removed. Because of the chemical reactions involved, sulfide-precipitated sludges are easier to manage than hydroxide sludges.

-3-

RESIDUALS MANAGEMENT TECHNOLOGIES

INTRODUCTION

Through the pretreatment of industrial wastewater, each industrial pretreater will generate residuals classified as hazardous, toxic, or nonhazardous. These residuals must be disposed of through environmentally sound and cost effective practices.

Prior to initiating the use of the system selection approach, the classification of treatment and disposal options must be undertaken. The criteria used and information gathered will vary depending on site specific cases. Classification of treatment and disposal options is shown in Figure 48.

The general sequence of events in system selection is:

- Selection of relevant criteria.
- Identification of options.
- Narrowing the list of candidate systems.
- Selection of a system.

In short, the selection mechanism used is the "Principle of Successive Elimination". Treatment and disposal systems which are compatible and appear to satisfy local relevant criteria are selected. Then the best system or systems are chosen by progressive elimination of weaker candidates.

The purpose of this section is to acquaint the reader with technical information relating to the treatment, transportation, storage, and final disposal of residuals. Technology descriptions, application, and advantages and disadvantages will also be discussed in order to aid in the selection of the most appropriate residuals management system(s).

This approach uses a matrix of information to aid in the selection of the best suitable technologies. The system approach includes economic, energy, and environmental impacts; system effectiveness; system implementation and compatibility; and administrative burdens.

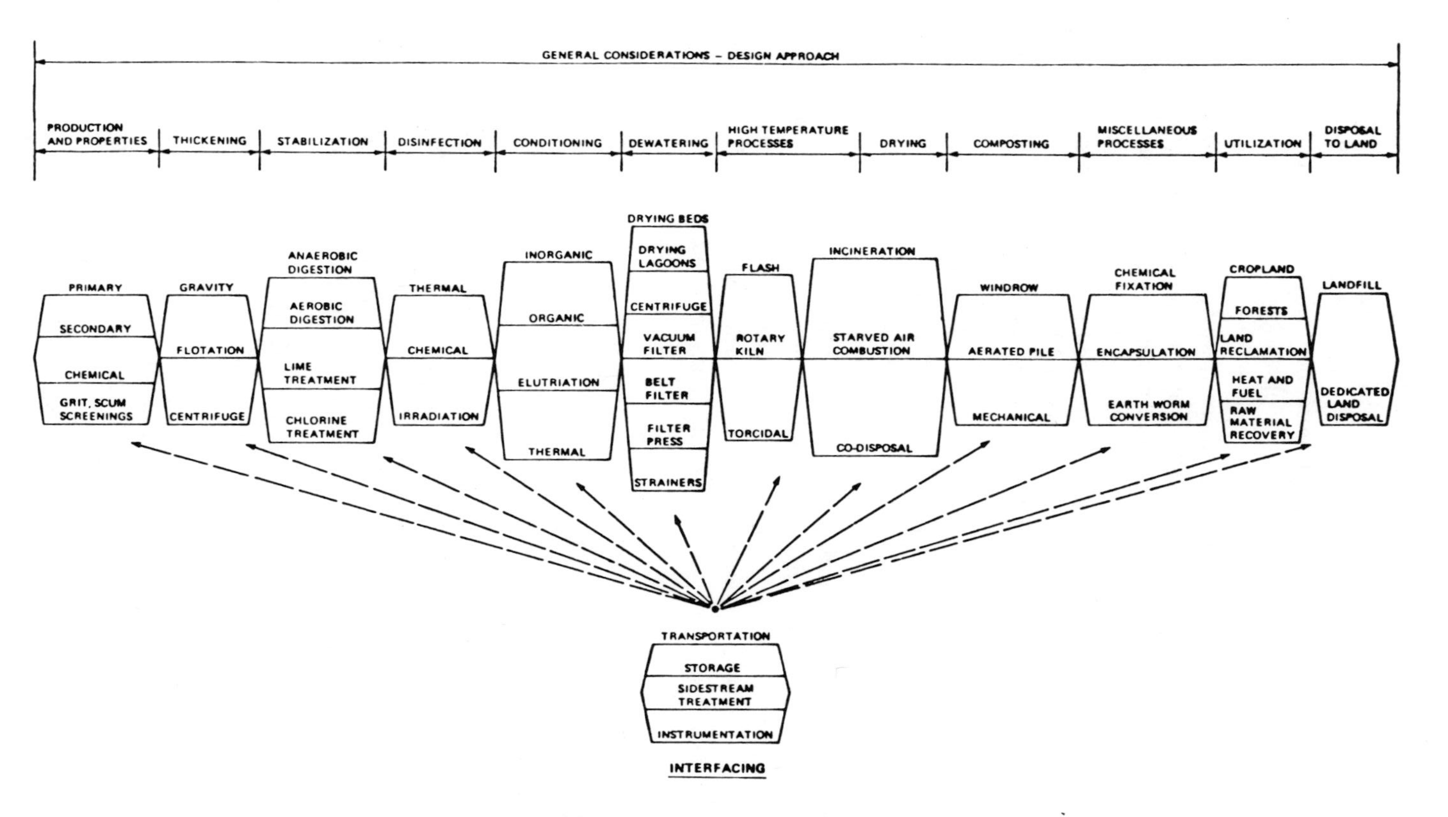

Figure 48: Classification of treatment and disposal options.

The primary goal of residuals management is complete reuse, however, very few of the numerous industrial subcategories offer any potential for the use of this type of residuals management approach. Besides the various methods of reuse, proper disposal could involve one or several of more than 30 residual management technologies. The tasks associated with properly managing residuals are neither simple nor cheap. In providing higher levels and additional treatment of wastewaters, greater volumes of residuals are produced. Through greater residual generation, and increasingly restrictive management requirements, options for the design of residuals management systems have become complicated. Therefore, there is a significant need for sound wastewater and residuals management. In satisfying this need serious consideration should be given to the interdependence of both the liquids and solids portion of the treatment facility. Costs for disposal and solids processing account for up to 40 percent of the total operating and maintenance cost of a facility. To alleviate these problems one must evaluate waste streams, residuals generated, and treatment and disposal methods, in order to utilize the most appropriate and cost effective alternative available.

To assure efficient utilization of resources for wastewater treatment, wastewater residuals, and disposal systems, a procedure for system selection is needed. A list of criteria used in the system selection procedures is shown in Figure 49. For additional information refer to EPA 625/1-79-011, *Process Design Manual for Sludge Treatment and Disposal*, USEPA, September, 1979.

RESIDUALS TREATMENT

Thickening

Thickening is defined in this manual as removal of water from sludge to achieve a volume reduction. The resulting material is still fluid.

Sludges are thickened primarily to decrease the capital and operating costs of subsequent sludge processing steps by substantially reducing the volume. Thickening from one to two percent solids concentration, for example, halves the sludge volume.

Depending on the process selected, thickening may also provide the following benefits: sludge blending, sludge flow equalization, sludge storage, grit removal, gas stripping, and clarification.

Although it is good design practice to pilot test thickening equipment before designing a facility, pilot testing does not guarantee a successful full-scale system. Designers must be cognizant of the difficulties involved in scale-up and allow for them in design.

The main design variables of any thickening process are:

- Solids concentration and volumetric flow rate of the feed system.

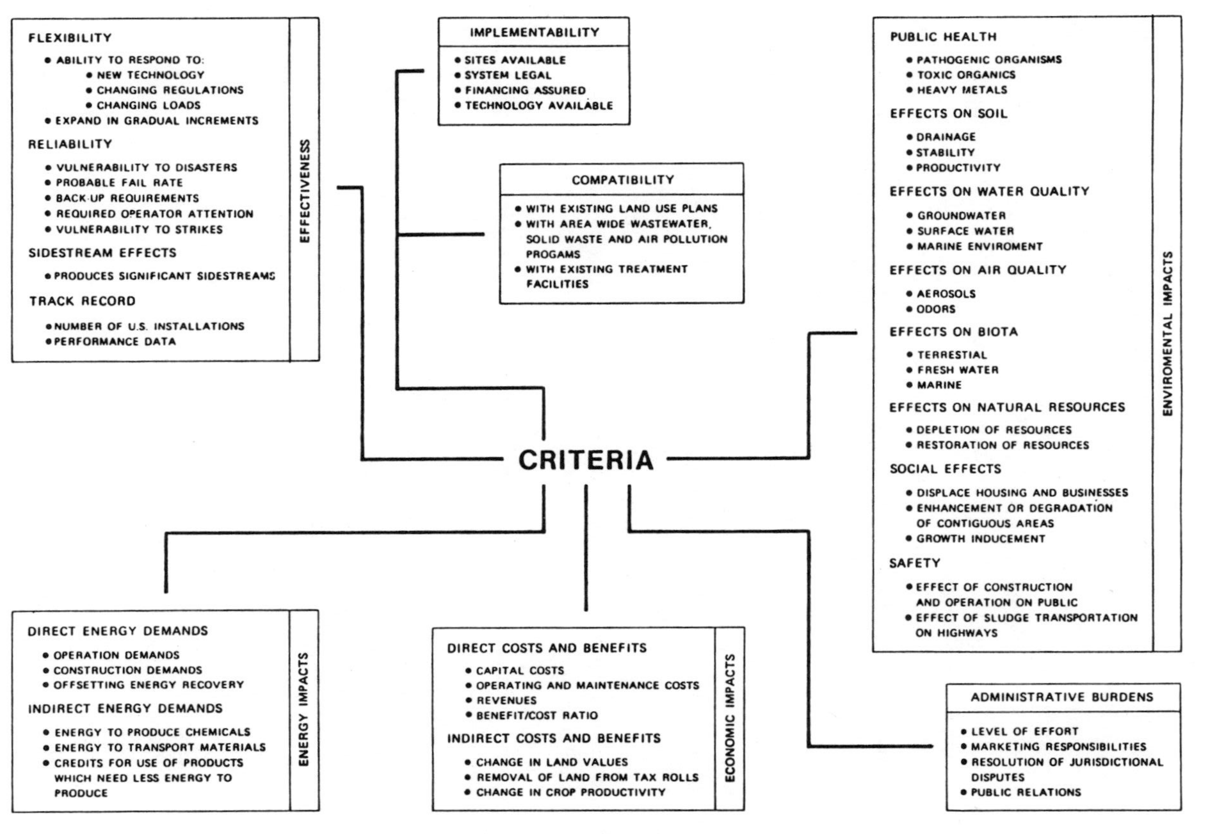

Figure 49: Criteria for system selection.

- Chemical demand and cost if chemicals are employed.
- Suspended and dissolved solids concentrations and volumetric flow rate of the clarified stream.
- Solids concentration and volumetric flow rate of the thickened sludge.

Other variables which impact the thickening process selected are: subsequent processing steps; operation and maintenance (O/M) cost and the variables affecting it; and the reliability required to meet successful operational requirements.

Thickening is generally subdivided into three general subcategories: gravity thickening, flotation thickening, and centrifugal thickening.

Gravity Thickening:

Description — Gravity thickening is a process which concentrates sludge particles in a solids/liquid mixture by utilizing the natural gravitational field and differences in specific gravity between sludge particles and water. The process can be either batch or continuous, with the latter being the most popular. In this process, thickened sludge is removed from the bottom of the tank and supernatant liquid is removed from the top. Figure 50 shows a typical gravity thickener.

Gravity thickeners are typically circular concrete tanks, though circular steel tanks and square or rectangular concrete tanks are also utilized. In most cases, tanks are provided with equipment to help uniformly distribute, concentrate, and remove sludge from the system. In designing a gravity thickener, the following variables are typically considered: minimum surface area; hydraulic loading; drive torque requirements (circular tank only); total tank depth; floor slope; lifting device (circular tank only); thickener supernatant; thickener feed pump and piping; and thickener underflow pump and piping.

Application — Gravity thickeners are excellent for sludges which act as particulate (nonflocculant) suspensions. Some flocculant sludges can also be gravity thickened when tanks are provided with low overflow rates, large surface areas, and/or treated with a coagulant.

Advantages and Disadvantages — Gravity thickening offers the following advantages over other thickening processes:

- Provides greatest sludge storage capabilities.
- Requires the least operational skill.
- Provides lowest unit operation (especially power) and maintenance cost.

Gravity thickening has the following disadvantages when compared to other thickening processes:

- Requires largest land area.
- Can be a source of odor.
- For some sludges, solids/liquid separation can be erratic, and for some sludges it will produce the thinnest, least concentrated sludge.

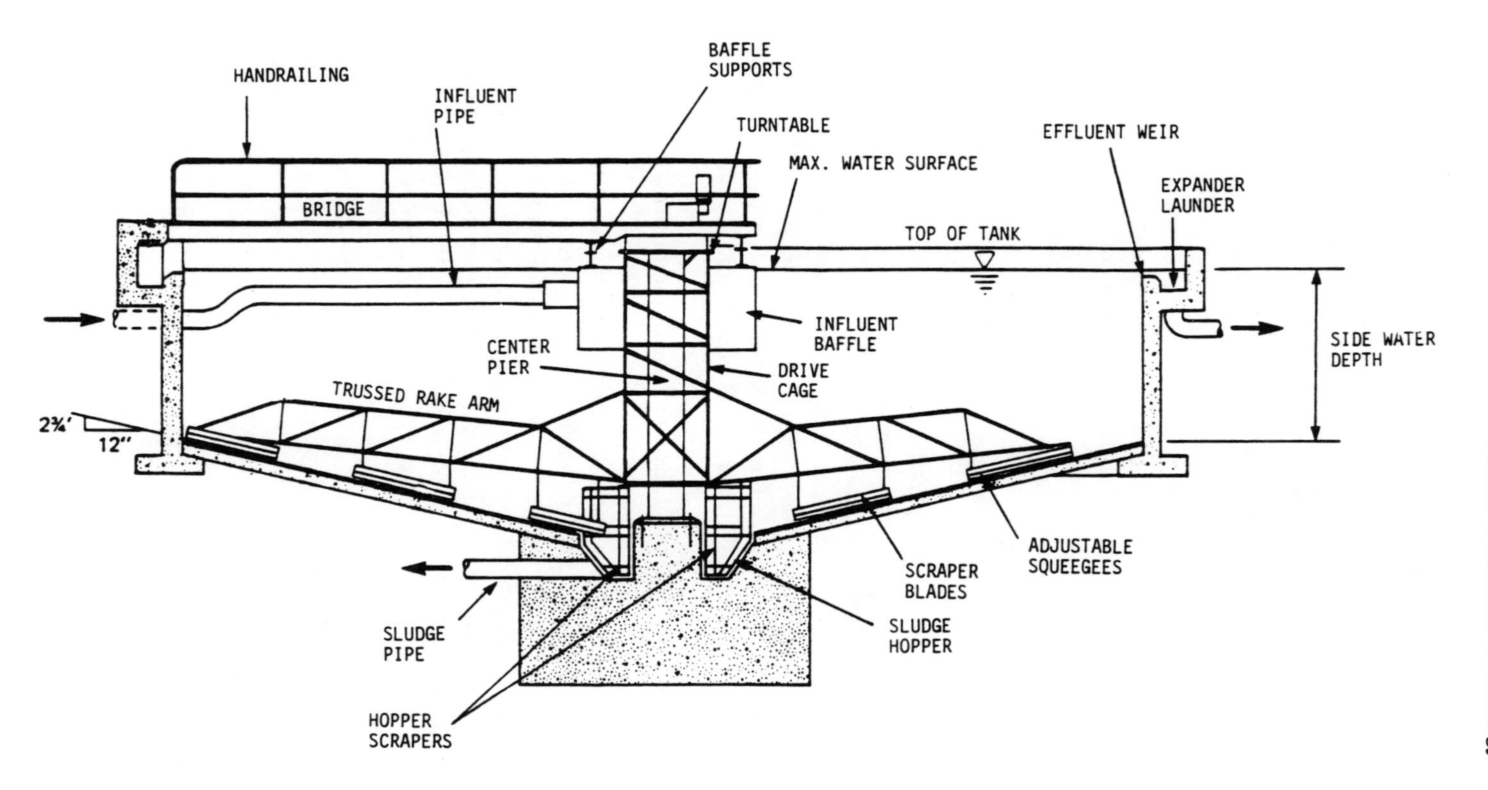

Figure 50: Typical gravity thickener.

Flotation Thickening

Description — Flotation is a process for separating solid particles from a liquid phase. Flotation of solids is usually created by the introduction of air into the system. Fine bubbles either adhere to, or are absorbed by the solids, which are then lifted to the surface.

In one flotation method, dissolved air flotation, small gas bubbles (50 to 100 μm) are generated as a result of the precipitation of a gas from a solution supersaturated with that gas. Supersaturation occurs when air is dispersed through the sludge in a closed, high pressure tank. When the sludge is removed from the tank and exposed to atmospheric pressure, the previously dissolved air leaves solution in the form of fine bubbles.

In a second method, dispersed air flotation, relatively large gas bubbles (500 to 1000 μm) are generated when gas is introduced through a revolving impeller or through porous media.

In biological flotation, the gases formed by natural biological activity are used to float solids.

In vacuum flotation, supersaturation occurs when the sludge is subjected initially at atmospheric pressure, to a vacuum of approximately nine inches of mercury in a closed tank.

Although all four methods have been used in industrial sludge residue management schemes, the dissolved air flotation process is the dominant method used in the United States. In this process, thickened sludge is removed from the surface of the tank and clear liquid is removed from the middle of the tank.

Figure 51 shows a typical dissolved air flotation thickener. Flotation thickener tanks can be either rectangular or circular in shape and constructed of either concrete or steel. Tanks are provided with equipment to provide uniform flow distribution at entrance, pressurize gas, skim off thickened sludge, and remove heavy nonfloatable solids. In designing a dissolved air flotation thickener, the following variables are typically considered: full, partial, or recycled pressurization systems; feed characteristics; surface area; net solids load; solids loading rate, hydraulic loading; air-to-solids ratio; polymer usage; type of pressurization equipment; and operating pressure.

Application — Dissolved air flotation thickeners are excellent for the biological sludges, especially those generated from the activated sludge process. Other difficult-to-thicken sludges, such as hydroxide sludges or fibrous sludges, have also been successfully thickened. Heavy coarse sludges, i.e. lime, will not float.

Advantages and Disadvantages — Dissolved air flotation thickening offers the following advantages:

- Provides better solids-liquid separation than a gravity thickener.
- Yields higher solids concentration for many sludges than a gravity thickener.

- Requires less land area than a gravity thickener.
- Has less chance of odor problems than a gravity thickener.

Dissolved air flotation thickening has the following disadvantages when compared to other thickening processes:

- Operating cost of dissolved air flotation thickener is higher than for a gravity thickener.
- Thickened sludge concentration is less than in a centrifuge.
- Requires more land than a centrifuge.
- Has very little sludge storage capacity.

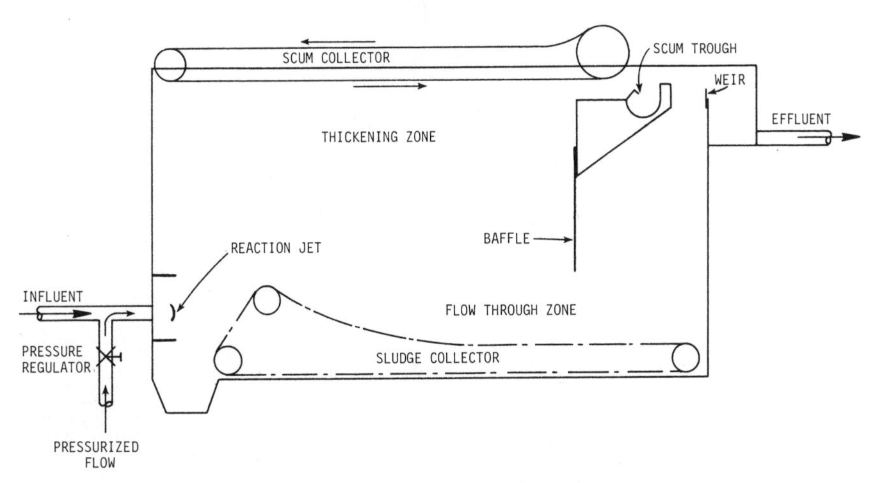

Figure 51: Typical flotation thickener.

Centrifugal Thickening: Centrifugal thickening is a process which separates and concentrates sludge by accelerating the solids/liquid mixture to several hundred or thousands of gravitational forces in a centrifuge. In thickening sludge residues, three different types of centrifuges are commonly used: disc nozzle, imperforate basket, and solid bowl decanter.

Disc Nozzle —

Description. Figure 52 features a cutaway view of a disc nozzle centrifuge. Feed normally enters through the top (bottom feed is also possible) and passes down through a feedwell in the center of the rotor. An impeller within the rotor accelerates and distributes the feed slurry, filling the rotor interior. Heavier solids settle outward toward the circumference of the rotor under increasingly greater centrifugal force. Liquid and the lighter solids flow inward through the cone-shaped disc stack. These lighter particles are settled out on the underside of the discs, where they agglomerate, slide down the discs, and migrate out to the nozzle region. The gap of 0.05 inches between the discs means that the particles have a short distance to travel before settling on the disc surface. Clarified liquid passes on through the disc stack into the overflow chamber and is then discharged through the effluent line.

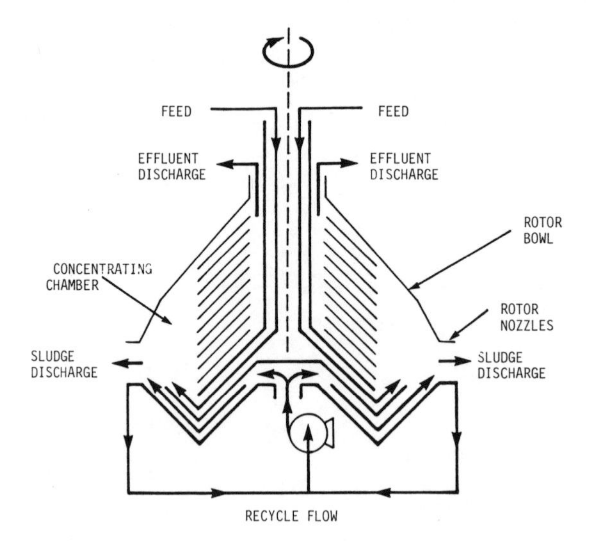

Figure 52: Schematic of a disc nozzle centrifuge.

Centrifugal action causes the solids to concentrate as they settle outward. At the outer rim of the rotor bowl, the high energy imparted to the fluid forces the concentrated material through the rotor nozzles. One part of this concentrated sludge is drawn off as the thickened product and another is recycled back to the base of the rotor and pumped back into the concentrating chamber. There, it is subjected to additional centrifugal force and is further concentrated before it is once again discharged through the nozzles. This recirculation is advantageous because it increases the overall underflow concentration; minimizes particle accumulation inside the rotor by flushing action; allows the use of larger nozzles, thus decreasing the potential for nozzle plugging; and helps to achieve a stable separation equilibrium that lends itself to precise adjustment and control.

Application. Disc nozzle centrifuges are normally applied to sludges having a majority of particles under 400 μm and void of fibrous materials. Even then, pretreatment systems, as shown in Figure 53 are recommended.

Sludge is pumped to a strainer in order to remove any large solids and fibrous materials. After screening, the flow may also go to a degritter. Under the velocities generated in a disc nozzle, any grit becomes abrasive and causes nozzle deterioration. The degritter does not

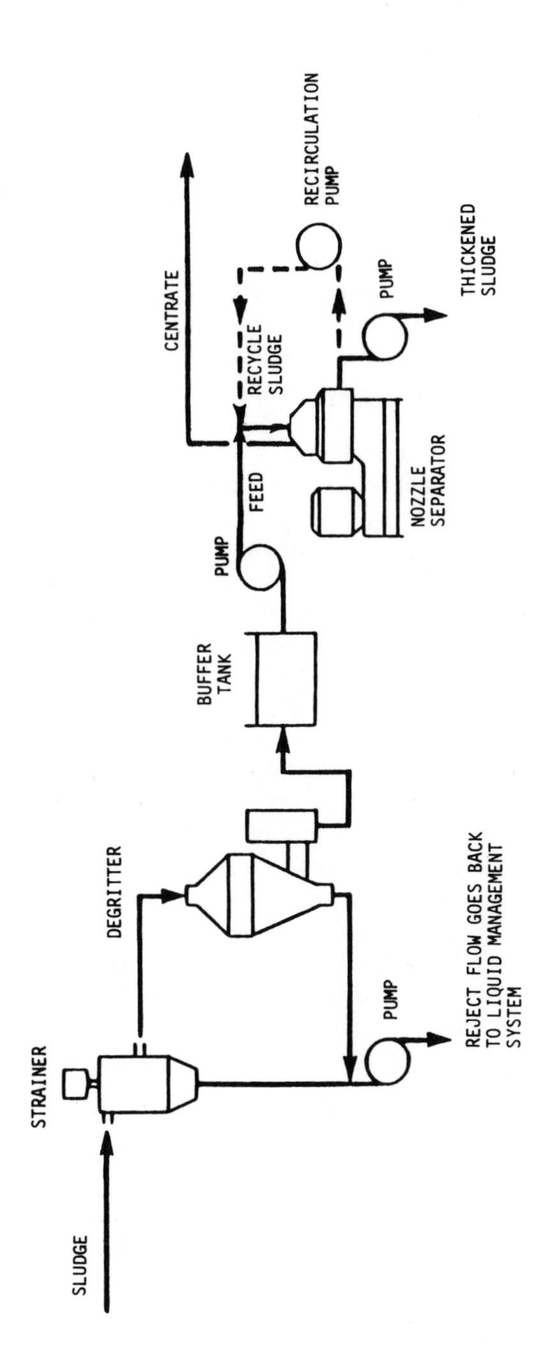

Figure 53: Typical disc nozzle with pretreatment system.

eliminate the problem completely, but it does increase the running time between nozzle replacements. Disc nozzles have been successfully applied to hydroxide and waste activated sludge.

Advantages and Disadvantages. Disc nozzle centrifuges for thickening offer the following advantages:

- Yields highly clarified centrate without the use of chemicals.
- Has large liquid and solids handling capacity in a very small space.
- Produces little or no odors.

Disadvantages of a disc nozzle system are as follows:

- Can only be used on sludges with particle sizes of 400 μm or less.
- May require extensive pretreatment system.
- May be a relatively high maintenance system.
- Requires skilled maintenance personnel.

Imperforate Basket —

Description. Figure 54 is a schematic of a top feed imperforate basket centrifuge illustrating general location of sludge inlet, polymer feed, and centrate piping and location of cake discharge.

The following describes one complete batch operating cycle of a basket centrifuge. When the "cycle start" button is pushed, the centrifuge begins to accelerate. After approximately 30 seconds, the feed pump is started through a timer relay. Depending on the feed pump rate, it will take one to three minutes for the bowl to reach operating speed. Sludge enters the unit through a stationary feed pipe mounted through the curb cap. This pipe extends to the bottom portion of the basket and ends at an angle just above the floor in order to impart a tangential velocity to the input stream. The duration of the feed time is controlled by either a pre-set timer or a centrate monitor that shuts the feed pump off when a certain level of suspended solids appears in the centrate. The centrate is normally returned to the liquid treatment system.

Deterioration in the centrate indicates that the centrifuge bowl is filled with solids, and separation can no longer take place. At this point, the sludge feed pump is turned off.

Turning off or diverting the feed pump decelerates the centrifuge. When the centrifuge has decelerated to 70 rpm, a plow (located by the center spindle shaft) is activated and starts to travel horizontally into the bowl where the solids have accumulated.

When the plow blade reaches the bowl wall, a dwell timer is activated to keep the plow in the same position for approximately 5 to 15 seconds until all the solids have been discharged. When the plow retracts, a cycle has been completed and the machine will automatically begin to accelerate, starting a new cycle.

Application. Imperforate basket centrifuges can be used on any type of sludge but is particularly applicable to soft or gelatinous-type sludges.

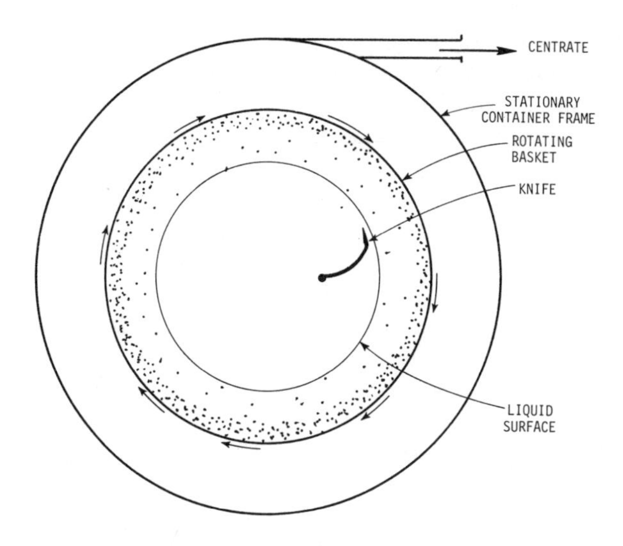

(a) Cross sectional top view

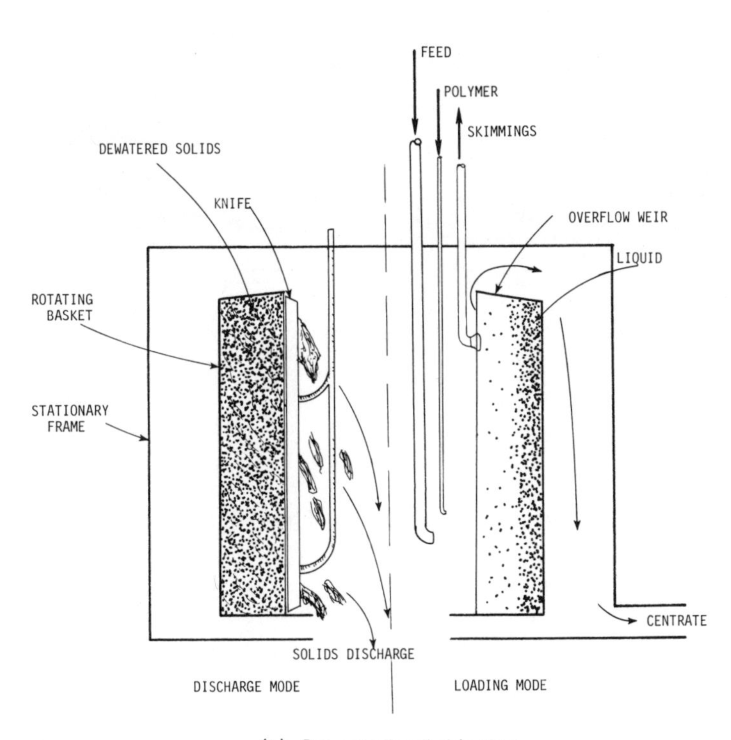

(b) Cross sectional side view

Figure 54: General schematic of imperforate basket centrifuge.

Advantages and Disadvantages. Imperforate basket centrifuges for thickening offer the following advantages:

- Is very flexible in meeting process requirements.
- Is not affected by coarse or gritty material.
- Of all centrifuges, has the lowest operation and maintenance requirements.
- Is excellent thickener for hard-to-handle sludges.
- Facility can be designed so that same machine can be used both for thickening and dewatering.

Disadvantages of an imperforate basket centrifuge are as follows:

- Unit is not continuous feed and discharge.
- Requires special structural support.
- Hydraulic capacity very limited.
- Has the highest ratio of capital cost to capacity.

Solid Bowl Centrifuge —

Description. Figure 55 is a schematic of a solid bowl decanter centrifuge. The sludge stream enters the bowl through a feed pipe mounted at one end of the centrifuge.

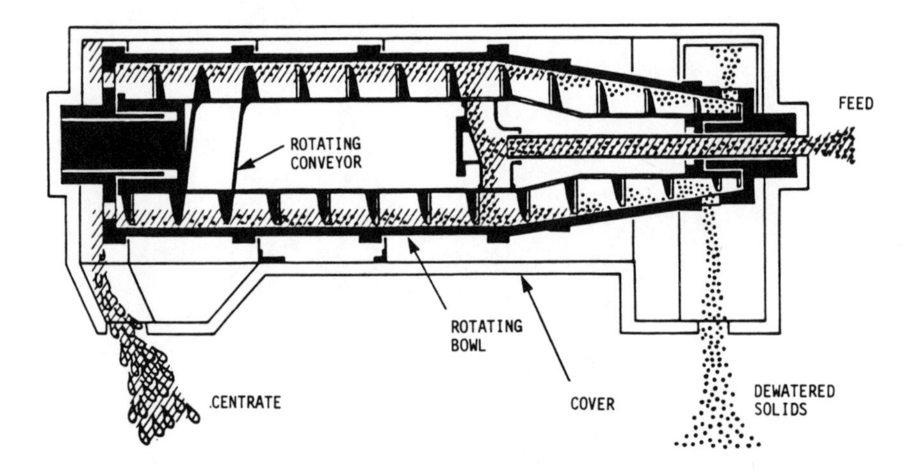

Figure 55: Schematic of a typical solid bowl decanter centrifuge.

As soon as the sludge particles are exposed to the gravitational field, they start to settle out on the inner surface of the rotating bowl. The lighter liquid, or centrate, pools above the sludge layer and flows towards the centrate outlet ports located at the large end of the machine.

The settled sludge on the inner surface of the rotating bowl is transported by the rotating conveyor towards the conical section (small end) of the bowl. In a decanter designed for dewatering, the sludge, having reached the conical section, is normally conveyed up an incline to the sludge outlet. Many soft sludges are too "slimy" to be conveyed upward without large doses of polyelectrolyte. In the newly designed machines, maximum pool depths are maintained; in addition, a specially designed baffle is located at the beginning of the conical section. This baffle, working in conjunction with the deep liquid pool, allows hydrostatic pressure to force the thickened sludge out of the machine independent of the rotating conveyor. This design eliminates the need for polymer addition to aid in conveying thickened sludge up the incline towards the sludge discharge and allows only the thickest cake at the bowl wall to be removed.

Application. Solid bowl decanters can be used on just about any type of sludge. Abrasive sludges are the exception, as they cause excessive wear on the machine.

Advantages and Disadvantages. Solid bowl decanter centrifuges for thickening offer the following advantages:

- Yields high throughput in a small area.
- Is easy to install.
- Is quiet.
- Has low capital cost for installation.
- Is a clean looking installation.

Disadvantages of a solid bowl decanter centrifuge are as follows:

- Is potentially a high maintenance item.
- May require polymers in order to operate successfully.
- Requires removal of gritty material in feed stream.
- Requires skilled maintenance personnel.

Stabilization

Stabilization is a process, biological or chemical, which will make a residue less odorous and putrescible and reduce the pathogenic organism content. In addition, some stabilization processes can also result in other basic changes in the residue.

The main design considerations in selecting a stabilization process are:

- Final disposal procedure planned for the residue.
- Concentration levels of various pollutants to be expected in the residual material to be stabilized.
- Amount of material that can be biologically or chemically oxidized.

This section will discuss five stabilization processes which can be applied to industrial residues: anaerobic digestion; aerobic digestion; composting; lime stabilization; and chlorine oxidation.

Anaerobic Digestion:

Description — Anaerobic digestion is the biological degradation of complex organic substances in the absence of free oxygen. During these reactions, energy is released and much of the organic matter is converted to methane, carbon dioxide, and water. Since little carbon and energy remain available to sustain further biological activity, the remaining solids are rendered stable.

Over the years, several basic process variations have been developed: low rate; high rate (single and two-stage); anaerobic contact; and phase separation.

Low-rate Digestion. Figure 56 shows the basic features of this simplest and oldest type of process variation. Essentially, a low-rate digester is a large storage tank. With the possible exception of heating, no attempt is made to accelerate the process by controlling the environment. Raw sludge is fed into the tank intermittently. Bubbles of sludge gas are generated soon after sludge is fed to the digester, and their rise to the surface provides the only mixing. As a result, contents of the tank stratify, forming three distinct zones: a floating layer of scum; a middle level of supernatant; and a lower zone of sludge. Essentially, all decomposition is restricted to the lower zone. Stabilized sludge, which accumulates and thickens at the bottom of the tank, is periodically drawn off from the center of the floor. Supernatant is removed from the side of the tank and recycled back to the liquid treatment system. Sludge gas collects above the liquid surface and is drawn off through the cover.

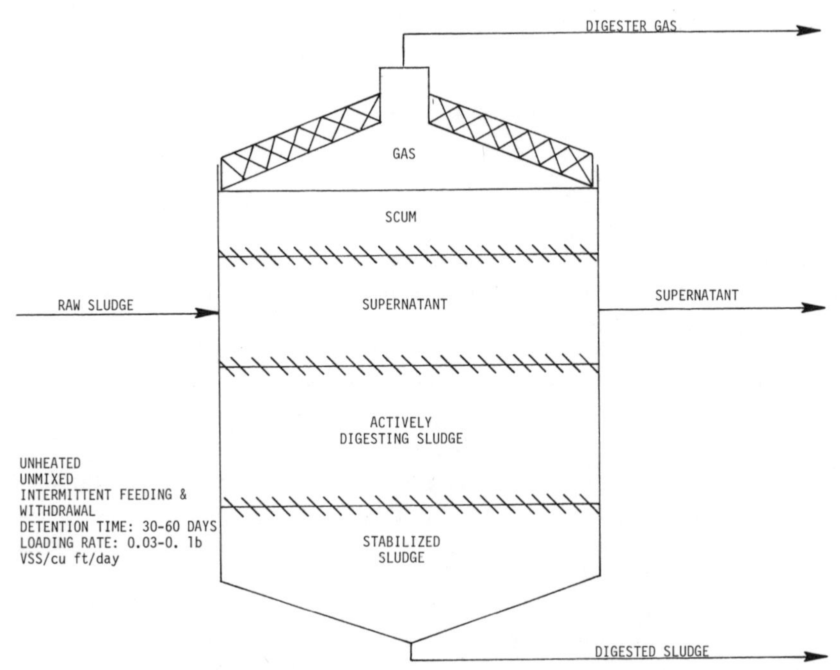

Figure 56: Low-rate anaerobic digestion system.

High-rate Digestion. Early research into ways to accelerate the anaerobic digestion process led to the development of the high-rate process variation. In this variation, thickening of the raw sludge, increasing the temperature of the liquid within the tank, auxiliary mixing of the tank contents, and uniform feeding of the raw sludge act together to create a steady and uniform environment, the best conditions for the biological process. The net result is that volume requirements are reduced and process stability is enhanced.

Figure 57 shows the basic layout of a single-stage, high-rate system. In some applications, another tank has been added to allow gravity separation of digested solids and decanting of supernatant liquor. This is called a two-stage system and is shown in Figure 58.

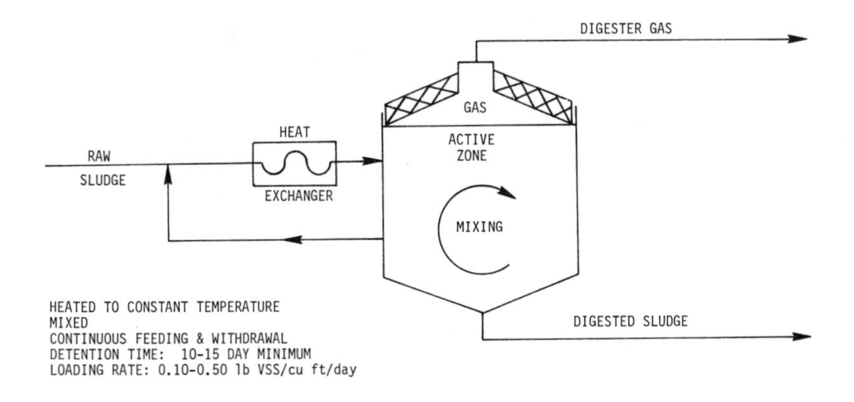

Figure 57: Single-stage, high-rate anaerobic digestion system.

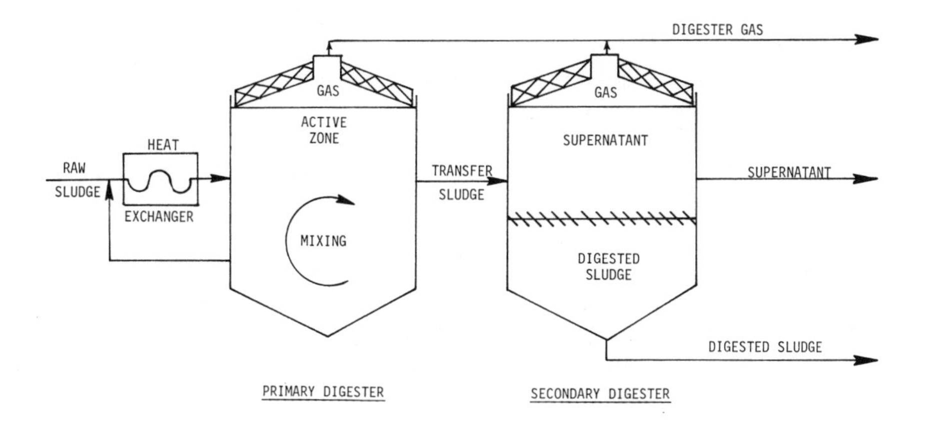

Figure 58: Two-stage, high-rate anaerobic digestion system.

Anaerobic Contact. In an effort to increase process stability of the high-rate system at low hydraulic residence times, the anaerobic contact process was developed. As shown in Figure 59, in this process variation, a portion of the active biomass leaving the digester is concentrated and then mixed with the raw sludge feed. This recycling allows for adequate cell retention to meet kinetic requirements while operating at a significantly reduced hydraulic detention time.

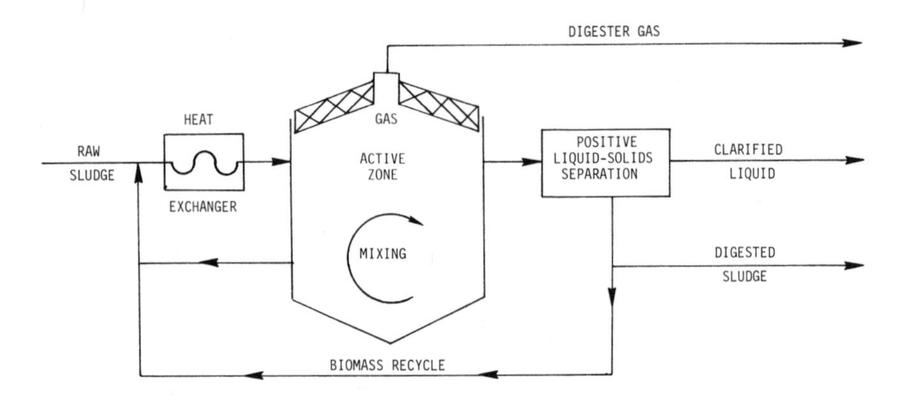

Figure 59: Anaerobic contact process.

Positive solids-liquid separation is essential to the operation of the anaerobic contact process. To gain any of the benefits from recycling, the return stream must be more concentrated than the contents of the digester.

Phase Separation. Anaerobic digestion involves two general phases: acid formation and methane production. In the three previous process variations, both phases take place in a single reactor. In the phase separation process, these two phases are physically separated. An effective means of separating the two phases is essential to the operation of this process variation. Possible separation techniques include dialysis, addition of chemical inhibitors, adjustment of the redox potential, and kinetic control by regulating the detention time and recycle ratio for each reactor. At this time, the latter seems the most practical, and a schematic of such a control method is shown in Figure 60.

Application — Anaerobic digestion is a process which has been extensively utilized in stabilizing municipal wastewater residuals but has only limited application on industrial residues — primarily residuals generated by food related industries. Because methane is a byproduct of anaerobic digestion, industries which generate significant amounts of organic residuals (gum and wood chemicals, petroleum refining, pharmaceuticals, pulp and paper, etc.) are now evaluating process applications, not only to stabilize and reduce the organic content of the waste, but to supply gas for operating internal plant operations.

Advantages and Disadvantages — Anaerobic digestion offers the following advantages over other stabilization processes:

- Produces methane, a usable source of energy.
- Reduces total residual mass through the conversion of organic matter to primarily methane, carbon dioxide, and water.
- Inactivates pathogens
- Yields a solids residue which may be suitable for use as a soil conditioner.

Anaerobic digestion has the following disadvantages when compared to other stabilization processes:

- Is susceptible to upsets.
- Has a high capital cost.
- Produces a poor quality sidestream.

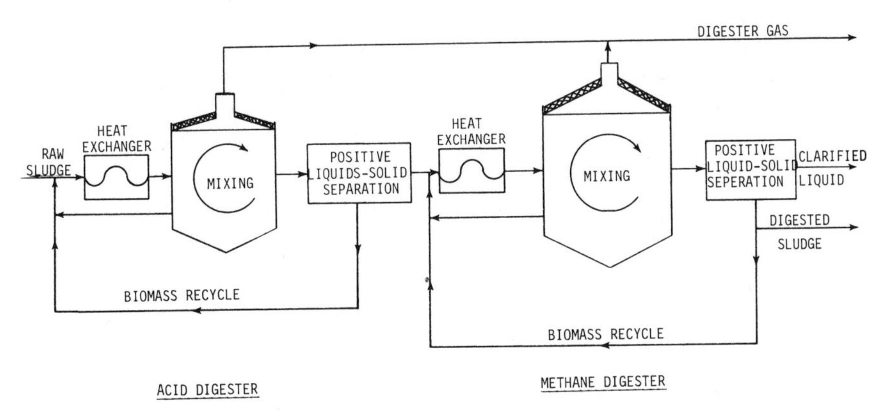

Figure 60: Two-phase anaerobic digestion process.

Aerobic Digestion

Description — Aerobic digestion is the biochemical oxidative stabilization of industrial wastewater sludges in open or closed tanks that are separate from the liquid process system. There are currently three process variations which can be used: conventional semi-batch; conventional continuous; and auto-heated mode of operation.

Conventional semi-batch operation. Originally, aerobic digestion was designed as a semi-batch process, and this concept is still a functional design alternative. Solids are pumped directly from the clarifiers into the aerobic digester. The time required for filling the digester depends on available tank volume, volume of waste sludge, precipitation, and evaporation. During the filling operation, sludge undergoing digestion is continually aerated. When the tank is full, aeration continues for two to three weeks to assure that the solids are thoroughly stabilized. Aeration is then discontinued and the stabilized solids settled. Clarified liquid is decanted, and the thickened solids are removed. When a sufficient amount of stabilized sludge and/or supernatant have been removed, the cycle is repeated. Between cycles, it is customary to leave some stabilized sludge in the aerator to provide the necessary microbial population for degrading the wastewater solids. The aeration device need not operate for several days, provided no raw sludge is added.

Many engineers have tried to make the semi-batch process more continuous by installing stilling wells to act as clarifiers in part of the digester. This has not proven effective.

Conventional Continuous Operation. Figure 61 shows the process flow diagram for this process variation which closely resembles the activated sludge process. As in the semi-batch process, solids are pumped directly from clarifiers into the aerobic digester. The aerator operates at a fixed level, with the overflow going to a solids-liquid separator. Thickened and stabilized solids are either recycled back to the digestion tank or removed for further processing.

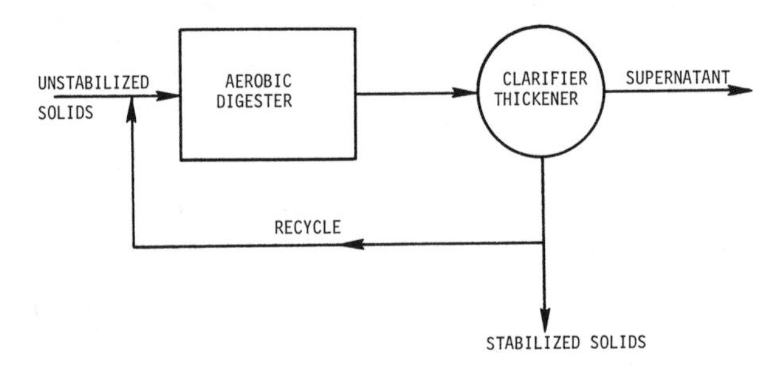

Figure 61: Process flow diagram for a conventional continuously operated aerobic digester.

Auto-heated Mode of Operation. A new concept in aerobic digestion is the auto-heated thermophilic aerobic digestion process. In this process, sludge from the clarifiers is usually thickened to provide a digester feed solids concentration of greater than four percent. Heat liberated in the biological degradation of the organic solids is sufficient to raise the liquid temperature in the digester to as high as 140°F. Advantages claimed for this mode of operation are: higher rates of organic solids destruction, hence smaller volume requirements; production of a pasteurized sludge; destruction of all weed seeds; 30 to 40 percent less oxygen requirement than for the mesophilic process, since few, if any, nitrifying bacteria exist in this temperature range; and improved solids-liquid separation through decreased liquid viscosity.

Disadvantages cited for this process are that it must incorporate a thickening operation, that mixing requirements are higher because of the higher solids content, and that nonoxygen aerated systems require extremely efficient aeration and insulated tanks.

Application — When residual stabilization has been required in the industrial sector (for organic sludges produced from biological treatment of industrial wastewaters), aerobic digestion has been the predominant process selected. This is primarily due to its low capital cost.

Advantages and Disadvantages — Aerobic digestion offers the following advantages over other stabilization processes:

- Has low capital cost requirements.
- Is relatively easy to operate.
- Will produce a supernatant low in BOD_5, suspended solids, and ammonia nitrogen.

Aerobic digestion has the following disadvantages when compared to other stabilization processes:

- Has high operating costs due to power costs to supply oxygen.
- Usually produces a digested sludge with very poor mechanically dewatering characteristics.
- Is significantly influenced in performance by temperature, location, and type of tank material.

Composting:

Description — Composting is the aerobic thermophilic decomposition of organic constituents to a relatively stable humus-like material. Environmental factors influence the activities of the bacteria, fungi, and actinomycetes in this oxidation decomposition process and affects the speed and course of composting cycles. Volatility and type of material, moisture content, oxygen concentration, carbon/nitrogen ratio, temperature, and pH are key determinants in the process. The composting process is considered complete when the product can be stored without giving rise to nuisances such as odors, and when pathogenic organisms have been reduced to a level such that the material can be handled with minimum risk.

Composting processes can be classified as either unconfined or confined systems. The unconfined systems can be further subdivided into windrow and aerated static pile processes.

Unconfined Process – Windrow Systems. The windrow process relies on natural ventilation with frequent mechanical mixing of the piles to maintain aerobic conditions. In areas of significant rainfall, it may be desirable for operational reasons to provide a roofed structure to cover the windrows for composting.

In the windrow composting process, the mixture to be composted is stacked in long parallel rows or windrows. The cross section of the windrows may be trapezoidal or triangular, depending largely on the characteristics of the mobile equipment used for mixing and turning the piles. The width of a typical windrow is 15 feet and the height is 3 to 7 feet.

Bulking agents such as wood chips, saw dust, straw, rice hulls, or composted material, is often added to increase the structural integrity of the mixture and thus, its ability to maintain a properly shaped windrow. Porosity of the mixed material is greatly improved, which in turn improves the aeration characteristics. External bulking agents can also provide a source of carbon for the composting process.

Convective air movement within windrows is essential for providing oxygen for the microorganisms. The aerobic reaction provides heat for warming the windrows. This causes the air to rise, producing a natural chimney effect.

The rate of air exchange can be regulated by controlling the porosity and size of the windrow. Turning of the windrow also introduces oxygen to the microorganisms. This method of aeration can be expensive if used excessively to obtain high oxygen concentrations and may reduce the temperature within the windrow.

As a result of the biological decay process, temperatures in the central portion of the windrow reach as high as 150°F. Operating temperatures of about 140°F may be maintained in the central portion of the windrow for as long as ten days. Temperatures in the outer layers are considerably cooler and may approach atmospheric conditions. During wet periods and winter conditions, maximum temperatures may only be 130°F to 140°F. A high temperature maintained throughout the pile for a sufficient period of time is important to the control of pathogens. A satisfactory degree of stabilization is indicated by a decline in temperature, usually to about 113°F to 122°F.

Unconfined Process – Aerated Static Pile. The aerated static pile system was developed in order to eliminate many of the land requirements and other problems associated with the windrow composting process and to allow composting of raw sludge. This system consists of the following steps: mixing of sludge with a bulking agent; construction of the composting pile; composting; screening of the composted mixture; curing; and storage. A diagram of an aerated pile for composting sludge is shown in Figure 62.

The forced air method provides for more flexible operation and more precise control of oxygen and temperature conditions in the pile than would be obtained with a windrow system. Since composting times tend to be slightly shorter and anaerobic conditions can be more readily prevented, the risk of odors is reduced.

Two distinct aerated static pile methods have been developed, the individual aerated pile and the extended aerated pile.

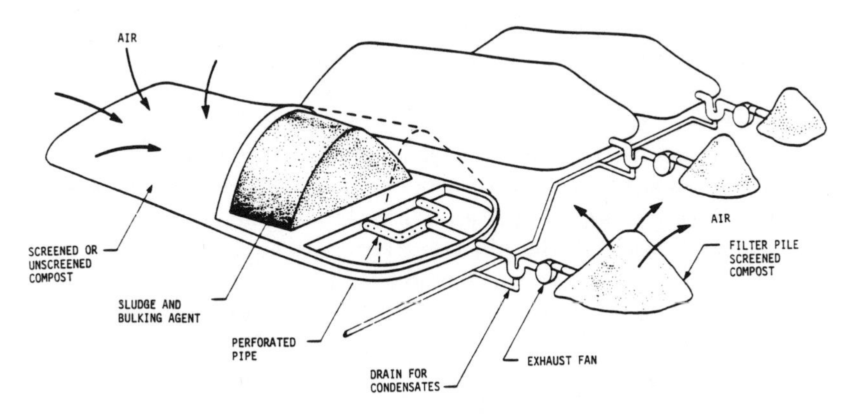

Figure 62: Configuration of individual aerated piles.

In the individual aerated pile method, a loop of perforated plastic pipe, 4 to 6 inches in diameter is placed on the composting pad, oriented longitudinally, and centered under the ridge of the pile under construction. In order to avoid short circuiting of air, the perforated pipe is terminated at least 8 to 10 feet inside the ends of the pile. A nonperforated pipe that extends beyond the pile base is used to connect the loop of perforated pipe to the blower. Figure 63 shows a diagram for an individual aerated pile.

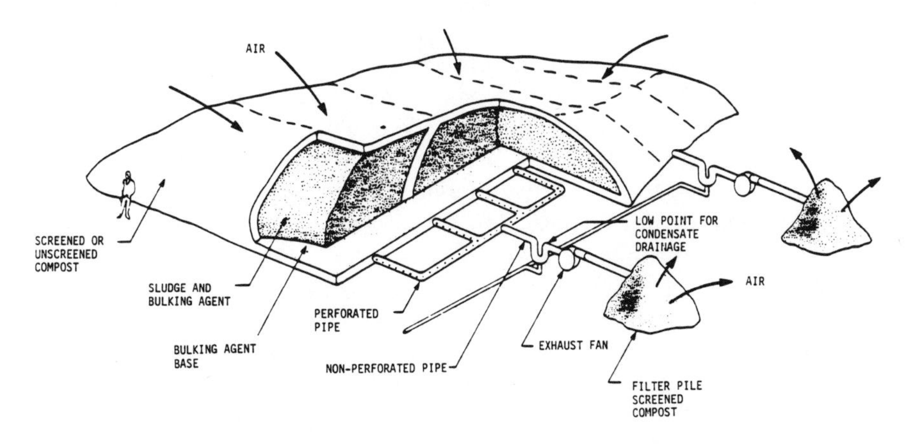

Figure 63: Aeration pipe set-up for individual aerated piles.

A 6 to 8 inch layer of bulking agent is placed over both the pipes and the area to be covered by the pile. This base facilitates the movement and even the distribution of air during composting and absorbs excessive moisture that may otherwise condense and drain from the pile.

A mixer or front end loader is used to mix the sludge residue and bulking agent. The resulting mixture is placed loosely upon the prepared base to form a pile with a triangular cross section.

The pile is then completely covered with a 12-inch layer of cured, screened compost or an 18-inch layer of unscreened compost. This outer blanket of compost provides insulation and prevents escape of odors during composting. Unstabilized sludge can generate odors during dumping and initial pile construction. The nonperforated pipe is connected to a blower that is controlled by a timer. Aerobic composting conditions are maintained if air is intermittently drawn through the pile. If the aeration rate is too high or the blower remains on too long, the pile will cool, and the thermophilic process will be inhibited.

The effluent air from the compost pile is conducted into a small, cone-shaped filter pile of cured, screened compost where malodorous gases are absorbed. The odor retention capacity of these piles is inhibited if their moisture content is greater than 50 percent. The odor filter pile should contain one cubic yard of screened compost for each four dry tons of sludge in the compost pile. Filter piles are sometimes constructed with a 4-inch base layer of wood chips to prevent high back pressures on the blower.

To make more effective use of available space, another static pile configuration called the extended aerated pile has been developed. An initial pile is constructed with a triangular cross section utilizing one day's sludge production. Only one side and the ends of this pile are blanketed with cured, screened compost. The remaining side is dusted with only about an inch of compost for overnight odor control. The next day, additional aeration pipe is placed on the pad parallel to the dusted side of the initial pile. The pile bed is extended by covering the additional pipe with more bulking agent and sludge-bulking agent mixture so as to form a continuous or extended pile. This process is repeated daily for 28 days. The first section is removed after 21 days. After seven sections are removed in sequence, there is sufficient space for operating the equipment so that a new extended pile can be started. Figure 64 shows such a system. The area requirement of an extended pile system is about 50 percent less than that for individual piles. The amount of recycled bulking agent required for covering the pile and bulking agent used in the construction of the base is also reduced by about 50 percent.

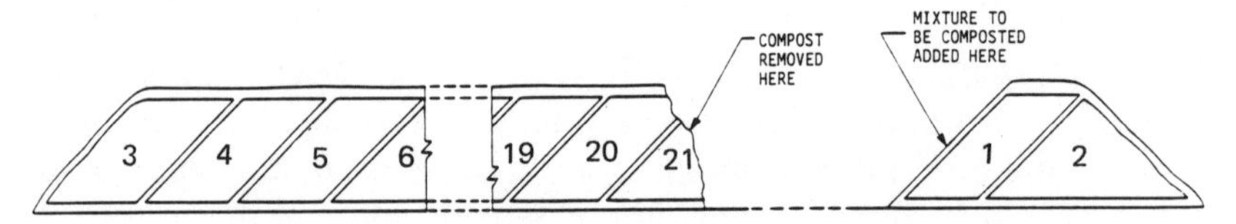

Figure 64: Configuration of extended aerated pile.

Confined Processes. In these processes composting is accomplished inside an enclosed container or basin. These processes are often called "mechanical" composting because mechanical action is generally used to transport or mix the compost. Mechanical systems are designed to minimize odors and process time by controlling environmental conditions such as air flow, temperature, and oxygen requirements.

The primary differences among mechanical composting systems are in the methods of process control. Some provide aeration by tumbling or dropping the material from one floor to the next. Others use devices which stir the composting mass. Tumbling the compost in a rotating cylinder is another approach. In addition, an endless belt is used to combine forced bottom aeration and stirring. Water is added to the composting mass at critical times to increase biological activity in some mechanical systems. Also, some mechanical composters can introduce heat to the composting mass to keep the composting reaction continuing at the optimum rate during cool weather.

The brief detention times which equipment manufacturers specify for mechanical composters do not allow adequate stabilization of the sludge. If shorter detention times are provided, a two- to three-month maturation period will be necessary to reduce the remaining volatile matter. Thus, the amount of time and total area required for mechanical processes approach that for unconfined processes. Mechanical processes are more capital-intensive than unconfined processes. Currently, only a few mechanical composting processes are operating in the United States and these are generally used to compost a mixture of refuse and wastewater sludge. A schematic of a typical confined composting process is shown on Figure 65.

Application — Data on composting of industrial residues generated from the categorical industries are very limited. Successful applications of composting to residuals generated from pulp and paper, pharmaceutical, petroleum refining, and timber products have been demonstrated. Research also indicates that some pesticides and TNT residuals can also be composted.

Advantages and Disadvantages — Composting offers the following advantages over other stabilization processes:

- Produces a material which can significantly improve the physical properties of many soils.
- Requires a low capital investment.
- Increases the attractiveness of the material for reuse.

Composting has the following disadvantages when compared to other stabilization processes:

- Process is labor intensive.
- Process has a high operating cost per unit of material processed.
- Poor management will create odor and dust problems.

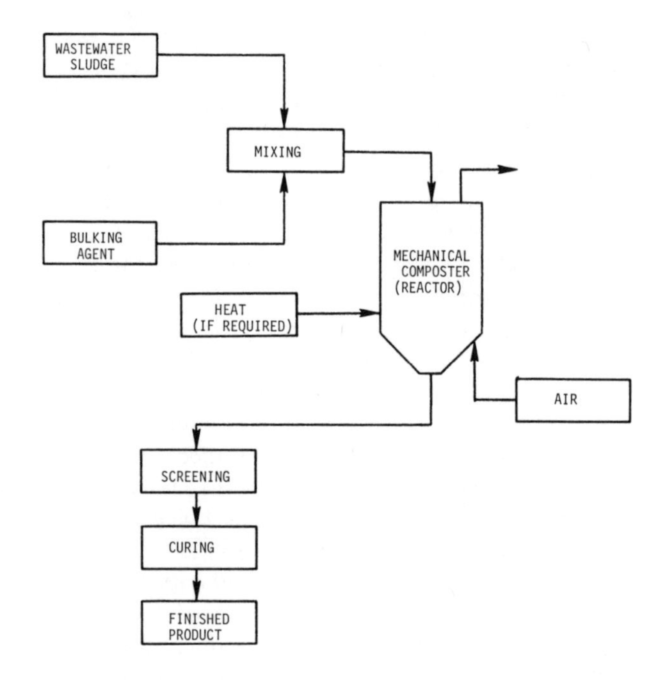

Figure 65: Typical process flow schematic confined composting system.

Lime Stabilization:

Description — Sufficient lime is added to the industrial residue to maintain the mixture at a pH of 12 for several hours. Unlike the previous three stabilization processes, no direct reduction of organic matter occurs during the process. In addition, lime does not make sludge chemically stable; if the pH drops below 11, biological decomposition will resume, producing noxious odors.

Application — No known application of lime stabilization to residuals generated from the categorical industries could be found. This stabilization process could be used for any of the industries who generate organic sludges and will be using land utilization for final disposal.

Advantages and Disadvantages — Lime stabilization offers the following advantages over other stabilization processes:

- Very simple process to operate.
- Very low capital cost.
- Very easy to dewater residual on mechanical dewatering equipment.

Lime stabilization has the following disadvantages when compared to other stabilization processes:

- Total mass of residual to be disposed of is significantly increased.
- Transport and ultimate disposal costs are high.

Chlorine Stabilization:

Description — Stabilization by chlorine addition is a proprietary process and is marketed under the registered trademark "Purifax." The process can be used to reduce putrescibility and pathogen concentration of organic residuals and also possibly improve dewaterability of the residual.

Chlorine treatment stabilizes sludge by both reducing the number of organisms available to create unpleasant or malodorous conditions and making organic substrates less suitable for bacterial metabolism and growth. Some of the mechanisms responsible are oxidation, addition of chlorine to unsaturated compounds, and displacement of hydrogen by chlorine.

In the chlorine stabilization process, sufficient acid is produced to reduce the pH of the sludge to a range of 2 to 3. Dissociation of HOCl to H^+ and OCl^- is suppressed by low pH and therefore is not significant. Cl_2 and HOCl are highly reactive and powerful bactericides and viricides. The chloride ion has no disinfection capability.

The process stream immediately following the chlorine addition is substantially a chlorine solution containing sludge. The solution contains (in molecular form) as much as ten percent of the total chlorine species present. The predominant species in solution is undissociated HOCl. HOCl and Cl_2 react with sludge to oxidize ammonia to chloramines and organic nitrogen to organic chloramines. Other reduced ions, such as Fe^{2+} and S^{2-}, are oxidized at the same time. Some of the oxidized end products, such as chloramines and organic chloramines, are germicidal and viricidal.

The chlorine stabilization unit consists of a disintegrator, a recirculation pump, two reaction tanks, a chlorine eductor, and a pressure control pump. A chlorine evaporator and/or a chlorinator, feed pump, and inlet flow meter can be purchased with the unit or separately. The unit is often supplied by the manufacturer as a complete package mounted on a skid plate and ready for installation. A detailed diagram of the unit is shown on Figure 66.

In the first operating step, sludge is pumped through a disintegrator which reduces particle size and therefore provides greater sludge surface area for contact with the chlorine. Chlorinated sludge from the first reactor is mixed with raw sludge just prior to reaching the recirculation pump. The combined flow then passes through the first reaction tank. Chlorine is added via an eductor located in the recirculating loop. Recirculation aids mixing and efficient chlorine use.

The ratio of recirculated reacted product to raw sludge at design capacity is about 7 to 1. System pressure is maintained in the 30 to 35 psi range, by a pressure control pump located at the discharge of the second reactor. The pressure provides a driving force to ensure penetration of chlorine into the sludge particles. The second reactor tank increases system detention time, allowing a more complete reaction between the sludge and the chlorine.

Flow patterns within the two reactor tanks are high, in the form of velocity spirals, with tangential discharges. The tanks are oriented with the spiral axis of the first in a horizontal plane and the second in a vertical plane. Solids that settle during periods of nonoperation are easily resuspended when the process is started again. The system is neither drained nor cleaned between operating periods. A holding tank should be provided for feed storage and for flow equalization. Blending done in the tank also helps to maintain feed uniformity, thus providing sludge of uniform chlorine demand and minimizing the need to frequently adjust chlorine dose. Sludge blending is particularly valuable for processing of primary sludges, which tend to be more concentrated when initially pumped from the sedimentation tank than at the end of the pumping cycle. Similarly, where primary and secondary sludges are treated together, blending can be accomplished in the holding tank. Continuously wasted activated sludge, however, may be adequately treated without prior blending, provided that solids concentration is nearly constant with time. Mixing is usually done by mechanical or air agitation. Air mixing is preferable, because it enhances aerobic conditions, reduces odors, and averts problems with fouling of the impellers by rags and strings. Odor can be controlled in the holding tank if a portion of the filtrate or supernatant from the dewatering process is returned to it.

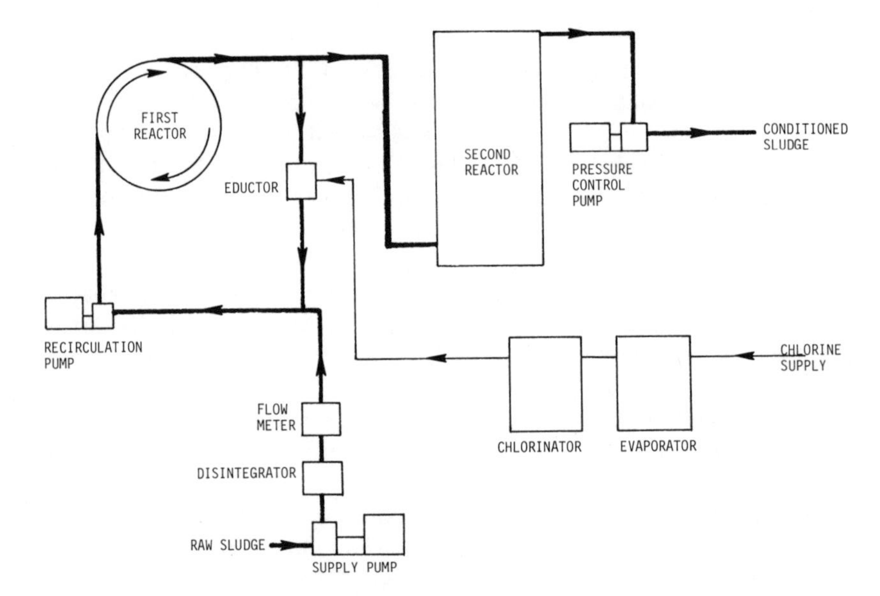

Figure 66: Schematic diagram of a chlorine oxidation system.

If the chlorine demand of the liquid fraction of the sludge is high, separation of some of the liquid from solids by thickening prior to chlorination may substantially reduce total sludge chlorine demand. If, however, chlorine demand is low, thickening will not be beneficial. Solids concentrations above certain defined limits should not be exceeded, because the diffusion rate of chlorine through the sludge is hindered and processing rates must be reduced to provide additional time for the chlorine to reach reaction sites.

Use of a holding tank downstream of the chlorine oxidation process allows subsequent processes to run independently and at their own best rate. Solids settling may occur in the tank after an initial period of flotation. The tank can, therefore, be used to separate the solid and liquid fractions of the stabilized product.

Application — No known application of chlorine stabilization to residuals generated from the categorical industries could be found. This stabilization process could be used for any of the industries which generate organic sludges.

Advantages and Disadvantages — Chlorine stabilization offers the following advantages over other stabilization processes:

- Process can be operated intermittently for only several hours per day. Therefore, operating costs are directly dependent upon production rates, and costs attributable to overcapacity are eliminated.
- It is a chemical process and is operationally insensitive to factors such as toxic materials in the feed stream.
- Process can treat a feed stream of widely varying character.

Chlorine stabilization has the following disadvantages when compared to other stabilization processes:

- Low pH of chlorine-stabilized sludge may require the sludge to be partially neutralized prior to further processing.
- Process does not reduce sludge mass.
- Process consumes relatively large amounts of chlorine, which is not only a significant cost, but special safety and handling precautions must be used.
- It could result in increased levels of toxic chlorinated organics in the treated materials.

Disinfection

Disinfection of industrial residuals, the destruction or inactivation of pathogenic organisms in the residuals, is carried out principally to minimize public health concerns. Destruction is the physical disruption or disintegration of a pathogenic organism, while inactivation, as used here, is the removal of a pathogen's ability to infect. An important but secondary concern may be to minimize the exposure of domestic animals to pathogens in the sludge. At the present time in the United States, use of procedures to reduce the number of pathogenic organisms is a requirement before sale of sludge or sludge-containing products to the public as a soil amendment, or before recycling sludge directly to croplands, forests, or parks. Since final use or disposal of sludge may differ greatly with respect to public health concerns, and since a great number of treatment options effecting various degrees of pathogen reduction are available, the system chosen for reduction of pathogens should be tailored to the demands of the particular situation.

Although there are many residual management technologies which reduce pathogens as part of the process (for example the stabilization processes), there are two types of processes designed specifically for the reduction of pathogenic organisms in sludge: pasteurization and high energy irradiation.

Pasteurization:

Description — The critical requirement for pasteurization is that all sludge be held above a predetermined temperature for a minimum time period. Heat transfer can be accomplished by steam injection or with external or internal heat exchangers. Steam injection is preferred because heat transfer through the sludge slurry is slow and undependable. Incomplete mixing will either increase heating time, reduce process effectiveness, or both. Overheating or extra detention are not desirable, however, because trace metal mobilization may be increased, odor problems will be exacerbated, and unneeded energy will be expended. Batch processing is preferable to avoid reinoculations if short circuiting occurs.

Figure 67 shows the flow scheme for a sludge pasteurization system utilizing a one-stage heat recuperation system to conserve energy. Principal system components include steam boiler, preheater, sludge heater, high-temperature holding tank, blow-off tanks, and storage basins for untreated and treated sludge. Sludge for pasteurization enters the preheater where the temperature is raised from 64°F to 100°F by vapors from the blow-off tank; 30 to 40 percent of the total required heat is thus provided by recovery. Next, direct steam injection raises the temperature to 157°F in the pasteurizer where the sludge resides for at least 30 minutes. Finally, the sludge is transferred to the blow-off tanks, where it is cooled first to 113°F at 1.45 pounds per square inch, and then to 98°F at 0.73 pounds per square inch.

Application — No known application of pasteurization to residuals generated from categorical industries could be found. This process could be used for any of the industries who generate organic sludges.

Advantages and Disadvantages — Pasteurization offers the following advantages over other sludge residual disinfection processes:

- It is a proven technology — in European countries — requiring basic skills in boiler operation and high temperature and pressure processes.

- Minimal pretreatment of the sludge is required.

Pasteurization has the following disadvantages when compared to other sludge residual disinfection processes:

- It consumes significant amounts of energy.
- Odor problems could develop.
- High operating cost is involved.

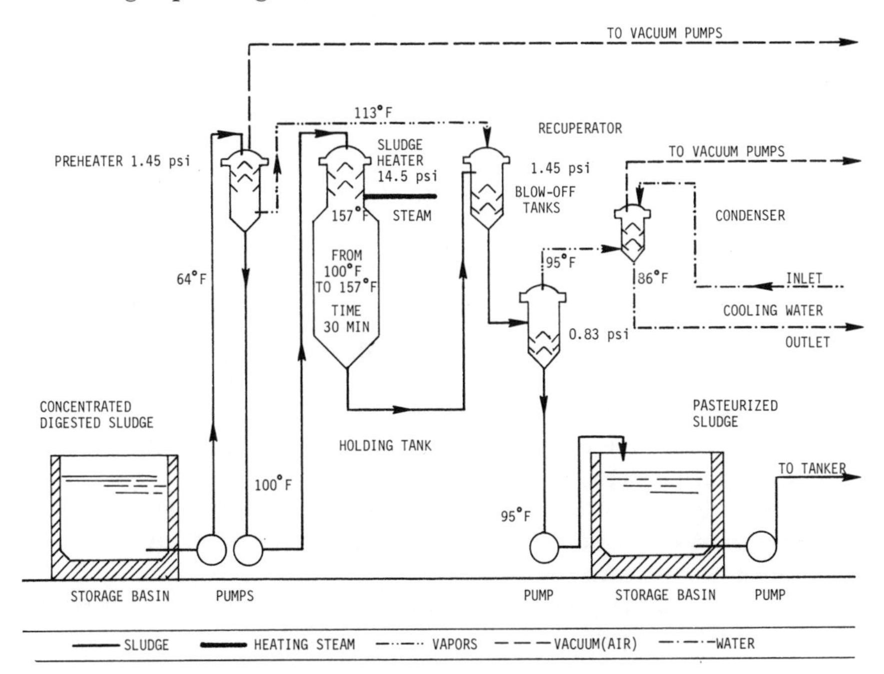

Figure 67: Flow scheme for sludge pasteurization with single stage heat recuperation.

High Energy Irradiation:

Description — Sludge disinfection with high energy electron beam has been considered for several decades but only recently has development work been conducted. A typical system would include a sludge screener, sludge grinder (optional), sludge feed pump, sludge spreader, electron beam power supply, electron accelerator, electron beam scanner, and sludge removal pump. A concrete vault would house the electron beam providing shielding for the workers from stray irradiation. In actual operation, the electron beams are first accelerated then leave the accelerator in a continuous beam that is scanned across the sludge several hundred times per second as the sludge flows by in a thin film. Figure 68 shows the general layout.

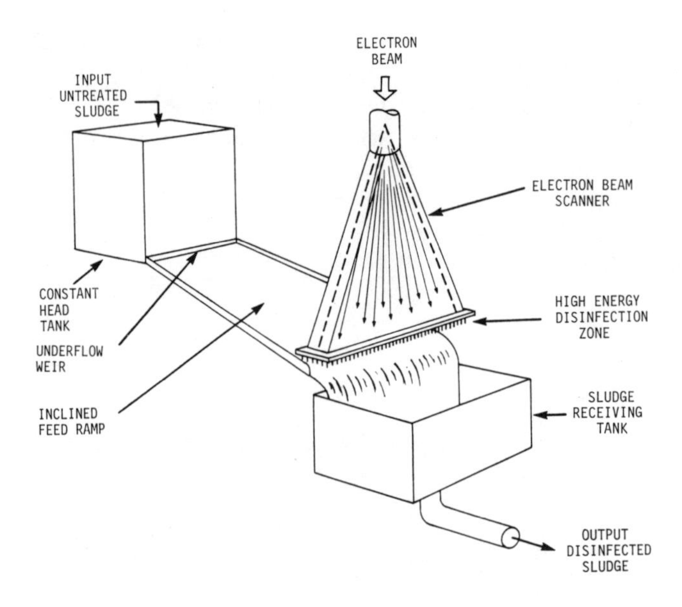

Figure 68: General electron beam disinfection operation.

Application — No commercial applications are in existence at this time — only research installations. This process may be of benefit to any industry which generates organic sludges.

Advantages and Disadvantages — The basic advantage high energy electron irradiation offers is that total energy requirements are only five to ten percent of pasteurization systems.

At this time the primary disadvantage is the lack of any operating facilities of commercial use to allow design evaluation.

Conditioning

Conditioning involves the chemical and/or physical treatment of a sludge stream to enhance water removal. In addition, some conditioning processes also provide disinfection, affect odor generation, alter the residue physically, provide limited solids destruction or addition, and improve residual capture.

Conditioning always has an effect on the efficiency of the thickening or dewatering process that follows. Any evaluation of the conditioning process must therefore take into consideration capital, operating, and maintenance costs for the entire pretreatment system and the impact of sidestreams on other pretreatment system processes, pretreatment system effluent, and resultant air quality. This type of analysis is necessary because conditioning processes differ and, therefore, produce differing consequences for the total system.

The main design considerations in selecting a conditioning process are:

- Sludge residual characteristics such as particle size and distribution, surface charge and degree of hydration, and particle interaction.
- Physical factors such as prior processing conditions (shear, heat, storage) and a method of conditioner application.

Numerous methods of residuals conditioning have been tried and are listed in Table 2. Of the methods listed only the first two, inorganic chemical addition and organic chemical addition are used to any great extent in industrial applications and therefore only they will be further discussed.

Table 2: Methods of Residuals Conditioning*

Method	Process
Inorganic chemical addition	Elutriation
Organic chemical addition	Electrical
Nonchemical addition	Ultrasonic
Thermal treatment	Bacterial
Mechanical screening and grinding	Solvent extraction
Freeze-thaw (direct, indirect, natural)	

*Filter aid not considered a conditioning process and is discussed separately under "Dewatering."

Inorganic Chemical Conditioning

Description — Inorganic chemical conditioning typically consists of mixing ferric chloride and lime with the sludge mixture.

Ferric chloride is added first. It hydroylzes in water, forming positively charged soluble iron complexes which neutralize the normally negatively charge sludge solids, thus causing them to aggregate. Ferric chloride also reacts with the bicarbonate alkalinity in the sludge to form hydroxides that act as flocculants. The following equation shows the reaction of ferric chloride with bicarbonate alkalinity:

$$2FeCl_3 + 3Ca(HCO_3)_2 \rightarrow 2Fe(OH)_3 + 3CaCl_2 + 6CO_2$$

Dehydrated lime is usually used in conjunction with ferric iron salts. Although lime has some slight dehydration effects on colloids, it is chosen for conditioning principally because it provides pH control and the $CaCO_3$, formed by the reaction of lime and bicarbonate, provides a granular structure which increases sludge porosity and reduces sludge compressibility.

The chemical dosage required for any sludge is determined in the laboratory. Filter-leaf test kits are used to determine chemical doses, filter yields, and suitability of various filtering media. These kits have several advantages over the Buchner funnel procedure. In general, it has been observed that the type of sludge has the greatest impact on the quantity of chemical required. Difficult-to-dewater sludges require larger doses of chemicals and generally do not yield as dry a cake.

Intimate admixing of sludge and coagulant is essential for proper conditioning. Mixing must not break the floc after it has formed, and detention is kept to a minimum so that sludge reaches the filter as soon after conditioning as possible. Mixing tanks are generally of the vertical type for small plants and of horizontal type for large plants. They are ordinarily built of welded steel and lined with rubber or other acid-proof coating. A typical layout for a mixing or conditioning tank has a horizontal agitator driven by a variable-speed motor to provide a shaft speed of 4 to 10 revolutions per minute. Overflow from the tank is adjustable to vary the detention period. Vertical cylindrical tanks with propeller mixers are also used.

Application — Inorganic chemical conditioning is associated principally with mechanical sludge dewatering with vacuum filters and recessed plate pressure filters being the major users. Technology in this method is well developed.

Advantages and Disadvantages — Inorganic chemical conditioning offers the following advantages over other conditioning processes:

- Technology is very well developed.
- Operator skill requirements are minimal.
- This process is extremely flexible in adapting to a wide range of changes in residual characteristics.

Inorganic chemical conditioning has the following disadvantages when compared to other conditioning processes:

- There are numerous problems such as lime scaling and ferric chloride corrosion with in-plant storage, preparation, and application of both lime and ferric chloride.
- This process increases total residue mass approximately one pound per pound of ferric chloride and lime added. This increases the amount of total residue requiring ultimate disposal or in the case of incineration, lowers the fuel value.
- Installations are typically very dirty looking.
- Capital cost can be quite high especially for lime storage and feeding facilities.

Organic Chemical Conditioning:

Description — Organic chemical conditioning consists of adding a polyelectrolyte or polyelectrolytes to the sludge mass. Polyelectrolytes are long chain, water soluble, specialty chemicals. They can be either completely synthesized from individual monomers, (basic sub-unit), or they can be made by chemical addition of functional monomers, or groups, to naturally occurring polymers. As of 1980, the completely synthesized polymers are the most widely used.

Anionic-type flocculants carry a negative electrical charge in aqueous solutions. Cationic polyacrylamides carry a positive electrical charge in aqueous solution.

Cationic polyelectrolytes are the most widely used polymers for sludge conditioning, since most sludge solids carry a negative charge. The characteristics of the sludge to be processed and the type of thickening or dewatering device used will determine which of the cationic polyelectrolytes will work best and still be cost-effective. For example, an increasing degree of charge is required when sludge particles become finer, when hydration increases, and when relative surface charge increases.

Cationic polyelectrolytes are available as dry powders or liquids. The liquids come as water solutions or emulsions. The shelf life of the dry powders is usually several years, whereas most of the liquids have shelf lives of two to six months and must be protected from wide ambient temperature variations in storage.

Applications — During the past decade, important advances have been made in the manufacture of polyelectrolytes for use in sludge treatment. Polyelectrolytes are now widely used in sludge conditioning and a large variety are available.

Selection of the correct polyelectrolyte requires that the designer work with polyelectrolyte suppliers, equipment suppliers, and plant operating personnel. Evaluations should be made on site and with the sludges to be conditioned. Since new types and grades of polymers are continually being introduced, the evaluation process is an ongoing one.

Advantages and Disadvantages — Organic chemical conditioning offers the following advantages over inorganic chemical conditioning process:

- Little additional sludge mass is produced.
- If dewatered sludge is to be used as a fuel for incineration, organic chemicals do not lower the fuel value.
- They allow for cleaner material-handling operations.
- They reduce operation and maintenance problems.

Organic chemical conditioning has the following disadvantages when compared to other conditioning processes:

- Operating results are sensitive to dosage. Fluctuations in sludge characteristics can cause significant operating problems.
- Technology is still being developed and significant changes in type and grades of polymers are continually occurring.

Dewatering

Dewatering is defined in this manual as the removal of water from sludge to achieve a volume reduction greater than that achieved by thickening. The resulting material is nonfluid.

Sludges are dewatered primarily to decrease the capital and operating costs of the subsequent direct sludge disposal or conversion and disposal process. Dewatering sludge from a 2 to 20 percent solids concentration, for example, reduces volume by 90 percent.

The main design variables of any dewatering process are:

- Solids concentration and volumetric flow rate of the feed stream.
- Chemical demand and cost.
- Suspended and dissolved solids concentrations and volumetric flow rate of the sidestream.
- Solids concentration and volumetric flow rate of the dewatered sludge.

Other variables which impact the dewatering process selected are: requirements of subsequent processing steps; operation and maintenance (O&M) cost and the variables affecting it; and the reliability required to meet successful operational requirements.

While numerous techniques fulfill the basic functional definition of dewatering, they do so to widely varying degrees. It is important to note these circumstances when comparing different devices. For example, drying beds can be used not only to dewater a sludge, but also to dry it to a solids concentration of greater than 50 to 60 percent. Depending on the circumstances and particular device involved, dewatered sludge from a mechanical device may vary from a wet, almost flowable form, to a harder and more friable form. Table 3 lists many of the dewatering processes available.

Table 3: Dewatering Processes or Equipment Available for Industrial Residual Management

Name	Name
Natural systems	Recessed plate pressure filters
Drying beds	Screw and roll press
Drying lagoons	Twin roll press
Centrifugal systems	Dual cell gravity filter
Imperforate basket	Tube filter
Solid bowl decanter	
Filtration systems	
Vacuum filters	
Belt filters	

Natural Sludge Dewatering Systems: When land is available, sludge dewatering by nature can be extremely attractive from both a capital and an operating cost viewpoint. Considering escalating electrical power costs, this method is even more attractive. Two types of systems can be categorized as natural: drying beds and drying lagoons.

Drying Beds —

Description. Drying beds may be classified as either conventional, paved, wedge wire, or vacuum assisted. In addition, drying beds may be open or enclosed.

Drying beds generally consist of a one to three foot high retaining wall enclosing a porous drainage media. This drainage media may be made up of various sandwiched layers of sand and gravel, combinations of sand and gravel with cement strips, slotted metal media, or a permanent porous media. Appurtenant equipment includes: sludge feed pipelines and flow meters; possible chemical application tanks, pipelines, and metering pumps; filtrate drainage and recirculation lines; possible mechanical sludge removal equipment; and a possible cover or enclosure.

Operational procedures common to all types of drying beds involve:

- Pump 8 to 12 inches of liquid sludge onto the drying bed surface.
- Add chemical conditioners continuously, if conditioners are used, by injection into the sludge as it is pumped onto the bed.
- Permit, when the bed is filled to the desired level, the sludge to dry to the desired final solids concentration. This concentration can vary from between 18 to 60 percent, depending on the type of sludge, processing rate needed, degree of dryness required for lifting, etc.
- Remove the dewatered sludge either mechanically or manually.
- Repeat the cycle.

Application. Greasy or slimy sludges do not dewater well on any type of drying bed. Conventional and paved beds are more applicable to sludges which will not produce odors upon being allowed to sit for several weeks to months.

Advantages and Disadvantages. Drying beds offer the following advantages over other dewatering processes:

- Higher dry cake solids contents than fully mechanical systems.
- Low total energy consumption.
- Small amount of operator attention and skill is required.

Drying beds have the following disadvantages when compared to other dewatering processes:

- Rational engineering design data are lacking to allow sound engineering economic analysis.
- Requires more land than fully mechanical methods.
- Labor intensive.

Drying Lagoons —

Description. Sludge drying lagoons consist of retaining walls which are normally earthen dikes two to four feet high. The earthen dikes normally enclose a rectangular space with a permeable surface. Appurtenant equipment includes: sludge feed lines and metering pumps; supernatant decant lines; and some type of mechanical sludge-removal equipment. The removal equipment can include a bulldozer, drag line or front-end loader. In areas where permeable soils are unavailable, underdrains and associated piping may be required.

Operating procedures common to all types of drying lagoons involve:

- Pumping liquid sludge, over a period of several months or more, into the lagoon. The pumped sludge is normally stabilized prior to application. The sludge is usually applied until a lagoon depth of 24 to 48 inches is achieved.
- Decanting supernatant, either continuously or intermitently, from the lagoon surface and returning it to the wastewater treatment plant.
- Filling the lagoon to a desired sludge depth and then permitting it to dewater. Depending on the climate and the depth of applied sludge, the time involved for dewatering to a final solids content of between 20 to 40 percent solids may be 3 to 12 months.
- Removing the dewatered sludge with some type of mechanical removal equipment.
- "Resting" the lagoon (adding no new sludge) for three to six months.
- Repeating the cycle.

Application. Applicable to any type of sludge which will not produce odors upon being allowed to sit for several months

Advantages and Disadvantages. Drying lagoons offer the following advantages over other dewatering processes:

- Low total energy consumption.
- Lagoons are not sensitive to sludge variability.
- Lagoons consume no chemicals.
- Small amount of operator attention and skill is required.

Drying lagoons have the following disadvantages when compared to other dewatering processes:

- Rational engineering design data are lacking to allow sound engineering economic analysis.
- Lagoons are more visible to the general public.
- Lagoons can create vector problems (for example, flies and mosquitos).
- Can be difficult to control at times.

Centrifugal Dewatering Systems: Centrifugal dewatering is a process which dewaters sludge by accelerating the sludge slurry to several hundred or thousands of gravitational forces in a centrifuge. In dewatering sludge residues, two different types of centrifuges are commonly used: imperforate basket and solid bowl decanter.

Imperforate Basket —

Description. The description is the same as described in centrifugal thickening. There is, however, one additional operation to be added to that discussion.

After the centrifuge bowl is filled with solids, the unit starts to decelerate. In the thickening mode, deceleration is at a speed of 70 rpm or lower before commencement of plowing. In the dewatering mode, another step called "skimming" takes place before the initiation of plowing. Skimming is the removal of soft sludge from the inner wall of sludge within the basket centrifuge. The skimmer moves from its position in the center of the basket towards the bowl wall. The amount of horizontal travel is set at the time of installation, and start-up depends on sludge type and downstream processing requirements. The skimming volume is normally 5 to 15 percent of the bowl volume per cycle. After the skimmer retracts, the centrifuge further decelerates to the 70 rpm level for plowing. Skimming streams are typically 6 to 18 gallons per cycle with a solids content of almost zero to eight percent. Treatment of this stream is typically by returning it either to the pretreatment system, or to some other pre-sludge handling step.

Application. Applicable to hydroxide or other soft sludges (biological, for example) or to sludges containing significant amounts of abrasive material.

Advantages and Disadvantages. Imperforate basket centrifuges offer the following advantages over other dewatering processes:

- Same machine can be used for both dewatering and thickening.
- It may not require chemical conditioning.
- Centrifuge has clean appearance, little-to-no odor problems, and fast start-up and shut-down capabilities.
- Basket centrifuge is very flexible in meeting process requirements.
- It is not affected by grit.
- It is an excellent dewatering machine for hard-to-dewater sludges.
- Does not require continuous operator attention.

Imperforate baskets have the following disadvantages when compared to other dewatering processes:

- Requires special structural support.
- Except for a vacuum filter, consumes most direct horsepower per unit of product processed.
- Skimming stream could produce significant recycle load.
- Limited size capacity.
- For easily dewatered sludges, has the highest capital cost versus capacity ratio.
- For most sludges, gives the lowest cake solids concentration.

Solid Bowl Decanter —

Description. The description is the same as described in centrifugal thickening.

Application. Solid bowl decanters can be used on most types of sludges. Advances in scroll design are even allowing units to dewater abrasive sludges over long periods of time before rebuilding of the scroll edge is required.

Advantages and Disadvantages. Solid bowl decanter centrifuges offer the following advantages over other dewatering processes:

- Centrifuges have clean appearance, little-to-no odor problems, and fast start-up and shut-down capabilities.
- It is easy to install.
- Provides a high throughput in a small surface area.
- Gives, for many sludges, a cake as dry as any other mechanical dewatering process except for pressure filtration systems.
- Has one of the lowest total capital cost versus capacity ratio.
- Does not require continuous operator attention.

Solid bowl decanters have the following disadvantages when compared to other dewatering processes:

- Scroll wear is potentially a high maintenance item.
- Requires skilled maintenance personnel.
- May require a grinder in the feed stream.

Filtration Dewatering Systems: Filtration can be defined as the removal of solids from a liquid stream by passing the stream through a porous medium which retains the solids. Figure 69 shows a flow diagram of a filtration system.

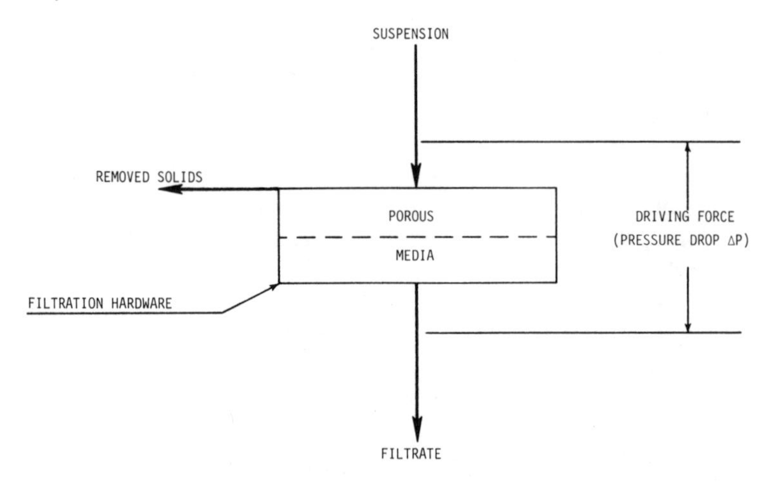

Figure 69: Flow diagram of a filtration system.

As indicated on Figure 69 a pressure drop is required in order for liquid to flow through the porous medium. This pressure drop can be achieved in four ways: by creating a vacuum on one side of the porous medium; by raising the pressure above atmospheric pressure on one side of the medium; by creating a centrifugal force on an area of the porous medium; and by designing to make use of gravitational force on the medium. Sludge filtration-dewatering processes use one or more of these driving forces and fall under the general filtration category of surface filters.

Some industrial sludge filtration dewatering systems — those employing vacuum or pressure filtration — may require a filter aid, especially for difficult-to-dewater sludges. Filter aid is any material — diatomite, perlite, cellulose, etc. — that serves to improve, or increase the filtration rate by physical means only. Filter aids are not added directly to the sludge body, as a conditioning agent is, but they are added in fixed amounts to the porous medium of the particular dewatering equipment. The amount of filter aid added is independent of sludge solids concentration. The filter aid literally becomes the "filtering surface" that achieves the liquid/solids separation, and the equipment functions as a filter holder.

Vacuum Filters —

Description. In vacuum filtration, atmospheric pressure, due to a vacuum applied downstream of the media, is the driving force on the liquid phase that moves it through the porous media.

There are approximately 20 different types of vacuum filters to select from with the most common in industrial waste sludge dewatering being the bottom feed, rotary drum type with belt (natural or synthetic fiber cloth, woven stainless mesh or coil spring media) and frequently with pre-coat. Figure 70 shows a cross sectional view of a fiber cloth-belt rotary vacuum filter.

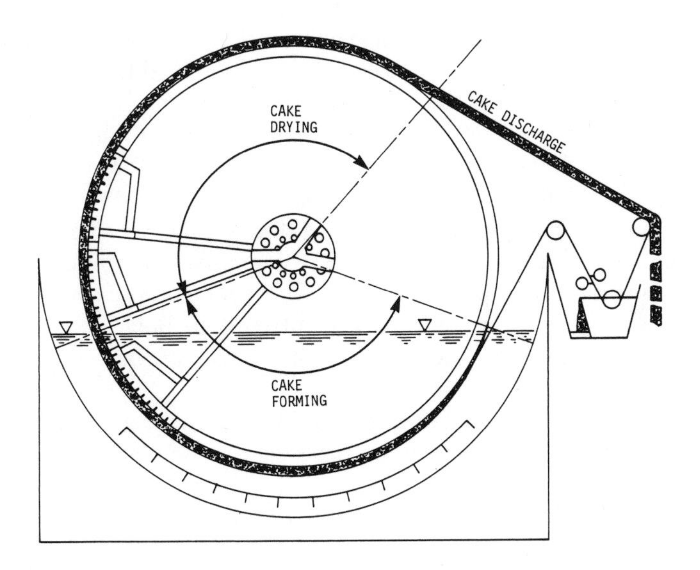

Figure 70: Cross sectional view of a fiber cloth-belt type-rotary vacuum filter.

A rotary drum-belt-type vacuum filter system normally consists of the following items: rotary drum mounted in a slurry vat with internal agitator and belt washing system, sludge conditioning system section on conditioning or pre-coat system, vacuum pump system and filtrate system.

In operation the rotary drum rotates, partially submerged, in a vat of conditioned sludge. The drum surface is divided into sections around its circumference. Each section is sealed from its adjacent section and the ends of the drum. A separate drain line connects each section to a rotary valve at the axis of the drum. The valve has "blocks" that divide it into zones corresponding to the parts of the filtering cycle. These zones are for cake forming, cake drying, and cake discharging (see Figure 70). A vacuum is applied to certain zones of the valve and subsequently to each of the drum sections through the drainlines as they pass through the different zones in the valve.

About 10 to 40 percent of the drum surface is submerged in a vat containing the sludge slurry. This portion of the drum is referred to as the cake-forming zone. Vacuum applied to a submerged drum section causes filtrate to pass through the cake-forming zone to the cake-drying or dewatering zone.

This zone is also under vacuum and begins where and when a drum section carries formed cake out of the sludge vat. The cake drying zone represents from 40 to 60 percent of the drum surface and terminates at the point where vacuum is shut off to each successive section. At this point, the sludge cake and drum section enter the cake discharge zone. In this final zone, cake is removed from the media.

The media leaves the drum surface at the end of the drying zone and passes over a small diameter discharge roll to facilitate cake discharge. Washing of the media occurs after discharge and before it returns to the drum for another cycle.

Application. Except for very fibrous sludges, dewatering of industrial sludge using vacuum filtration is very dependent on proper sludge conditioning or proper pre-coat. Even with proper conditioning and pre-coat, greasy or slimy sludges are difficult to dewater or to obtain release from the filter media. Vacuum filters have been used in almost all the categorical industries.

Advantages and Disadvantages. Vacuum filters offer the following advantages over other dewatering processes:

- Does not require skilled personnel.
- Has low maintenance requirements compared to continuous operating equipment.
- Provides a filtrate with a low suspended solids concentration.

Vacuum filters have the following disadvantages when compared to other dewatering processes:

- Requires continuous operator attention.
- Consumes the largest amount of energy per unit of sludge dewatered in most applications.
- Auxiliary equipment (vacuum pumps) are very loud.

Belt Filter Press —

Description. In belt filter press filtration, pressure is applied by tension to a continuous moving filtering belt by either another belt or rollers. This driving force moves the liquid through the porous belt material.

There are many various design alternatives with belt filter presses but all alternatives can be grouped into three basic stages: chemical conditioning of the incoming slurry; gravity drainage to a nonfluid consistency; and, compaction of the resulting material. Figure 71 depicts a simple belt press and shows the location of the three stages. Although present-day belt presses are more complex, they follow the same principles indicated in Figure 71.

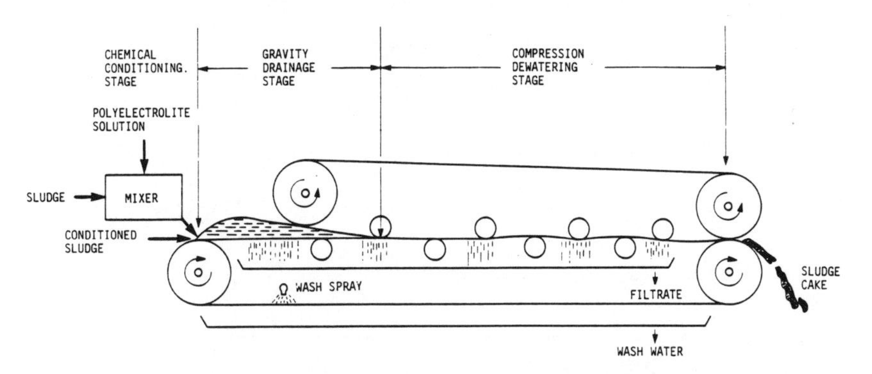

Figure 71: Basic three stages of a belt filter press.

Good chemical conditioning is the key to successful and consistent performance of the belt filter press. After conditioning, the readily drainable water is separated from the slurry by discharge of the conditioned material onto the moving belt in the gravity drainage section. The third stage of the process begins as soon as the sludge is subjected to an increase in pressure, due to either the compression of the sludge between the carrying belt or the application of a vacuum on the carrying belt. During pressure application, the sludge cake, squeezed between the two belts is subjected to flexing in opposite directions as it passes over the various rollers.

Application. Except for very fibrous sludges, dewatering of industrial sludges using belt filter presses is very dependent on proper sludge conditioning. Belt filter press are applicable to many of the sludges generated at categorical industries.

Advantages and Disadvantages. Belt filter presses offer the following advantages:

- High pressure machines are capable of producing very dry cake.
- Low power requirements.

Belt filter presses have the following disadvantages:

- Very sensitive to incoming feed characteristics.
- Machines hydraulically limited in throughput.
- Short media life as compared with other devices using cloth media.

Recessed Plate Pressure Filters —

Description. In recessed plate pressure filtration, fluid pressure generated by pumping slurry into the unit, is the driving force on the liquid phase that moves it through the porous media.

A recessed plate pressure filter system normally consists of the following items: the presses assembly, pressure feed pump, conditioning system, and sludge holding tank. Figure 72 shows a schematic side view of a recessed plate pressure filter. Figure 73 shows a cross section of a fixed volume recessed plate filter assembly.

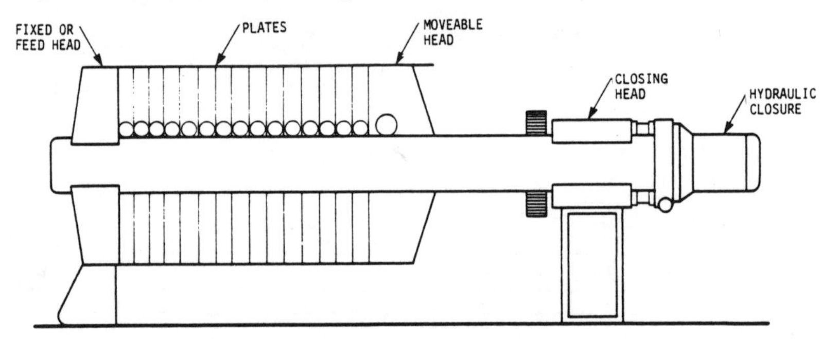

Figure 72: Schematic side view of a recessed plate pressure filter.

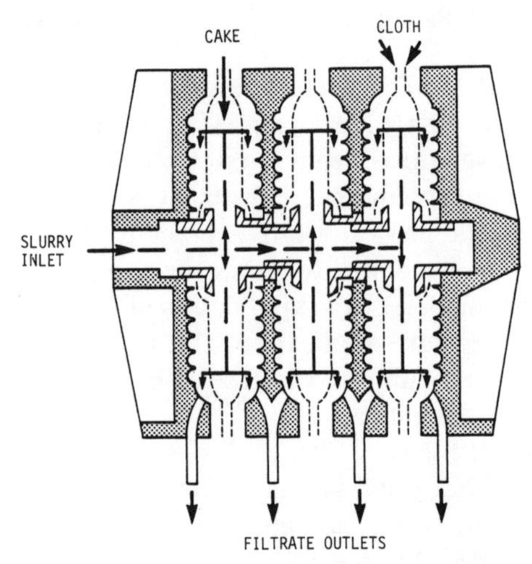

Figure 73: Cross section of a fixed-volume recessed plate filter assembly.

The operation of a recessed plate pressure filter is best explained by describing a typical pressure filtration cycle which begins with the closing of the press. Sludge is pumped for a 20- to 30-minute period until the press is effectively full of cake. The pressure at this point is generally the designed maximum and is maintained for a one- to four-hour period. The filter is then mechanically opened, and the dewatered cake drops from the chambers onto a conveyor belt for removal.

Application. Although recessed plate pressure filters can be used on a wide variety of waste sludge residuals, their high capital and operating cost normally prohibits their use to sludges which are difficult to dewater, or in those sludge management situations where sludge cake must be in excess of 30 to 40 percent solids. As with all filter dewatering equipment, sludge conditioning is very important. In addition, precoating may also be required.

Advantages and Disadvantages. Recessed plate pressure filters offer the following advantage over all other dewatering systems:

- Highest cake solids concentration.

Recessed plate pressure filters have the following disadvantages:

- Batch operation.
- High labor cost.
- High capital cost.
- Special support structure requirements.
- Large area requirements.

Screw Press —

Description. In screw press filtration, pressure generated both by pumping and mechanical squeezing is the driving force on the liquid phase that moves it through the porous media.

A screw press installation normally consists of the following items: screw press, filtrate pump, and sludge conditioning equipment. The dewatering device itself employs a screw surrounded by a perforated steel (screen) cylinder. Sludge is pumped inside the screen and is deposited against the screen wall by the rotating screw. The cake that forms acts as a continuous filter. The screw moves the progressively dewatered sludge against a containment at the outlet and further dewaters the sludge by pressure of the screw action against the restriction. Figure 74 shows a typical screw press layout.

Applications. Screw presses are, generally, only going to be applicable to fibrous type sludges. Gelatinous and slimy sludges, even with excellent sludge conditioning, will tend to ooze through the retaining media.

Advantages and Disadvantages. Screw presses offer the following advantages:

- Low power consumption.

- Minimum space requirements.
- Minimal maintenance requirements.

Screw presses have the following disadvantages:

- Limited operating experience upon which to base design.
- Not capable of dewatering gelatinous or slimy sludges.

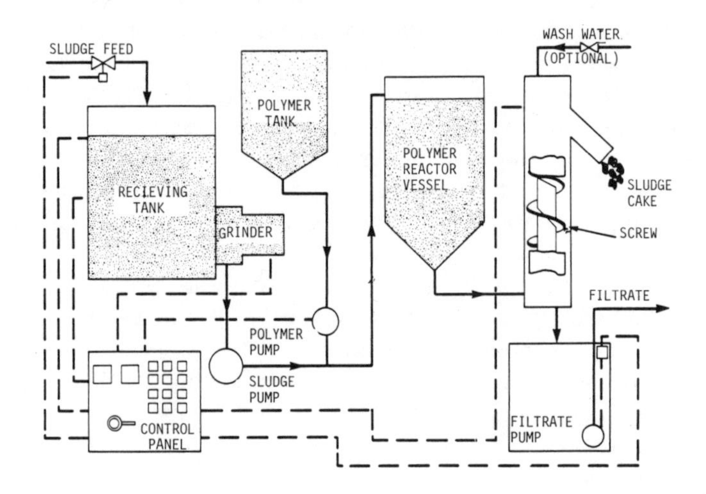

Figure 74: System schematic for one type of screw press system.

Twin Roll Press —

Description. In twin roll press filtration, pressure generated by mechanical squeezing of the two rollers, is the driving force on the liquid phase that moves it through the porous media.

The unit consists of a pair of perforated rolls, one roll fixed and the other moveable, so that the nip (or space) between the rolls can be varied. The horizontal rolls are mounted in a sealed vat. Sludge is pumped into the vat under a slight pressure of two to four pounds per square inch. This low vat pressure moves the sludge into the nip, where it is further dewatered by a nip pressure load of 100 to 400 pounds per linear inch of roll length. Filtrate passes from the sludge through the perforated rolls and discharges by gravity. Compressed cake is removed from the rolls with a doctor knife and discharged into a shredder and conveyor. Figure 75 shows a cross section view of a twin roll vari-nip press.

Application. Twin roll presses are applicable to fibrous or primary sludges. Gelatinous and slimy sludges, even with excellent sludge conditioning, will tend to ooze through the nip. Majority of experience has been in the pulp and paper industry.

Advantages and Disadvantages. Twin roll presses offer the following advantages:

- High solids concentration, 30 to 40 percent solids.
- Low power requirements.
- Has a clean appearance.
- Provides a high throughput in a small surface area.

Twin roll presses have the following disadvantages:

- Not capable of dewatering gelatinous or slimy sludges.
- Limited experience in dewatering many industrial residues.
- High capital cost.

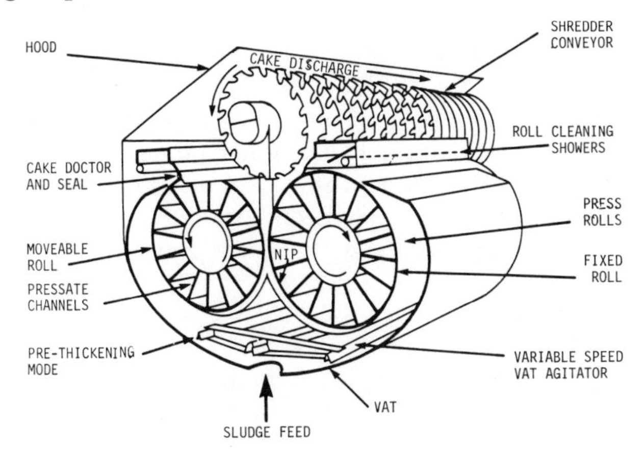

Figure 75: Cross section view of a twin-roll vari-nip press.

Dual-Cell Gravity (DCG) Filter —

Description. In the dual-cell gravity filter, pressure generated by the weight of the moving plug of solids is the driving force on the liquid phase that moves it through the porous media.

The unit consists of two independent cells formed by a nylon filter cloth. The cloth travels continuously over guide wheels and is rotated by a drive roll and sprocket assembly. Dewatering occurs in the first cell, and cake formation, in the second cell.

Sludge is introduced in the dewatering cell, where initial liquid/solids separation takes place. The dewatering solids are then carried over the drive roll separator into the second cell. Here, they are continuously rolled and formed into a cake of relatively low moisture content. The weight of this sludge cake presses additional water from the partially dewatered sludge carried over from the dewatering cell. When the cake of dewatered solids grows to a certain size, excess quantities are discharged over the rim of the second cell to a conveyor belt that moves the material out of the machine. Figure 76 shows a cross section view of a dual-cell gravity filter.

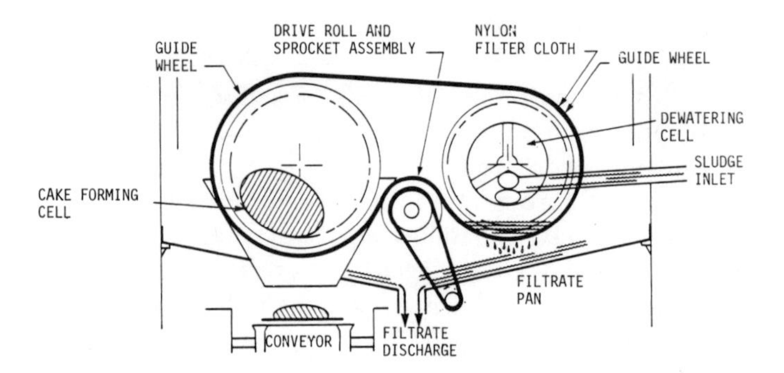

Figure 76: Cross section view of a dual cell gravity filter.

Applications. Has been used on a wide variety of waste residuals generated from pulp and paper, leather tanning, mechanical products, paint and ink and other categorical industries. It is capable of a decent performance in hydroxide sludges.

Advantages and Disadvantages. Dual-cell gravity filters offer the following advantages:

- Low capital cost.
- Low power requirements.
- No specialized operator skills required.

Dual-cell gravity filters have the following disadvantages:

- Limited throughput capacity.
- High conditioning requirements.
- Messy operation.
- Low solids content.

Tube Filters —

Description. Tube filters can be either of the pressure type or of the gravity type.

Pressure-type units are more commonly known as tube filter presses. Typically, this is a vertically-mounted device consisting of an outer cylinder which is covered with a filter media. Slurry is pumped into the annular space between the bladder and media-covered wall. When this area is full, the bladder is filled with liquid, and the slurry is compressed against the filter media. Filtrate flows through the media and is discharged. When the desired cake solids concentration has been obtained, liquid pressure is released and the cake is discharged with a blast of air.

In gravity-type units, sludge is mixed with polymer and then held in suspended porous bags. The weight of the sludge forces water out of the bag sides and bottom. Sludge is retained for a maximum of 24 hours, depending upon the desired dryness, and is then released through a bottom opening.

Application. Pressure-type systems have been used extensively in industrial applications dewatering a wide range of residuals. These units are especially attractive for difficult to dewater sludge since units can generate over 5,000 psi of pressure on the sludge mass. Experience with gravity-type units is almost nonexistent.

Advantages and Disadvantages. Tube filters offer the following advantages:

- Gravity units are very low capital cost.
- Pressure units can dewater almost any type sludge.
- Units have relatively low maintenance cost.
- Units are easy to operate and do not require skilled personnel.

Tube filters have the following disadvantages:

- Very low throughput capacity per unit floor area.
- Are batch processes.
- Can be a messy operation.

Thermal Drying

Description: Thermal or heat drying is an evaporative process in which the water content of the residual sludge is reduced to eight-to-ten percent without combusting the solids being dried. The heat for drying may either be applied directly — exhaust gases from a combustion source — or indirectly using steam. For economic reasons, the moisture content of the material to be dried is reduced as much as possible through mechanical dewatering.

There are basically five thermal drying techniques: flash, rotary, toroidal, multiple hearth, and atomizing spray. Other drying processes such as composting and air drying were discussed earlier in this chapter.

Flash Drying: Flash drying is the instantaneous vaporization of moisture from solids by introducing the sludge into a hot gas stream. The system is based on several distinct cycles that can be adjusted for different drying arrangements. The wet sludge cake is first blended with some previously dried sludge in a mixer to improve pneumatic conveyance. Blended sludge and hot gases from the furnace at about 1200°F to 1400°F are mixed and fed into a cage mill in which the mixture is agitated and the water vapor flashed. Residence time in the cage mill is only a matter of seconds. Dry sludge with eight-to-ten percent moisture is separated from the spent drying gases in a cyclone, with part of it recycled with incoming wet sludge cake and another part screened and sent to storage. Figure 77 shows a typical arrangement for a flash drying system.

Rotary Driers: Rotary driers can be of two general types: direct or indirect.

Direct rotary driers are the most common consisting of a cylinder that is slightly inclined from the horizontal and revolves at about five-to-eight revolutions per minute. The inside of the dryer is equipped usually with flights or baffles throughout its length to break up the sludge. Wet cake is mixed with previously heat dried sludge in a pug mill. The system may include cyclones for sludge and gas separation, dust collection scrubbers, and a gas incineration step. Figure 78 shows a typical process schematic for a direct drying rotary dryer system.

Indirect systems dry the sludge by bringing the residual sludge into contact with a warm surface. One example of this type of dryer is the jacketed and/or hollow-flight dryer and conveyor. Steam is forced through the jacket and/or hollow-flights while the flight is rotating to provide turbulence and mixing. Figure 79 shows a typical process schematic for an indirect drying rotary dryer system.

Toroidal Dryers: A toroidal dryer has no moving parts. Transport of solids within the drying zone is accomplished entirely by high-velocity air movement.

Dewatered sludge is pumped into a mixer where it is blended with previously dried sludge. Blended material is fed into a doughnut-shaped dryer, where it comes into contact with heated air at a temperature of 800°F to 1100°F. Particles are dried, broken up into fine pieces, and carried out of the dryer by the air stream. Dry sludge is separated from the drying gases in a cyclone, with part of it recycled with incoming wet sludge cake and the other part sent to storage. Figure 80 shows a general process flow schematic for a toroidal drying system.

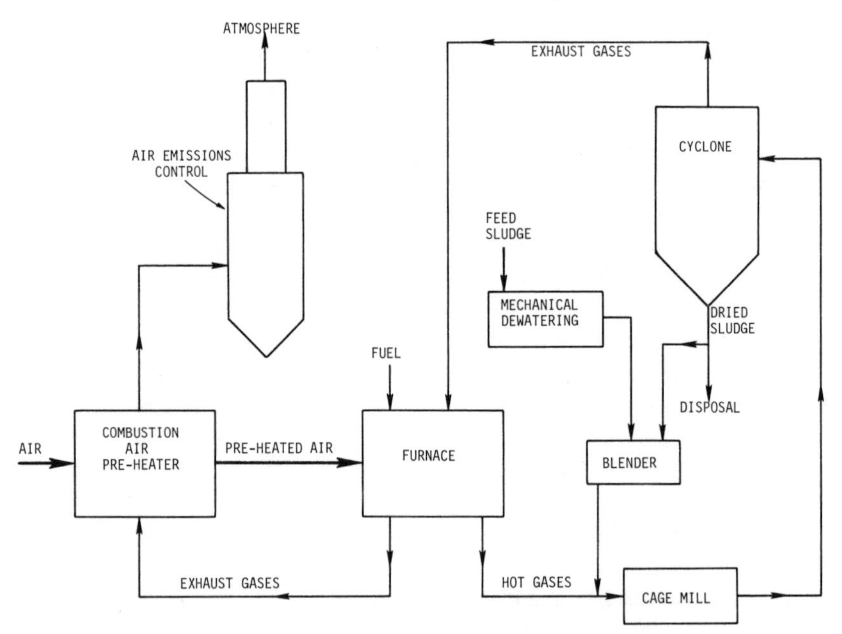

Figure 77: Typical arrangement of a sludge residual flash drying system.

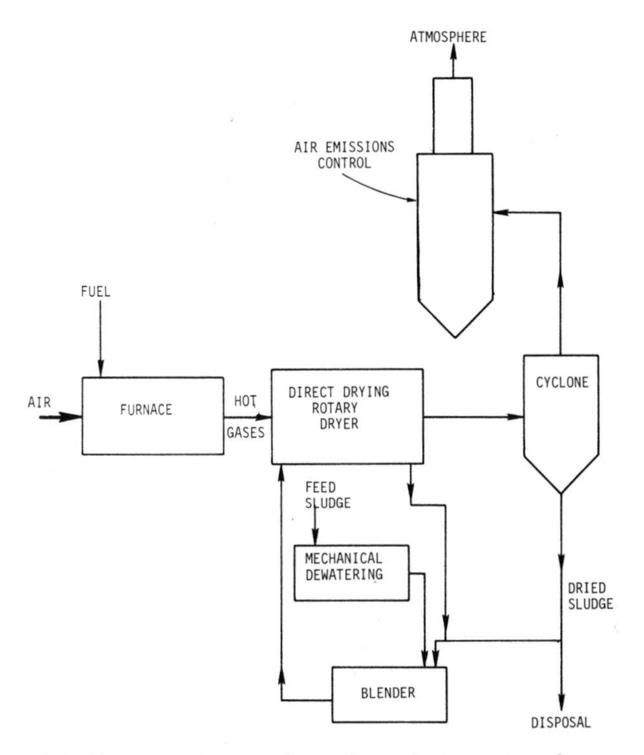

Figure 78: Process schematic for a direct drying rotary dryer system.

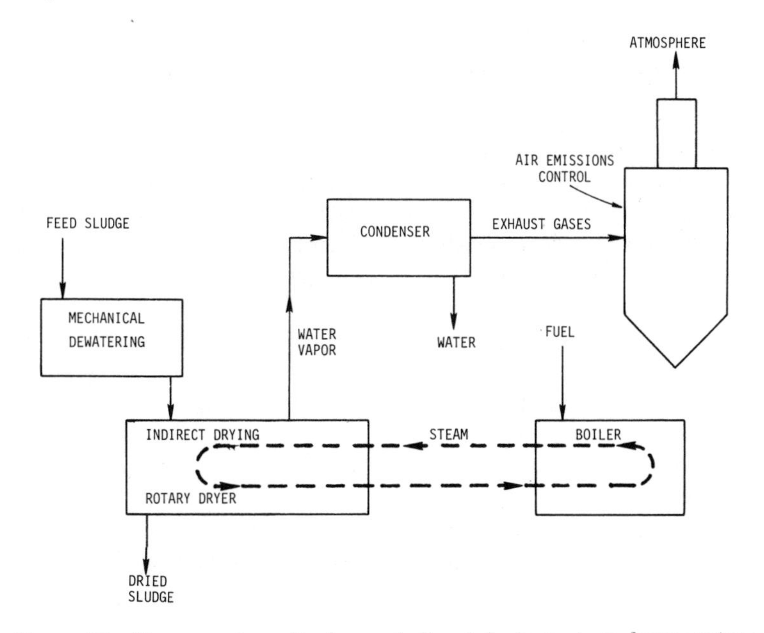

Figure 79: Process schematic for an indirect drying rotary dryer system.

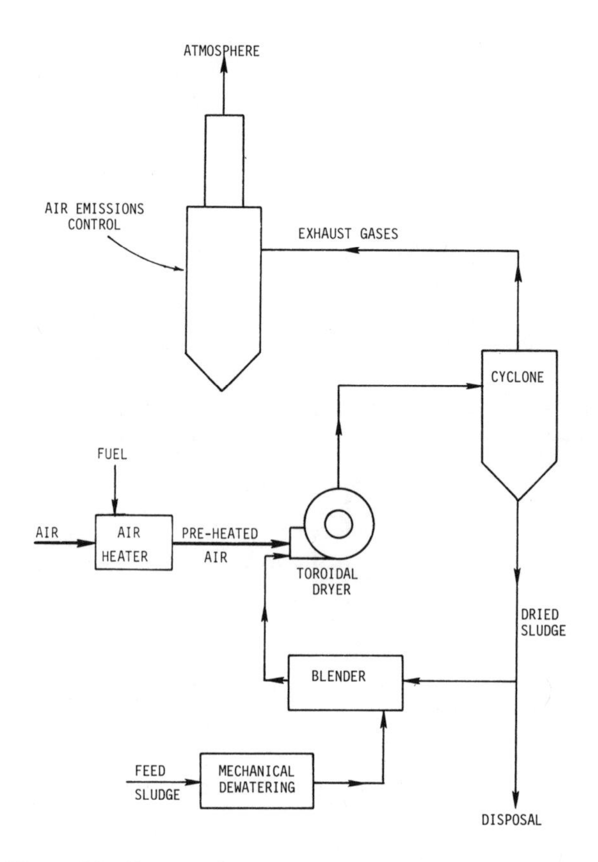

Figure 80: Process schematic for a toroidal drying system.

Multiple Hearth: Multiple hearth incinerators are discussed in the section on "Thermal Destruction." This type of incinerator can be adapted for heat drying of sludge by incorporating fuel burners at the top and bottom hearths plus down draft of the gases. Incoming sludge is dewatered and mixed in a pug mill with previously dried sludge before entering the furnace.

Atomizing Spray: These systems are similar to flash-drying systems in that almost instantaneous drying occurs in both. Atomizing drying involves spraying liquid sludge in a vertical tower through which hot gases pass.

Dust carried with hot gases is removed by a wet scrubber or dry dust collection.

Applications: Can be used in any type of sludge residual. Although thermal drying of sludges is over 50 years old, very few thermal drying systems presently exist.

Advantages and Disadvantages: Thermal drying offers the following advantages:

- Drying can produce a material that when coupled with nutrient supplement can be used as a fertilizer or directly as a fuel source.
- Adds flexibility to final disposal option.
- Reduces volume and weight.

Thermal drying has the following disadvantages:

- Drying can produce an odor problem.
- Drying has high energy costs.
- Drying systems require highly skilled operators.

Thermal Destruction

Description: Thermal destruction is a broad field, impossible to cover in the few pages allocated in this manual. In this section, a brief description of the three main thermal destruction processes — incineration, starved air combustion, and high pressure/high temperature oxidation — will be covered. A more complete discussion can be found in the USEPA *Process Design Manual For Sludge Treatment and Disposal,* ERIC, Cincinnati, Ohio, EPA 625/1-79-011.

Incineration — Incineration is a two-step oxidation process involving first drying and then combustion. Drying and combustion may be accomplished in separate units or successively in the same unit, depending upon temperature constraints and control parameters. The drying step should not be confused with preliminary dewatering, which is usually done mechanically prior to incineration. In all furnaces, the drying and combustion processes follow the same phases: raising the temperature of the feed sludge to 212°F, evaporating water from the sludge, increasing the temperature of the water vapor and air, and increasing the temperature of the dried sludge volatiles to the ignition point. Although presented in simplified form, incineration is a complex process involving thermal and chemical reactions which occur at varying times, temperatures, and locations in the furnace.

Several types of incinerators exist with the most common being the multiple hearth and fluid bed types. The units being utilized are the electric furnace, cyclonic furnace, and the modular furnace.

In addition to incineration, residuals containing high heat values can be utilized as fuel alternatives. By using residuals in this manner, energy recovery from these wastes could be harnessed for the production of steam electricity and a broad range of resource recovery alternatives.

Multiple Hearth Furnace. Due to furnace durability, simple operation, and potential of operating over a wide variation in feed quality and loading rate, the multiple hearth furnace (MHF) is the most widely used incinerator in the United States. An outstanding feature of the

MHF is that it can be divided into four zones. The first zone or upper hearth zone is where drying takes place through the evaporation of water. The second zone consists of the central hearths; here combustion takes place with temperatures reaching 1400°F to 1700°F. The third zone or the fixed carbon zone reduces volatile matter to carbon dioxide. The fourth is the cooling zone where ash is cooled by incoming combustion air.

Fluid Bed Furnace. The fluidized bed furnace is a vertically oriented, cylindrically shaped furnace containing a sand bed and fluidizing air diffusers. During operation, air is injected into the furnace at a pressure of 3 - 5 psig to fluidize the sand bed. At this time the bed expands to 100 percent of its normal volume. Temperature of the bed is maintained between 1400°F and 1500°F by burners above or below the sand bed. Drying and combustion occur in either the dense or dilute phases in the sand bed. After the combustion gases pass through the combustion zone, ash is carried out of the top of the furnace and is removed by air pollution devices. Sand losses do occur at approximately five percent of the bed volume for every 300 hours of operation.

Starved Air Combustion — Unlike incineration, which is a complete combustion process, starved air combustion is incomplete combustion and occurs when insufficient oxygen is provided to satisfy the combustion requirements. Under these conditions, the end products are combustible gases, tars, oils, and a solid char that can have an appreciable heating value. The relative proportion of each varies with the amount of heat applied and the feed moisture.

High Pressure – High Temperature Oxidation — Any burnable substance may be oxidized in the presence of water at a sufficiently high temperature (flameless combustion). In this process, residual sludge at four-to-six percent solids is pressurized and excess air is added for oxidation. The mixture is then pumped through a heat exchanger and heated to the desired reaction temperature and discharged into the reactor for oxidation. After an adequate length of time, the oxidized slurry is then cooled in the heat exchange, gases are removed and processed, and the slurry is removed from the system and dewatered before final disposal.

Application: There are numerous types of incinerators, one or more of which is capable of incinerating any particular sludge residual. For many of the hazardous and toxic wastes, incineration may be the only means of disposal, but great care will be required in designing a system to meet all environmental and legal constraints.

Advantages and Disadvantages: Thermal destruction processes offer the following advantages:

- Maximum volume reduction is obtained.
- When designed correctly, will destroy or reduce toxics that may otherwise create adverse environmental impacts.

- Potential to recover energy through combustion of waste products.

Thermal destruction has the following disadvantages:

- Capital costs are high.
- Operating and maintenance cost are high.
- High temperatures create high maintenance requirements and can reduce equipment reliability.
- Highly skilled and experienced operators are required.
- Potential for adverse environmental impacts may require significant cost increases for protection of environment.

RESIDUALS TRANSPORTATION OR CONVEYANCE

Description

Transportation, conveyance or the movement of residuals between residual treatment processes is also part of the overall residual management strategy. Processes for transportation or conveyance of industrial wastewater residuals can be grouped into two broad categories: those for movement of material within or within close proximity of residual processing (pumping and pipelines and conveyors); and, those for movement of material some distance from the residual or generating facility (long distance pipeline, truck, rail, or barge).

Pumping and Pipelines: In most cases, unless a sludge has been dewatered, sludge can be transported efficiently and economically by pumping through pipelines. The most important factor in designing a sludge pipeline, no matter what length, is estimating the friction loss to be used.

The correct application of a pump is a second important factor. Pumps which are currently utilized in sludge transport include: centrifugal, torque flow, plunger, piston, piston/hydraulic diaphragm, progressive cavity, rotary, diaphragm, ejector, and air lift pumps. Each has its advantages and disdisadvantages, and careful consideration must be given to selection.

Conveyors: Dewatered or dried residuals can be conveyed short distances by most forms of industrial materials handling equipment including: belt, tubular, and screw conveyors; slides and inclines; elevators; and pneumatic systems. Each may be used to advantage in certain applications. Because the consistency of residual solids is highly variable, and because the solids are often difficult to move and may tend to flow, the design of this equipment must consider the most severe conditions that may be expected.

Trucks: Trucking provides a viable option for transport of both liquid and dewatered material. Trucking provides flexibility not found in other modes of transport since terminal points and route can be changed readily at low cost. Truck hauling allows more flexibility than pumping or other transport modes. This flexibility is valuable since locations for disposal may change.

Rail: Rail transport is suitable for transporting sludges of any solids concentration. It is, however, not a commonly used method, and probably never will be, due to insufficient volume to allow for economical cost.

Barge: Barge transport for the ocean dumping of sludge has been practiced for many decades around the world. Recent decisions to limit ocean dumping, combined with rapidly escalating costs for dewatering or drying sludges, have led to more consideration of barge transport of liquid sludges between the residual management site and the land disposal site many miles distant. Barge transportation of sludges is generally only feasible for (1) liquid sludges (to the solids concentration limit at which it may be pumped); and (2) over longer distances, generally over 30 miles.

Sludge Applications

Any sludge can be transported by any of the methods described above. Economics and ease of operation will dictate selection.

Advantages and Disadvantages

One can not list advantages and disadvantages of each transportation mode as residual characteristics and properties play such a significant part in mode selection.

RESIDUALS CONTAINMENT TECHNIQUES

Residual containment techniques are associated with the final disposal option, but are significant enough to warrant a separate discussion in this manual. Two containment techniques will be discussed: linings and solidification.

Linings

Description: Development of lining materials since the mid-1960's has lead to a rapid growth in their use as a means of pollution control. Table 4 lists many of the liner materials available. Selection of the one to use will depend on waste characteristics, type of containment basin to be used, state requirements, and cost. For storage of residuals generated from categorical industries, liners should definitely be: impervious; durable; flexible over a wide range of temperatures; easy to repair and maintain; and, resistant to chemical, biological, mechanical, weathering and deterioration damage.

Table 4: Listing of Common Lining Materials

Asphalt concrete	Gunite
Asphalt panels	Plastics
Bentonite clays	Rubber latex
Chemical treatments	Soil asphalt
Compacted clays	Soil cement
Compacted solids	Sprayed asphalt membranes
Concrete	Steel
Elastomers	Waterborne treatments

Application: Lining containment is available for many types of waste residuals. Since the field is young, and new lining materials and experiences are continually being developed, residual managers should keep in touch with the knowledgable suppliers.

Advantages and Disadvantages: Containment with a properly applied liner offers the following advantages:

- Reduces the risk of groundwater contamination from impounded material to a very low level.
- Lined impoundment may be the only environmentally acceptable method available for residual disposal.

Containment within a lined basin has the following disadvantage:

- Possible break in the liner, allowing leakage of waste material into the environment — normally by groundwater contamination.

Solidification

Description: The term solidification is used to mean either the fixation or encapsulation of waste sludge residuals into a solid matrix and products. Fixation techniques chemically and physically mix the waste sludge residual with a solidification agent — normally a proprietary compound but also limestone, fly ash, lime, etc. Encapsulation methods physically surround the waste sludge, for example the encapsulation of a 55 gallon drum within polyethylene or asphalt. The intent of both technologies is to reduce the permeability of the sludge and produce an endproduct having a high shear strength. Table 5 lists the five basic solidification processes currently available.

Table 5: Basic Solidification Processes

Silicate and cement based
Lime based
Thermoplastic based
Organic polymer based
Encapsulation techniques

Applications: Although experience with both technologies is still very limited, the bulk of solidification work has been with the fixation processes. To date, fixation processes have been tried on sludges from the following categorical industries: petroleum refining, pulp and paper, organic chemicals, plastics, iron and steel, steam electric, electroplating, ore mining, paint and ink, and pesticides. Industrial-scale data for encapsulation could not be found at this time.

Advantages and Disadvantages: The advantages and disadvantages of the five solidification process are shown in Table 6.

RESIDUALS DISPOSAL AND/OR UTILIZATION

The most difficult part of residual management for the categorical industries will be finding the final resting place for the residuals. Since the 1970's, available options for final disposal have been increasingly restricted. This section of the manual will discuss the following available options: landing farming; land filling; and sale or give away. Ocean disposal is not mentioned since at the present time no new permits are being issued for ocean disposal of residues, and enforcement compliance schedules are being issued to phase out existing ocean disposal practices as soon as possible. Long-term storage is essentially covered in the section on "Residuals Containment Techniques". Deep well injection is not considered an option for residual management, except in the very special case where injection may be into a deep underground cavern.

Land Farming

Description: Land farming can be defined as the controlled addition of a waste residual into the upper horizons of the soil. This method utilizes plants and the soil matrix to remove residual waste constituents. Land farming is also based on the fertilizing and soil conditioning properties of the residuals. Reclamation of land stripped by mining, fertilization of food and non-food crop lands such as forests and grass farms have been primary areas for use of this method. Other names used to describe land farming are: land spreading, land application, sludge farming, and soil cultivation.

The amount of residuals deposited per unit area of ground depends on the characteristics of the residual and the type of soil involved. For liquid waste, below eight-to-ten percent solids, direct injection into the ground would be the most practical method of application. For residuals above ten percent solids application will require the mixing of the material into the top six to eight inches of earth using a disk harrow or rototiller. Periodic cultivation is also advisable depending on the residuals involved.

Table 6: Advantages and Disadvantages of the Five Solidification Processes

Process	Advantages	Disadvantages
Cement-based	1. Additives are available at a reasonable price. 2. Cement mixing and handling techniques are well developed. 3. Processing eqiupment is readily available. 4. Processing is reasonably tolerant of chemical variations in sludges. 5. The strength and permeability of the end-product can be varied by controlling the amount of cement added.	1. Low-strength cement-waste mixtures are often vulnerable to acidic leaching solutions. Extreme conditions can result in decomposition of the fixed material and acce rated leaching of the contaminants. 2. Pretreatment, more-expensive cement types, or costly additives may be necessary for stabilization of wastes containing impurities that affect the setting and curing of cement. 3. Cement and other additives add considerably to weight and bulk of wastes.
Lime-based	1. The additives are generally very inexpensive and widely available. 2. Equipment required for processing is simple to operate and widely available. 3. Chemistry of pozzolanic reactions is well known.	1. Lime and other additives add to weight and bulk of waste. 2. Stabilized sludges are vulnerable to acidic solutions and to curing and setting problems associated with inorganic contaminants in the waste.
Thermoplastic	1. Contaminant migration rates are generally lower than for most other techniques. 2. End-product is fairly resistant to most aqueous solutions. 3. Thermoplastic materials adhere well to incorporated materials.	1. Expensive eqiupment and skilled labor are generally required. 2. Sludges containing contaminants that volatilize at low temperatures must be processed carefully. 3. Thermoplastic materials are flammable. 4. Wet sludges must be dried before they can be mixed with the thermoplastic material.

(continued)

Table 6: (continued)

Process	Advantages	Disadvantages
Organic polymer	1. Only small quantities of additives are usually required to cause the mixture to set. 2. Techniques can be applied to either wet or dry sludges. 3. End-product has a low density as compared to other fixation techniques.	1. Contaminants are trapped in only a loose resin-matrix end-product. 2. Catalysts used in the urea-formaldehyde process are strongly acidic. Most metals are extremely soluble at low pH and can escape in water not trapped in the mass during the polymerization process. 3. Some organic polymers are biodegradable. 4. End-product is generally placed in a container before disposal.
Encapsulation	1. Very soluble contaminants are totally isolated from the environment. 2. Usually no secondary container is required, because the coating materials are strong and chemically inert.	1. Materials used are often expensive. 2. Techniques generally require specialized eqiupment and heat treatment to form the jackets. 3. The sludge has to be dried before the process can be applied. 4. Certain jacket materials are flammable.

Application: Land farming has been used for over 25 years in the petroleum industry to dispose of many of their oily sludges. Recent work indicates that some residuals generated from the pharmaceutical and organic chemicals industry are also amenable to land farming techniques. In addition, the application of land farming to sludges generated from scrubber operation used to control air pollution are also being studied.

Advantages and Disadvantages: Land farming offers the following advantages:

- The process is relatively environmentally safe, when properly designed and operated.
- The process will generally improve soil structure and fertility.
- Requires no dependence on high maintenance or failure-prone equipment.
- Uses a natural process that recycles the waste residual back into the environment.
- The process can be utilized with a wide range of solid concentrations.

Land farming has the following disadvantages:

- Except for the petroleum industry, there is very little operating experience on land farming of other categorical residues.
- The potential exists for developing odors.
- The potential exists for soil contamination due to salt or metal buildup over period of years.
- Air pollution from evaporation, sublimation, and wind erosion during initial application of material.
- Surface runoff can be a problem.

Landfilling

Description: Landfilling can generally be defined as a method of disposal in which waste is systematically deposited and covered with earth to control environmental impacts. In most cases landfill sites are situated in areas with low water tables or areas of least probable flooding.

Since the advent of RCRA regulations, costs for the landfilling of categorical industry residues will be significantly higher. RCRA requires that landfills be designed, constructed, and operated so that discharges (liquid or gas) are minimized or do not occur. To achieve this, liquid and gaseous discharges must be monitored and effluent and emission control limits must be met. New landfills are also required to line their disposal cells with Bentonite clay or plastic liners to prevent leaching to the groundwater. Also, the soil cover must be deeper than the plow layer, which is between eight and ten inches.

Co-disposal — Co-disposal is the mixing of residuals with solid municipal waste or soil thereby utilizing the natural water absorbing properties of the solid waste or soil.

Depending on residual quantity and quality and state and local waste disposal regulations, co-disposal may be one of the most environmentally sound methods of disposal.

The co-disposal method is recommended only for stabilized or dewatered sludges. Once sludges are dewatered or stabilized they can be disposed of through the use of sludge/refuse mixtures or sludge/soil mixtures. Application rates and procedures depend primarily on solids content.

Application: Under proper design and operation any residual can be safely disposed of in a landfill, though the cost of doing so may be extremely expensive. Many of the techniques discussed in the section on "Residuals Containment Techniques" would be applicable.

Advantages and Disadvantages: Landfilling offers the following advantages:

- It may be the only method of final disposal of a residue.
- It allows stockpiling of a residual that in the future could become a valuable resource of raw material if economics for recovery became attractive.
- It can be used for every type of residue generated.

Landfilling has the following disadvantages:

- The possibility of fire and explosions.
- Poisoning by direct contact with waste material.
- Air pollution due to evaporation, sublimation or wind erosion.
- Ground water contamination from leachate.
- Surface water contamination from runoff.

Sale or Give Away

Description and Application: Sale or give away offers great opportunity for complete reuse of residual wastes. It involves the transfer of residuals to reclaimers or industries who can use the residuals as a raw material. Solvents and waste oils can be upgraded and reused. Some sludges can be used directly as fertilizers. Metals can be reclaimed for value.

Advantages and Disadvantages: Sale or give away of sludge offers the following advantages:

- It uses a waste product beneficially by recycling back into the environment.

Sale or give away of sludge has the following disadvantages:

- Material will have a variable quality, and therefore an uncertain market.

APPENDIX

BIBLIOGRAPHY

The following general references were used for the pretreatment technologies discussed in this Volume.

1. USEPA. *Innovative and Alternative Technology Assessment Manual (draft).* Office of Water Program Operations, Washington, D.C. 20460, EPA 430/9-78-009, September 1978.
2. USEPA. *Process Design Manual-Wastewater Treatment Facilities for Sewered Small Communities.* USEPA Technology Transfer, Cincinnati, Ohio 45268, EPA 625/1-77-009, October 1977.
3. USEPA. *Process Design Manual-Upgrading Existing Wastewater Treatment Plants.* USEPA Technology Transfer, Cincinnati, Ohio 45268, EPA 625/1-71-004a, October 1974.
4. USEPA. *Development Document for Existing Source Pretreatment Standards for the Electroplating Point Source Category.* Effluent Guidelines Division, Washington, D.C. 20460, EPA 400/1-79/003, August 1979.
5. USEPA. *Lagoon Performance and the State of Lagoon Technology.* Environmental Protection Technology Series, EPA-R2-73-144, June 1973.
6. USEPA. *Upgrading Lagoons.* USEPA Technology Transfer, Cincinnati, Ohio 45268, EPA 625/4-73-001b, August 1973.
7. Metcalf and Eddy Inc. *Wastewater Engineering.* McGraw-Hill Book Co., New York, 1979.
8. USEPA. *Performance and Upgrading of Wastewater Stabilization Ponds.* MERL Cincinnati, Ohio 45268, EPA, 600/9-79-001, May 1979.

9. USEPA. *Areawide Assessment Procedures Manual Vol. III.* MERL Cincinnati, Ohio 45268, EPA 600/9-76-014, July 1976.
10. WPCF. *MOP 8-Wastewater Treatment Plant Design.* Water Pollution Control Federation 1977.
11. WPCF. *MOP 11-Operation of Wastewater Treatment Plants.* Water Pollution Control Federation, 1976.
12. USEPA. *Process Design Manual for Carbon Adsorption.* USEPA Technology Transfer, Cincinnati, Ohio 45268, EPA 625/1-71-002a, October 1973.
13. USEPA. *Process Design Manual for Suspended Solids Removal.* USEPA Technology Transfer, Cincinnati, Ohio 45268, EPA 625/1-75-003a, January 1975.
14. Shell, G.L. and R.C. Wendt "Spray Cooling." *Pollution Engineering,* July 1977.
15. USEPA. *A Survey of Alternate Methods for Cooling Condenser Discharge Water.* Water Pollution Control Research Series 16130 DHS 01/71, January 1971.
16. Hwang, S.T. and P. Fahrenthold. "Treatability of the Organic Priority Pollutants by Steam Stripping." Paper presented at the August 1979 meeting of A.I.C.H.E.
17. API. *Manual on Disposal of Refinery Wastes.* American Petroleum Institute, 1969.
18. USEPA. *Hazardous Material Incinerator Design Criteria.* IERL, Cincinnati, Ohio 45268, EPA 600/2-79-198.

The following general references were used for the sludge treatment technologies discussed in this Volume.

19. USEPA. *Process Design Manual for Sludge Treatment and Disposal.* MERL, Cincinnati, Ohio 45268, EPA 625/1-79-011, September 1979.
20. Epstein, E., J. Alpert, and W. Toffel. "Composting of Industrial Wastes." *Proceedings of National Conference on Municipal and Industrial Sludge Composting.* November 14-16, 1979.
21. Bell, R.G., J. Pos, and R.J. Lyon. "Production of Composts from Soft Wood Lumber Mill Wastes." *Compost Science.* March-April 1973.
22. Suler, D. "Composting Hazardous Industrial Wastes." *Compost Science/ Land Utilization.* July-August 1979.
23. Svarovsky, L. "Filtration Fundamentals." *Solid-Liquid Separation,* Butterworths, Inc., Ladislav Svarousky, editor, 1977.
24. Middlebrooks, E.J., C.D. Perman, and I.S. Dunn. *Wastewater Stabilization Pond Linings.* Special report 78-28 Department of the Army, CRREL, Corps of Engineers, Hanover, New Hampshire. November 1978.
25. USEPA. *State-of-the-Art Study of Land Impoundment Techniques.* MERL, Cincinnati, Ohio 45268, EPA 600/2-78-196, December 1978.

26. Kays, W.B. *Construction of Linings For Reservoirs, Tanks and Pollution Control Facilities*. John Wiley and Sons, 1977.
27. Pojasek, R.B. "Solid-Waste Disposal: Solidification." *Chemical Engineering*. August 13, 1979.
28. *Toxic and Hazardous Waste Disposal Volumes I and II*. Ann Arbor Science Publishers Inc., Ann Arbor Michigan, 1979.
29. Smith, D.D. and R.P. Brown. "Deep Sea Disposal of Liquid and Solid Wastes." *Industrial Water Engineering*. September, 1970.
30. Huddleston, R.L. "Solid Waste Disposal: Land Farming." *Chemical Engineering*. February 26, 1979.
31. Frank, A. "Oil Refinery Sludge — An Industry Looks To Land Cultivation." *Sludge Magazine*. May-June 1978.
32. Duvel, W.A. "Solid-Waste Disposal: Landfilling." *Chemical Engineering*, July 2, 1979.
33. Lazar, E.C. "Summary of Damage Incidents from Improper Land Disposal." *Management and Disposal of Residues from The Treatment of Industrial Wastewaters*. February 3-5, 1975.
34. Kohn, P. M. "Wastes Find Fertile Field As Low-Cost Plant Nutrients." *Chemical Engineering*. August 2, 1976.
35. *Applications of Sludges and Wastewaters on Agricultural Land: A Planning and Educational Guide*, edited by B.D. Knezek and R.H. Miller. USEPA reprint of NCR publication 235. USEPA Office of Water Program Operations, Washington, D.C., 20460, MCD-35, March, 1978.
36. USEPA. *Evaluation of Sludge Management Systems: Evaluation Checklist and Supporting Commentary*. Office of Water Program Operations, Washington, D.C., 20460, EPA 430/9-80-001, MCD-61, February, 1980.
37. USEPA. *A Guide to Regulations and Guidance for the Utilization and Disposal of Municipal Sludge*. Office of Water Program Operations, Washington, D.C., 20460, EPA 430/9-80-015, MCD-72, September, 1980.
38. USEPA. *Process Design Manual: Municipal Sludge Landfills*. ERIC. Cincinnati, Ohio, 45268, EPA 625/1-78-010, SW 705, October, 1978.